THE CONTINENTAL SHELF

The undersigned, Secretary General of the I n s t i t u t e of I n t e r n a t i o n a l L a w, declares herewith that the text of this work is in conformity with the one which has been submitted to the Jury appointed for the attribution of the G r o t i u s p r i z e 1952.

Footnotes and Appendix added to the original text are printed in italics.

Geneva, July 10, 1952.   (*Signé*) HANS WEHBERG

(Article 19 of the "Règlement des Prix institués par M. James Brown Scott en mémoire de sa mère et de sa sœur Jeannette Scott".)

THE
CONTINENTAL SHELF

BY

M. W. MOUTON
Doctor of Law
Captain Royal Netherlands Navy

*This work was awarded the Grotius Prize 1952
of the Institute of International Law*

SPRINGER-SCIENCE+BUSINESS MEDIA, B.V.
1952

© *Springer Science+Business Media Dordrecht 1952*
Originally published by Martinus Nijhoff, The Hague, Netherlands in 1952
Softcover reprint of the hardcover 1st edition 1952
All rights reserved, including the right to translate or to
reproduce this book or parts thereof in any form

Additional material to this book can be downloaded from http://extras.springer.com

ISBN 978-94-017-5671-6   ISBN 978-94-017-5966-3 (eBook)
DOI 10.1007/978-94-017-5966-3

TO MY WIFE

PREFACE

This study was submitted to "The Institute of International Law", for a competition. The subject was: "To make a critical study of the juridical position of the Continental Shelf and of the questions concerning the utilization of the sea covering it, of its soil and subsoil beyond the outside limits of the territorial waters".

The Introduction explains the author's conception of the subject.

The small figures in the text refer to the Bibliography at the end. French, Spanish, German and Dutch quotations, if not given in the original language, have been translated by the author.

The author is greatly indebted to all those who have given him their advice or have helped him in any other way.

Wassenaar, Holland
1 August 1952.

CONTENTS

CHAPTER III

THE REGIME OF THE HIGH SEAS AND NAVIGATION

CHAPTER IV

MINERAL RESOURCES

CHAPTER V

OFFSHORE DRILLING TECHNIQUE AND MINING

Chapter VI

SOME ASPECTS NOT YET DISCUSSED, SUMMARY AND PROPOSALS

Additional material from *The Continental Shelf,*
ISBN 978-94-017-5671-6, is available at http://extras.springer.com

INTRODUCTION

The continental shelf is the part of the sea-bottom and the soil underneath, which is covered by shallow waters, up to a depth where the slope of the sea-bottom increases noticeably in steepness, which fringes large parts of the continents, over varying distances from the coasts.

Although the continental shelf has had some attention of international lawyers at an earlier date, it came into the limelight of interest only recently. The first instrument concerning this recent development was the Treaty between the United Kingdom and Venezuela relating to the submarine areas of the Gulf of Paria, of February 26, 1942, but a real impetus to the development of a continental shelf theory was only given by the Proclamation of President Truman with respect to the natural resources of the subsoil and sea-bed of the continental shelf of September 28, 1945. This Proclamation was followed in a short time by declarations and decrees of other countries.

What is the problem involved?

Modern engineering skill has made it possible to exploit the minerals contained in the subsoil of the continental shelf. To extract oil, being the principle mineral concerned for the time being, installations have been built in the sea, carrying derricks. It is obvious that such installations are obstacles for navigation, particularly if built in considerable numbers and in places frequented by shipping.

Leakages, and worse still a blow-out, could cause damage to fisheries through pollution of the sea-water with oil.

As soon as these installations are being built outside the territorial sea of a particular State, the question arises, who has the right to do so, any State, or only the coastal State.

In some proclamations, declarations, and national decrees, referred to above, the mineral resources in the sea-bed or subsoil of the continental shelf have been claimed by the coastal State. In other instruments sovereignty over the continental shelf was claimed and even over the waters covering the shelf. Did these States have any right to do so, and what is the significance of such unilateral acts in International Law? The principle of the high seas is interfered with as far as navigation and fisheries are concerned. Here we are in the middle of a most fascinating subject, questions besiege us, problems dawn on our minds, the possibility of serious conflicts becomes obvious.

Oil reserves becoming smaller, exploitation of new fields is in the general interest of mankind, but so is an umhampered navigation on the high seas connecting the continents.

The growing world population needs an increasing food supply and one of the sources of this supply, fisheries, will be intensified. This will enlarge the danger of exhaustion and ways have to be found to prevent depletion of this valuable source of proteins.

Again President Truman took the lead with a Proclamation of the same date as the one referred to above, establishing conservation zones. Again his example was followed by other countries, although in many cases on a more drastic scale. Some instruments went as far as proclaiming the sovereignty of the coastal State over sea areas extending to a considerable distance from the coast irrespective of the existence of a continental shelf.

Can a State claim such rights, rights not only on the mineral resources outside its territory, including of course its territorial waters, but also on the biological resources, for instance in the waters covering its continental shelf? Here again questions of the utmost importance. The world needs these resources, and a solution has to be found to regulate exploitation activities in such a way, that a rush and grab policy will be forestalled, that each part of the world population gets its share, that exhaustion and conflicts will be prevented.

These problems, where often traditional rights are involved, are not easy to solve. Not a single person will be able to find a conclusive answer to all these questions. But each contribution of human thought may help. Several scientific associations and last but not least the International Law Commission, dedicated a lot of thought and time to the subject. There are enough sources of conflict already in the world. Let us try to tackle this young problem together and lead it into channels where a peaceful development of the exploitation of these so much needed resources will be possible.

In the following pages a number of aspects of the subject will be discussed.

This study starts from the premiss that the juridical position of the continental shelf can only be spoken of in the sense of "de jure constituendo" or in that of "de lege ferenda". The sequence we have chosen is as follows.

First of all we attempt to give a definition of the notion "continental shelf". Then we will investigate which subjects come under the continental shelf theory. As there appears to be difference of opinion whether Fisheries are related to the continental shelf we have thought fit to start with the Fisheries, fishery rights being one of the rights derived from the principle of the freedom of the seas. We will deal with navigation immediately afterwards because this subject comes under the same principle. After that we will deal with the main subject, the mineral resources, and the means of their exploitation.

The first Chapter deals with the continental shelf as a scientific notion and in the light of the use which is made at the moment of the shelf, we will investigate whether this scientific notion as understood by geologists, oceanographers and geographers, can be used in law. A description of the main features is given and the relevant paragraphs of several legal instruments relating to the description and delimitation of the continental shelf are critically analized.

The second Chapter, on Fisheries, starts with a biological discussion on the geographical distribution of fish, in order to answer the question of the relation between continental shelf and fisheries. Again the relevant legal instruments are critically discussed and an attempt is made to base legal provisions on facts, because as we will see, a certain discrepancy between facts and proposals made for regulations has existed in the past. Special attention has been given to bottom-fisheries and sedentary fisheries, and the possible consequences of oil-exploitation for the fisheries.

The next Chapter deals with the other right derived from the principle of the freedom of the seas, i.e. navigation. An attempt is made to reduce criticism relating to the interference with free navigation to the right proportions. Some deviations were made into the fields of the theory of territorial waters and contiguous zones, to elucidate the discussion on the high seas. Pipelines as compared to telegraph cables are viewed in the light of International Law.

The fourth Chapter deals with the declarations and decrees in so far as mineral resources are concerned. A great deal of attention is dedicated to the nature of the rights which States may claim on the soil and subsoil and especially on the minerals contained therein. In particular reference is made to mine law in different countries and certain suggestions as to the application of mine-law principles to the subject under discussion are made.

In order to understand what oil-exploitation in the sea involves a Chapter is dedicated to offshore drilling technique. Some other mining techniques not interfering with shipping and fisheries are mentioned.

In the last Chapter salient points of the previous Chapters are summarized. Some special aspects are discussed, for instance the consequences of war and the question whether international control is desirable and possible.

Some suggestions are made concerning ideas and proposals which may be of some use for future international deliberation. The trend of the study is directed towards international agreement as the only means to solve international problems.

Where the method is concerned, we are aware of two peculiarities which may give rise to criticism if not explained beforehand.

One is, that we have introduced more references to other sciences than usually will be found in a treatise on a chapter of International Law. We believed, however, that in a new field of law, it was justified to investigate the facts on which the whole conception of the continental shelf-theory rests. We felt that it would serve a good purpose to dedicate some space to the experts in other branches of science and to listen to what they said on these facts, preferably in their own words in order to avoid misunderstanding and misinterpretation of these facts, which unfortunately has already taken place. Our investigation led us as we have said above to the conclusion that there existed a certain amount of discrepancy, for instance between biological facts and theories advanced in International Conferences concerning fishery rights.

The other is, the great number of quotations. This, however, has been done consciously. We know that in International Law books or articles often too much copying takes place. However, in a new field of law, opinions of authors are more valuable than in older fields, where the practice of States and the Treaties are more important. We prefer to quote as fully as possible. The reader can see for himself, instead of either having to look it up in the source or believing us for having rightly given the gist of the author's idea.

We quoted extensively also to avoid the passages to be wrenched from their context. One cannot solve a new problem alone. One has to put the opinions next to each other, in their original wording, in order to be able to attain a certain amount of progression in thought and give the reader the chance, without forcing him to go through all the sources, to compare the arguments and judge whether he can or cannot agree with the conclusions we have produced.

A scientific work should be a most honest and constructively critical contribution. Only then it has value in the solving of a problem. This we have tried.

The first duty of all who work in International Law is to promote world-peace. We believe that particularly when working in a field of International Law so strewn with possible conflicts, one should be aware of this great responsibility.

CHAPTER I

ATTEMPT TO DEFINE THE CONTINENTAL SHELF

SECTION 1. GENERAL IDEA OF THE NOTION
"CONTINENTAL SHELF"

The notion "continental shelf" is of purely geological, geographical and oceanographical origin. What do geologists and other scientists mean when they speak of "the continental shelf" (le plateau continental, der Kontinentalsockel)? From the middle of the last century onwards an increasing number of soundings in the seas and oceans have been carried out for scientific as well as practical purposes (shipping and telegraph cables).

One of the outstanding features of the sea-bottom relief was the fact that going from the shore towards the sea, the soundings showed that the depth increased slowly to a certain figure, not everywhere the same, but let us say for the time being about 200 metres, whereafter the depth increased more rapidly. In other words, generally speaking, it was found that the land shelves away to the sea with a small angle or gradient to an average depth of 200 metres, after which the gradient increases rather rapidly to a steeper slope going down to ocean-depth. The isobath of 200 metres forms in this simplified picture an edge.

The part of the sea-bottom between the shore and this edge is called the continental shelf. To quote Krümmel [1], I, p. 103–104: "The need to give a special name to the framelike rim or margin of the continents with their long, sometimes narrow, sometimes wide but hardly ever completely failing shallow-water banks, was first (1887) felt by Hugh Robert Mill, and for that purpose, the notion *continental shelf* was used".

The part between the edge and the ocean-bottom is called the continental slope. To give an idea of the extent and the geopgraphical form of the continental shelf a chart is given, which shows clearly the great differences in width of the continental shelf, for instance between the east and the west coast of South America, or between the coast of China and that of Spain and Portugal. In figures, quoted by Umbgrove [2], p. 99, the shelves all over the world cover 27,500,000 square kilometres, or 7.6 per cent of the surface of the oceans. He makes however a distinction between inner shelves and outer shelves, the latter covering a surface of 9,900,000 square kilometres, or 3.1 per cent of the ocean surface.

He writes, p. 98,: "Generally the shallow platforms bordering the continents are classified in two categories, viz. the inner and the outer shelves. The first group comprises such regions as the shelf of the North Sea and the Sunda Sea. Bathymetric charts show them to be furrowed by river-like trenches. And indeed, in the two examples just mentioned their sub-aerial origin as rivers extending over these regions during low stands of the sea-level in Pleistocene times was clearly demonstrated by data furnishing converging evidence. Their course can be followed towards the debouchement of present-day rivers. Moreover, the frequent finding of the remains of large vertebrates as far as the Dogger Bank shows the North Sea to have been a land area in the near past.

In the case of the Sunda-shelf the congruence of such animals as fresh-water fishes clearly proves e.g. the rivers of western Borneo and eastern Sumatra to have been connected in sub-recent times. The shelf of the Barents Sea is considered to belong to the European continent ....". Concerning the Sunda Sea Molengraaff [3] wrote earlier, p. 274: "By the beginning of the ice-period the sea regressed as a result of the fixing of great quantities of water in the pleistocene ice-caps and its level was then at least 40 fathoms (72 metres) lower than nowadays". (In a note he remarks that 40 fathoms is a conservative estimate, probably too low. Penck's estimate was 100–150 metres. We add that Umbgrove [2], p. 105, estimates "the amount of the lowering of the sea-level at the time the ice-sheets reached maximum size .... at 75 metres, possibly as a lower limit 90

to 100 metres"). "A big area became dry, Sumatra, Borneo and
Java were connected with each other and with Malacca into one
united country". Molengraaff then describes how in that plain,
water was collected from the mountain ranges bordering it.
The rivers of Sumatra and Borneo were branches of a big river
which debouched in the China Sea. The former course of these
rivers, partly as continuation of the existing one's, can be followed
by soundings showing the drowned river valleys, or is demonstrated by tin-ore deposits in old river beds outside the isle of
Sinkep, as we will see later.

After the ice-period, the sea-level started to rise as the ice-caps
on higher latitudes melted. The Sunda-plain was inundated and
forms the so-called Sunda-plat [a] with an average depth of 50
metres.

This, a little more detailed, description of one of the "inner
shelves" was necessary as we will see. The distinction between
inner shelves and outer shelves, made by Umbgrove [b] is, in our
opinion, of great importance for the subject we are dealing
with. He leaves, in his article on the origin of continental shelves [4],
p. 250, and in his book [2], p. 99, "the origin and history of the
inner shelf regions .... out of consideration .... *since they do
not belong to the marginal zone proper of the continents*".

It seems that he does not comprise the inner shelves under the
notion "continental shelf". In other words, certain shallow
terraces, belong geologically speaking, rather to the continental
masses proper than to the part which geologists call the continental shelf. In that train of thought the actual continental shelf
outside the North Sea would start somewhere above the Dogger
Bank and finish at the edge of the 100-fathom line running
along the depression off the Norwegian coast and further round
the Shetland Islands, whereas the actual continental shelf outside
the Sunda Sea would start somewhere on the border of the
South China Sea. Kuenen [5] does not seem to make this distinction.

---

[a] Sunda-plat = Sunda-shelf.
[b] Also Krümmel [1], p. 105, makes a similar distinction when he speaks of "echte
oder Randschelfe" (real or margin-shelves) and „falsche Schelfe" (false shelves)
such as "Binnenschelfe" (inner shelves) of which he mentions those in the Baltic,
Persian Gulf and the Hudson Bay, and the Aussenschelfe (outer shelves), but here
in the sense of "loose" from the continents such as Faröer- or Iceland-shelves or
freely situated banks like the Seychellen-Bank.

He speaks, however, p. 104, about insular shelves, referring to the belts of shallow water, which skirt the larger islands.

Must we assume that there is a continental shelf in a wide sense of all undersea terraces to a maximum depth of 200 metres, and a continental shelf stricto sensu, i.e. the margins of the continents directly facing the ocean? Has any distinction been made by Governments or lawyers when the continental shelf became a subject of governmental proclamations or decrees or juridical publications or discussions? To mention a few examples: First of all, in which way did proclamations refer to the continental shelf? The Proclamation by President Truman of September 28, 1945, (for Proclamations see Chapter IV) only used the term "continental shelf". In a Press release of 28 September, 1945 [6], a depth limit, namely the 100-fathom isobath, was mentioned. The Declaration of the President of Mexico of 29 October, 1945, mentions "the continental shelf, which is delimited by a two-hundred-metre isobath". The Decree Law No. 803 of 2 November, 1949, of Costa Rica refers to the submarine platform or the continental and insular shelves adjacent to the continental and insular coasts of the national territory, at whatever depth it is found. The Congressional Decree No. 102 of 17 March, 1950, of Honduras uses the same wording, adding however "and whatever its extent". The Royal Pronouncement of the Kingdom of Saudi Arabia with respect to the subsoil and sea-bed of areas in the Persian Gulf of May 29, 1949, speaks of those areas of the Persian Gulf seaward from the coastal sea of Saudi Arabia but contiguous to its coasts. The boundaries of such areas will be determined in accordance with equitable principles by our Government in agreement with other States having jurisdiction and control over the subsoil and sea-bed of adjoining areas.

The Proclamation of the Bahrain Government of June 5, 1949, promises a similar limitation.

We notice that in the instruments of the Western Hemisphere the continental shelf is mentioned with or without further delimitation. In the Decrees of Costa Rica and Honduras the term "insular shelf" is used. In the Saudi Arabian Pronouncement, however, no continental shelf is mentioned. Turning to discussions and publications we find the following explanation.

Hudson said in the 66th meeting of July 12, 1950, of the International Law Commission [7], p. 24: In the Persian Gulf "there was no continental shelf". He then went on advocating the exploitation of mineral reserves in the Persian Gulf. The Chairman pointed out that the scope of Mr. Hudson's statement went far beyond the continental shelf in the strict sense of the word. Young [8], p. 531, in his article on Saudi Arabia offshore legislation remarks: "As a factual matter, no continental shelf exists in the Persian Gulf, which is merely a basin much less than 100 fathoms deep on the Asian continental mass" and on p. 532: ".... the Saudi Arabian Pronouncement exemplifies one approach to the difficult problem of how to divide amicably submarine areas of narrow seas where the continental shelf doctrine is not applicable". In a later article [9], p. 790, he is more precise in saying: "In the Persian Gulf, in which there is, *strictly speaking*, no continental shelf ....". These statements are referred to by François, in his first Report to the International Law Commission [10], p. 36: "In point of fact there is no continental shelf in the Persian Gulf". J.Y.B(rinton) [11], p. 133, also states: "The theory of jurisdiction on which the Arabian claims are based is necessarily somewhat different from that supporting the American policy due to the fact that there is no continental shelf in the Persian Gulf".

In Art. 1 of the Iranian Bill of 19 May, 1949, relating to Persian Gulf subsea resources, however, we read: "The natural resources existing at the bottom of the sea or under the bottom of the sea up to the limits of the continental shelf of the Iranian coasts in the Persian Gulf and Oman Sea belong to the Iranian Government" and Art. 2: "Should the continental shelf of Iran extend to the coasts of another country or be common with another adjacent country, the limits of the interested countries will be fixed equitably between the interested governments with respect to the natural resources of the continental shelf".

From these few utterances a picture of the continental shelf as a legal notion evolves, which seems somewhat blurred. It has been said that in the Persian Gulf there is no continental shelf, or, there is, *strictly speaking*, no continental shelf. In the Iranian Bill we read of the continental shelf of the Iranian coasts in the Persian Gulf. Is the denial of the existence of a continental shelf in the Persian Gulf based on the distinction made by Umbgrove? Or must we classify the bottom and subsoil of the Persian Gulf under the group "sea-bed and subsoil beneath shallow seas and gulfs, in which no shelf edge exists" as Young [12] does (p. 227)? It is true that there is no edge in the Persian Gulf, but the edge of the continental shelf of the Asian continental mass lays just outside the Persian Gulf [a], in the Gulf of Oman. Can we speak about a continental shelf when the coast of a continent is dented into bays or more so into gulfs? Geologists speak about the continental shelf in the Gulf of Mexico. Or is the criterion the age of the formation? Umbgrove [2], describing the inner shelves, mentions the sub-aerial origin of the shelves of the North Sea and Sunda Sea and he and Molengraaff explain the North Sea-bottom and the Sunda-"plat" to have been land area in the near past. But if geological age is a criterion, the part of the continental shelf of an open straight continental coast, the origin of which could be explained by the post glacial rise of the sea-level, should not be called continental shelf stricto sensu either. What we mean is this. Umbgrove [2], p. 101, states that: "a glacial lowering of sea-level to an amount of 100 meters is inadequate to explain the features of the shelf surface ....". Several geologists believe that the history of the continental shelf has partly to be explained by subsidence of the layers underneath. Umbgrove [4], p. 249, summarizes his theory by saying: "The history of the shelf was rather complicated .... The area was subjected to changes of sea-level *and* movements of the bottom ...." and "In order to explain the common occurrence of a shelf edge at the isobath of 200 meters a widespread subsidence of the shelf area in the order of 100 meters has to be assumed".

It does, however, not seem to serve any purpose to borrow a name of another branch of science which does not fit the thing for which we want to use it. The "fact" that we can speak about the continental shelf in the Mexican Gulf and not about the continental shelf in the Persian Gulf, does not elucidate our

---

[a] We may add that the distinction is certainly not a general one. We mentioned already Kuenen in this connection, and Dacqué [18], p. 48, considers the Persian Gulf as lying on the continental shelf: "der Persische Golf liegt noch auf dem Kontinentalschelf".

problem in the least. A rather drastic solution has been suggested by François in his Report [10], p. 44: "On se demandera donc si une meilleure solution ne serait pas d'écarter l'idée du plateau continental ...." [a], which was actually taken over in the Report of the International Law Commission covering its second session, 5 June — 29 July, 1950 [14], in so far as the areas in question are not described as continental shelf, but as (p. 22): "submarine areas situated outside the territorial waters" (of a littoral State). The Report goes on: "... where the depth of the waters permitted exploitation it (the area) should not necessarily depend on the existence of a continental shelf". The heading of this part of the Report reads nevertheless: "The Continental Shelf".

Young [12] also suggests, p. 227, the term "submarine areas", which term was employed in the British-Venezuelan Treaty of February 26, 1942. Apart from the reason already mentioned above, (places where no edge exists) he gives as further reasons that this term would include "that landward portion of the continental shelf lying within territorial waters and subject to their regime", and "insular shelves".

SECTION 2. DELIMITATION OF THE CONTINENTAL SHELF

But apart from the name, there are other difficulties. The notion "continental shelf" has also been used in proclamations or decrees to circumscribe the extent of the claim, to delimit the area of the control and jurisdiction or of the sovereignty of the littoral State. In some cases, only the term "continental shelf" is used, and in the Decrees of Costa Rica and Honduras for example with the addition "at whatever depth it is found". Obviously the pure oceanographical or geological notion is meant. In the first Report of François to the International Law Commission [10], p. 40, he puts the question: "... how should this continental shelf be defined? Should an ocean depth of 200 metres (100 fathoms) be adopted as the outer limit?". The Report of the Committee on Coastal Waters and Appurtenant Subsoil for the 3rd Conference of the International Bar Associ-

---

[a] We have taken here the original French text because the English translation does not appear to us to be quite exact.

ation [15], p. 10, gives the following answer: "The definition of the continental shelf should express the geographical-geological conception of this formation. Although the end of the shelf .... generally appears to coincide with the 200 m. isobath this naturally is not an exact figure. The geographical-geological limit would therefore seem preferable". However, other solutions, which may be more practical have been suggested. In other proclamations the extent of the continental shelf was further specified by a maximum depth of 100 fathoms or elsewhere by 200 metres.

We have to go back to the geologists. The picture of the continental shelf, given in the beginning of this Chapter, was rather a simplified one. The real features of the continental shelf are not nearly as simple. Some geologists agree to an average depth of the edge of 200 metres, like Kossinna [16], p. 875, some to 100 fathoms. Apart from the fact, that the edge is not a sharp line between the shelf and the slope, but rather a rounded area in which the gradient increases more rapidly, several geologists mention quite different figures for its depth. Paul Weaver [17], p. 352, says: "In many areas the outer edge of the continental shelf is at a depth less than 200 meters and as shallow as 130 meters, but the slope outside the edge is so steep that this edge is horizontally very near the 200 meter contour". Carsey [18], p. 361–385,: ".... the shelf is a compound feature with a gentle slope to about the 50-fathom [a] line (91.45 metres), then a slight steepening to the 70-fathom depth (128.03 metres). Beginning at this depth of approximately 70 fathoms, the slope steepens markedly, but there is a slight further increase at the 100-fathom depth (182.9 metres) into the gulf basin. As many authors have used the 100-fathom depth in their definition for the continental shelf, this is the popular concept today. However, a close study of the detailed map by Veatch and Smith, which extends 530 miles along the east coast of the United States and is contoured in 5-fathom intervals, shows that the break is near the 60- to

---

[a] One has to be careful with measures. There are five different fathoms. We have taken here the English fathom: 1.829 metres. However there are Danish and Norwegian fathoms (favn): 1.883 metres; a Swedish fathom (famn): 1.781 metres; a Spanish fathom (braza): 1.672 metres and a Dutch fathom (vadem or vaam): 1.699 metres. The last one is no longer used. Morales [19], p. 145, gives for "cien brazas" (100 fathoms) 188.88 metres. He must have used still another fathom.

80-fathom depth (109.74 to 146.32 metres). A very detailed account on this break in various parts of the world is given by Shepard in his Submarine Geology (1948) and although he found some variance, the average depth at which the greatest change in slope occurs is 72 fathoms" (131.688 metres) [a]. Kuenen [20], p. 76,: "The continental terrace underlies the shallow shelf with its slight seaward inclination and the continental slope, starting at about 130 meters below sea-level and continuing to the depths of a few thousand meters where it merges imperceptibly in the ocean floor", and his book [5], p. 154,: ".... the depth at the edge averages 130 m. but may attain 500 m.". Krümmel [1], p. 105, mentions 500 metres deep edges of the Barents-Norwegian-Labrador- and Arctic shelves. Bourcart [21], p. 126,: "Il n'est donc pas facile de préciser le lieu exact du changement de pente. Pour G. de Joly par exemple, le Plateau continental, sur la côte française du Golfe de Gascogne, serait limité à la cote- 300...

Dans certains cas, il est difficile de trouver le rebord (edge). Le très large plateau continental, qui borde l'Irlande au Sud-ouest, ne se termine pas brutalement par une fosse, mais il se prolonge vers le banc du Porcupine, dont les pentes sont très douces et convexes". Finally we quote from the Memorandum on the Regime of the High Seas, prepared by the Secretariat of the United Nations [22], p. 49: "This emergence in law of a concept derived from other sciences raises a delicate question of definition. It cannot be denied that the definition of the seaward limit of the continental shelf as being the 200 metre bathymetric contour line (which is the line roughly corresponding to a depth of 100 fathoms, i.e. 182 metres 90 centimetres) is not geographically satisfactory. As has been stated by Mr. Emm. de Martonne, the 200 isobath is merely 'a rough approximation of the idea of the continental shelf'. (Géographie Physique, p. 252) .... To define the seaward limit of the continental shelf satisfactorily from the geographical and oceanographical standpoint, the depth line of the rim should be used; but that varies quite considerably", and, on p. 50, quoting Bourcart, who "seeking

---

[a] The figures in metres, between brackets, have been inserted by the author. We apologize for the mixture in spelling of the word metre and meter. Metre is the official spelling of the Oxford Dictionary, but we left the word meter, in quotations, where this (American) spelling was used.

the 'real' definition of the continental shelf states (p. 130): 'The only accurate method of defining the continental shelf is to consider it as lying between the shore and the first substantial fall-off -on the seaward side- *whatever its depth*' (underlining by Secretariat) ..." and finally on p. 102, the memorandum suggests "... Generally speaking, the current concept of the continental shelf as a plateau with a fall-off at about 200 metres, although open to criticism geographically, should be accepted for legal purposes". Young [12], p. 234, proposes to select .... an appropriate standard for fixing a "legal edge".

If depth is to be a criterion [a] for the delimitation of the continental shelf, and there is far more reason to do so than to use it for fixing the width of territorial waters as was suggested by Valin [23], p. 38 [b], we have to chose from the range of figures supplied by the geologists. If the choice of a "legal edge-depth" would fall, for instance, on the 200 metre-, or 100-fathom line as has been suggested by Young [12], p. 235; :"even when the shelf in fact terminates at a lesser depth", we shall have to decide which of the two measures is preferable. The difference being 17.1 metres may not seem important, the more so as the average depth of the real edge is smaller than 200 metres (i.e. 130 m) so that this difference on a steep slope will not amount too much in a horizontal direction. If the depth of the edge is greater than 200 metres, a difference of 17.1 metres in depth would, measured horizontally on the gently sloping shelf, be of more importance. However, as most hydrographic charts are still giving soundings in fathoms, the 100-fathom line would be preferable (that is to say using fathoms of 1.829 metres!). For practical reasons this figure seems acceptable. The 100-

---

[a] Other examples where depth is used as a criterion are the regulations relating to sedentary fisheries, which will be discussed later.

[b] He suggested that the sea up to the point at which the seabottom ceased to be reached by a sounding line pertained to the adjoining coast. If we realise that *wire*-soundings have been performed of over 10,000 metres, (Mindanao Trough) which of course Valin could not know, we see that his criterion would nowadays mean that all coastal States could claim all the oceans as territorial waters, so that the seas would become condominium. (We have mentioned wiresoundings to be fair to Valin, echo-soundings would even give higher figures). The text (Liv. V, Tit. I) reads: "... il auroit été mieux peut-être de juger du domaine sur la mer voisine de la côte, par la sonde, & d'en assigner les bornes précisément à l'endroit où la sonde auroit cessé de prendre fonds; de manière que hors la portée de la sonde, la mer eût été reconnue libre, pour la navigation & la pêche, comme ne pouvant être du domaine de personne".

fathom line is indicated on most sea-charts. If the real edge is less deep, then it will be horizontally near the 100-fathom line, because of the steepness of the slope. In the case that the real edge is deeper, not much harm is done either, because the depth at which it is technically possible to work is, for the time being, not nearly as great as 100 fathoms.

The legal edge will not always be parallel to the real edge. As the real edge varies in depth, it may dip, as it actually does, near the "Franklin shore" on the Atlantic shelf of the United States from East of Cape Henry to East of the Atlantic City from 60 to 80 fathoms. (Umbgrove [2], p. 113). Theoretically we have to be prepared for a legal edge even crossing the real one if the latter, for instance, dips from 90 fathoms to 110 fathoms [a].

SECTION 3. THE SURFACE OF THE CONTINENTAL SHELF

The next difficulty supplied by the geologists is the fact that the surface of the continental shelf is not so smooth as the ideal picture might have made us believe.

In the first place we mention the submarine canyons (Kuenen [5], p. 485 et seq. and Umbgrove [2], p. 120 et seq.). Kuenen describes this geomorphological feature as follows, p. 487,: "A typical canyon starts as a steep, narrow gorge cutting across the continental shelf for a few dozen kilometers and running straight down the continental slope to great depths. At the edge of the shelf the bottom may lie many hundreds to a thousand meters below the adjoining sea-floor .... The transverse section is V-shaped, and in ground plan a moderately sinuous course is followed .... Tributaries, generally heading well into the shelf or beginning at the top of the continental slope, come in, forming a dendritic pattern". Umbgrove, p. 122, distinguishes 3 types: "1) As a first group submarine gorges originating near the edge of the shelf and running downward to great depth have to be mentioned. In many places they were found crowded together in great numbers .... They are cut back only a short distance

---

[a] This would not be so extraordinary. It means that an artificial borderline would be fixed independent of a natural borderline. Something comparable we find in the Treaty of 17 May, 1939, (Netherlands Statute book No. 30) between the Netherlands and Germany, whereby a borderline between coal deposits was recognized independently from the (artificial) frontiers between the Netherlands and Germany.

into the platform of the shelf, their headward extensions being seldom more than 5 to 10 miles"; p. 123: "2) A few gorges of the type mentioned just now extend across the continental platform, their much shallower headward extensions reaching the vicinity of the shore near or at the debouchement of a large river" (Congo, Indus, Hudson, etc.), and p. 124: "The Hudson canyon is cut 1,120 meters below the rim near the edge of the shelf". Finally, p. 125: "3) Another type of submarine canyon was revealed by sonic soundings off the coast of California. They are characterized by a dendritic river-like pattern, deeply incised in the surface of the continental shelf and thence continuing towards the great depth of the Pacific Ocean". These quotations will suffice to show that the shelf edge is notched in many places and the shelf surface may be furrowed especially near the edge.

Apart from these furrows, some shelves have an irregular surface. Kuenen [5] (quoting Shepard), p. 154,: "Hills of 20 m(etres) or more occur on 60% of the profiles, hollows of 20 m(etres) or more on 35% .... Off formerly glaciated coasts the shelf is wide and deep with a very irregular surface, due to glacial troughs and morainic deposits .... In areas of active coral growth the shelves are shallow and strewn with irregular shoals and banks".

Carsey [18] speaks of "the irregularities in the shelf depth including in particular such extreme depths as 100–400 fathoms noted on the shelf in areas formerly covered by glaciers. The weight of the ice could have depressed these areas; also the glaciers probably scoured much material from them". Carsey also "counted 164 topographic features which occur along the shelff off the coast of Texas and Louisiana. These topographic features continue southward along the shelf off the east coast of Mexico. Because of the great number of salt domes in the Louisiana-Texas land area, one may speculate that the topographic domes on the continental shelf may easily be underlain by salt plugs. Two clusters of topographic features have a relief of 200–300 feet above the ocean-floor".

These facts will make it difficult in some cases to delimit the continental shelf and ways have to be found to overcome these difficulties.

Young [12], p. 235, suggests: "A possible boundary line should not be regarded as discontinuous merely because it may be interrupted by submarine canyons, running out from land. On a principle somewhat analogous to the headland theory for bays, such canyons should be spanned by straight lines connecting the 100-fathom contours. By the same analogy, the permissible length of such lines might be limited to that applied by the coastal state to its bays".

If a principle of International Law is going to be established on the lines of the original idea that the coastal State can claim the natural resources of the soil and the subsoil of the continental shelf and subject them to control and jurisdiction, these claims should be limited to the depth where it is technically possible to extract these resources.

We have to distinguish here between mineral and biological resources. In the canyons, which are, as we have seen, on the edge very deep, extraction of mineral resources is, for the time being, out of the question. For this purpose canyons should not be included in the area in which these claims are applicable. However, if the continental shelf is going to be the area for exclusive rights of the littoral State on fisheries, or for conservation areas, the cutting of the edge by a canyon does not play a role. Later we will deal in detail with this question, in the Chapter on Fisheries.

Strictly speaking, we should not allow basins inside the continental shelf deeper than 100 fathoms to be included in the area for claims either. We could, however, agree with Young, loc.cit., where he suggests: "Isolated patches of limited size which are over 100 fathoms in depth should be disregarded and absorbed into the shelf area", because these basins will, as data which are available so far show, not exceed in considerable measure the depth of 100 fathoms, as the canyons do. Under "isolated patches", however, we have to understand, as Young does, those which do not connect with the ocean depths.

We believe that the condition that the resources should be workable is a reasonable one, which corresponds with the aims of the majority of the proclamations and decrees. Consequently, as will be further explained in the Chapter dealing with Offshore drilling technique, a proviso as suggested by the Committee in its Report to the International Law Association at Copenhagen in 1950 [24], p. 15, that it should be open to a coastal State to prove that its continental shelf, as a result of exceptional geological conditions, underlies greater depths, should not be accepted.

Young [12], p. 235, speaks of "depriving a State arbitrarily of so obvious a part of its continental shelf", if, in the case we are discussing, a claim to deeper areas of the continental shelf would not be recognized. It seems to us that a State cannot be deprived of something it has not got so far on any legal principle. The question is: What can, on a new principle of International Law, yet to be accepted generally, be recognized as a legal claim? Young reverses the question.

The criterion of technical possibilities has been put forward by Hudson in the International Law Commission in the 67th meeting [25], p. 7: "Control and jurisdiction over the sea-bed and subsoil of submarine areas outside the marginal sea may be exercised by a littoral State for the exploration and exploitation of the natural resources therein contained to the extent to which such exploitation is feasible". In Hudson's view, exploitation becomes impossible beyond a depth of 200 metres (p. 9). We find the criterion back in the Report of the International Law Commission [14], p. 22, as we have seen already: "Where the depth of the waters permitted exploitation" and in the Second Report of the Rapporteur François to the International Law Commission [26], p. 67: "The rights might be attributed as far as the line which, because of the depth of the sea overlying the shelf, constitutes the extreme limit of possible working".

We can close this discussion by suggesting that delimitation, in a legal sense, between neighbouring States bordering the same shelf, and between opposite States on the same inner shelf, should be left to the States concerned by way of bilateral treaties on the lines of the British-Venezuelan Treaty of 26th February, 1942, and on certain principles (see below), granted, we repeat, that the principle mentioned above is to be established in International Law, a question which we have to discuss presently.

SECTION 4. THE CONTINENTAL SLOPE

As we have given some attention to the condition of the practical exploitation possibilities, it may be useful to have a glance at the continental slope. As will be shown in the Chapter on Offshore drilling technique, the 100-fathom line is a very liberal limit of the possibilities to work. As a matter of fact no greater depth than 30 metres has yet been used to build constructions. (See Chapter on Offshore drilling technique). The 200 metre limit will therefore leave us a reasonable length of time to settle the problem internationally without the danger of being caught up by the drilling technique.

But, if the edge lies at a depth considerably less than 100 fathoms or 200 metres, for instance at the average depth of 130 metres, and the technique has developed so far, that drilling at a depth of 130 metres and more has become possible, would it be feasible to work on the slope between the 130- and the 200 metre line? Weaver [17], p. 353–355, writes: "There are two types of continental slope. In one type the continental slope has a single steep slope .... In the other it is compound .... so that it consists of a steeper upper segment and a gentler lower one". The last type is far from smooth: ".... Where soundings are numerous, the slope can be seen to have much local relief" or, as he says a few lines further: "with compounding and local roughness". He also draws attention to the relation of the land profile, p. 353: "The narrower the continental shelf, the steeper both the continental slope off it seaward, and the landrise inland from it". He gives three profiles in the Gulf of Mexico to demonstrate the difference. The lower profile has a gradient of 1 : 21, the two upper profiles 1 : 90. Kuenen [5], p. 155, quoting Shepard: "The average slope is $4\frac{1}{4}°$ for the first 2,000 meters. Off large delta's the slopes are gentle and smooth, averaging $1\frac{1}{3}°$ to 2,000 meters, but with numerous irregularities. Off fault coasts the average slope is $5\frac{2}{3}°$. Along the Pacific coasts the slopes are steeper than the average ....".

Table 8 on p. 106 gives for the average slope of the shelf 1 : 540 and for the average slope of the continental slope 1 : 15. Carsey [18] compares the two figures of the gradient of the shelf and that of the continental slope: "The order of magnitude of the average gradient on top of the shelf is approximately 1 to 400 while the slope beyond the brink of the continent may be as much as 1 to 9". Umbgrove [2], p. 98, gives the following figures for the outer slope of the Atlantic shelf of the United States: "In a few sections the slope is as steep as $7\frac{1}{2}°$ (about 13%) or 700 feet per statute mile".

Another phenomenon characteristic for the continental slope has to be mentioned, namely the occurrence of slumping of the sediments.

Kuenen [5], p. 241, mentions: "The sudden slumping of larger masses which one might term submarine landslides. Both rock falls and mud slumps are known to occur. A slow slipping en masse of unconsolidated sediment can also be imagined, more akin to the creep on valley slopes" and on p. 244: "In deeper water, beyond the 200 meter depth line, the material can drop to the bottom undisturbed, and where a marked declivity occurs, the deposits are originally laid down with a considerable slope. Here, however, the fine composition and high percentage of clay, coupled with the considerable water content must render the muds highly mobile" and, p. 500: "The sliding of sedimentary deposits on steep submarine slopes is proved by fossil evidence, by observations in the Californian Canyon heads and also by dredgings and soundings showing hard bottom in regions where deposition takes place on adjoining flatter portions of the sea-floor" [a].

When we remember the frequent occurrence of canyons on the continental slopes and think of the steepness of some slopes and the rough profile of the ones which have a lesser gradient but are still much steeper than the shelf, and finally the occurrence of slumping, we realize that the continental slope does not offer ideal conditions to build installations and we believe that it will last a long time before the technique will have developed so much as to be able to tackle these difficulties, if ever . . . .

One way to meet the supposed situation, mentioned in the beginning of this section, namely, the exploitation of resources underneath the continental slope between the 130- and 200 metre

---

[a] See for canyons and submarine landslides the recent article by Shepard [27], p. 7–9.

lines is the technique of directional drilling, i.e. not vertically, but at an angle. If, we repeat *if*, the technique has developed so far that building a drilling platform in as much as 130 metres of water has become possible, then it would become feasible to extract oil from underneath the slope. One may ask, would not the same solution be possible to exploit the slope beyond the 200 metre line if the edge lies at 200 metres? The answer would be in the affirmative, but it seems, as we have said already, for the time being out of the question to build installations in 200 metres of water. But, when in the future that has been attained, the lawyers of that time may further develop the system which will then have worked, we hope, satisfactorily for a long time. Besides, if a legal system is adopted whereby installations may be built on the shelf to a depth of 200 metres, directional drilling from such an installation underneath the continental slope, would not interfere with navigation and nobody would care, like nobody cares about coal-mines under the high seas.

SECTION 5. THE WIDTH OF THE CONTINENTAL SHELF

A glance at the chart has shown us immediately the vast differences in width of the continental shelf in different areas. Kuenen [5], p. 105: "The breadth of the shelf varies from less than 1 to more than 1,000 km." (kilometres). Carsey [18] says that the width may vary from 0 to 800 miles (about 1,480 kilometres). The most extensive shelf on earth (says Kuenen [5], p. 105, who, as we have said above, does not make a distinction between inner and outer shelves) "is that connecting Java, Sumatra, Malacca and Borneo. It has an area of 2 million km². Other wide shelves are found south of the Behring Strait, north of Siberia, between Australia and New Guinea, to the East of the Argentine, off Korea and in the North Sea, all of about 1 million km²". Evans [28], p. 80: "It (the shelf) is generally narrow along the west sides of both North and South America, but on the opposite side of the Pacific along the Asiatic coast, it is the widest in the world, reaching 750 to 800 miles in the Yellow Sea and in the Gulf of Siam". And finally, Umbgrove [2], p. 104: "Along the coast of Angola and off the Ivory coast the isobath of 1,500 to 2,000 meters approaches the coast so nearly that no shelf at all is present".

This wide divergence in width has given rise to some remarkable thoughts. Quoting in chronological sequence, François, in his first Report to the International Law Commission [10], p. 39–40: "It must now be considered whether the grant of special rights in respect of the mineral resources of the subsoil as well as marine resources should be made conditional upon the existence of a continental shelf. It is undeniable that this will lead to discrimination against those countries which have no continental shelf or whose continental shelf does not stretch beyond the limits of their territorial waters", and a few lines further on: ".... it must be borne in mind that the adoption of a depth line of 100 fathoms as the outer limit of the continental shelf is likely to allot to the various States portions of the high seas varying greatly in extent. This would establish an unjustifiable inequality between States".

We do agree that it would be a discrimination against countries which have no continental shelf in the strict sense of the word, for instance those situated on the Persian Gulf, to adhere to the notion "continental shelf" stricto sensu [a]). We have, however, some doubts whether this sort of discrimination is meant, or that the Rapporteur had in mind the differences in width we have described above. Besides, it seems a rather strong wording to speak about allotting ("attribuer" in the French text) portions of the high sea. This expression could, in our opinion, only be used in the cases like Chile and Peru, who claim a part of the ocean. Most other countries, however, only claim certain rights concerning the mineral or marine resources, or on the shelf itself.

However, let us see the opinion of other lawyers. Brierly said in the 67th meeting of the International Law Commission [25], p. 20: "that if the Commission was of the opinion that the right of control and jurisdiction depended on the presence of the continental shelf, it was committing an injustice towards certain countries, such as Chile, that possessed no continental shelf". Here is no doubt possible. Brierly obviously refers here to the geo-

---

[a] Compare the opinion of Hudson in the Summary Record of the 66th meeting of the International Law Commission [7], p. 24,: "He felt that lawyers had no right to prevent the exploitation of the resources for the benefit of mankind".

graphical difference in width of continental shelves. He seems, however, to have corrected his views, when he explains his thoughts,
p. 21: "Certain countries not possessing a continental shelf might
desire to exploit the marine subsoil and the number of such
countries was fairly large. If, for technical reasons, they were
unable to exploit that subsoil there was nothing to be done about
it. On the other hand, there were regions, where such exploitation
was feasible and should be permitted to coastal States. Those
areas required definition but not necessarily one based on the
presence of a continental shelf". Here he refers obviously to the
countries on the Persian Gulf and he uses the word continental
shelf in the strict sense. Sandström (p. 22): "wondered why Sweden, for example, a country without a continental shelf, should
not have the right of exploring and exploiting the subsoil of the
Baltic ...". He also uses the term in the strict sense.

In the same meeting Amado, Summary Record [25], p. 26, draws
the attention of the Rapporteur to a text submitted to the International Law Association by Mr. Govare c.s. The passage in
question reads as follows: "In the event of the break in the continental shelf occurring at a distance less than 20 sea-miles from
the coast, sovereignty, together with control and the exclusive
right of exploitation would be prolonged to 20 sea-miles from that
coast". Amado had quoted the passage "because it could be
applied to countries without a continental shelf and because the
rights of such States would be unlimited unless a limit were
prescribed to them".

In the Report of the Commission of which Govare was Char-
man [29], it is said on the 7th page that the word "souveraineté"
should not be translated in English by "sovereignty" but by
"autorité et compétence comportant le droit de concession et de
contrôle", which we may translate by the term "jurisdiction and
supervision". Besides, in the translation read by Amado, it
seems that the French word "contrôle" has been translated
wrongly by "control". The translation should be "supervision".
However, what did Govare c.s. mean with his suggestion? Govare
seems also to be worried by the unequal distribution of continental shelves in the world, 7th page: "La nature des masses conti-
nentales, dont la structure est loin d'être identique pour toutes les
parties du globe, est fort différente d'un continent à l'autre et

suivant les faces envisagées d'un même continent. Il en résulte que l'attribution de la souveraineté sur le sous-sol marin et le contrôle de l'exploitation de la mer territoriale devraient être régis par deux considérations et non pas par une seule norme". It is not clear how the territorial sea comes into the picture. However he goes on explaining that France "rests" on a large continental shelf, on the Atlantic coast, which widens towards the north, England resting also on this shelf. In the Mediterranean the situation is different. Whereas in the Gulf of Lion the isobath of 200 metres is at an average distance of 50 km. off the coast, it is quite different more to the East, where depths of 2,000 metres are found very near the coast of the "Massif des Maures". He then continues: "En conséquence, et afin de préserver les droits légitimes de tous les Etats, quelles que soient leurs positions eu égard à l'étendue de leur plateau continental, il convient d'établir le double principe ci-après. Premièrement: La souveraineté de l'Etat riverain doit être prolongée sur le plateau continental avec le droit de contrôle et l'exclusivité d'exploitation de son sous-sol marin. Deuxièmement: ...." follows the passage which Amado quoted. He ends by saying: "Cette limite est largement suffisante pour couvrir toutes les possibilités d'exploitation qui se présenteront, même dans un avenir assez éloigné, dans le cas où le plateau continental n'existe pas ou ne s'étend pas loin des côtes".

As he gives also a suggestion to deviate from these principles by bilateral agreements between States, in case the distance between their coasts is less than 40 miles, and gives as an example the Channel and the "Pas de Calais", where France and England should agree on a demarcation line, he obviously halts between two opinions in this second rule, namely the non-existence of a continental shelf in the sense of the North Sea, where no edge exists, and the non-existence in the case where the coast slopes steeply into the ocean. It is difficult to understand why it should not be allowed to divide the Channel in places where it is wider than 40 miles and where the depth is still less than 100 fathoms or 200 metres.

Continuing with the 67th Summary Record [25], p. 27, Brierly suggested the following formula: "When a continental shelf exists, its limit is also the limit of the zone in which the State has the right to explore and exploit the marine subsoil". François and Cordova "thought that account should be taken of the fact that in certain cases the continental shelf was very narrow. Mr. Brierly's formula might, accordingly, lead to an inequality in favour of States possessing no continental shelf". Here the last speakers must have thought that Brierly meant a continental shelf stricto sensu, and that his proposal would therefore not supply a limit for countries on the Gulf of Persia, where no continental shelf stricto sensu existed. These countries would be in an advantageous position without having a real continental shelf compared with those who had a real one, albeit a narrow one.

But Spiropoulos definitely referred to the difference in width when he expressed his thought (same page) "That if it adopted Mr. Brierly's proposal, the Commission might also commit an injustice towards States possessing perhaps only 20 metres of continental shelf, whereas others had a shelf extending for 200 kilometres".

The same thought has been expressed in the Memorandum on the Regime of the High Seas [22], p. 105, discussing the differences in width: "Consideration will probably have to be given to the question whether such inequalities should not be rectified by providing States without a continental shelf with special contiguous zones. Conversely, where the continental shelf is very extensive, it may be considered advisable not to grant special rights to the littoral State over its continental shelf beyond a certain distance from the coast. The curtailment of the area in question might be effected either by a given bathymetric curve or by deciding upon a horizontal distance from the shore".

What becomes clear from these discussions is, that some speakers based their misgivings about inequality on the fact, that some countries have no continental shelf stricto sensu, for instance those situated on the Persian Gulf. They very rightly thought that it would be most unfair and ridiculous to exclude those countries from the legal regulations to be proposed. The solution for their worries has already been suggested above: i.e. to comprise under the "legal" notion of the continental shelf also the "inner shelves", the submarine areas covered by water not exceeding 100 fathoms or 200 metres depth.

But this "having no shelf" thought has been mixed in the minds of some speakers with the fact that some countries have neither an outer nor an inner shelf, their coast just going straight down into the ocean, so that the 100-fathom line is very near their shores. Here the difference in width plays a role.

Although it is perfectly understandable that feelings of justice moved these speakers, one could doubt whether we ought to carry our justice so far. In the first place François suggests an answer in saying in his Report [10], p. 40: " .... it could perhaps be argued that where there is no continental shelf the grant of the rights in question would not be of great practical value". We are also inclined to agree with the remarks made in this respect in the Report of the Committee on coastal waters etc. of the International-al Bar Association [15], p. 8: "This leads to a question of justicia distributiva. It may be hard on Switzerland that it has no sea-coast to base a fishing industry upon, as well as it may be hard on Holland that it has no Alps to boost its tourist trade; do the lack of such natural assets in itself justify the consideration of compensatory measures and even if one should feel differently would in the case under discussion the allotment of a few miles of deep sea be of any actual use to the coastal nation that has no continental shelf?". In an even less high-brow way we feel tempt-ed to say, that it is not the question of dividing a cake amongst hungry children, who are watching jealously whether the portions meeted out are exactly equal and of compensating one or two who missed the cherry with an extra almond. It is purely a question of geography, as Brierly said in the 66th meeting [7], p. 29, whilst Cordova thought, p. 67 of the Summary Record of the 67th meeting [25], "that any injustice there was, was at the most a geographical one". The idea is actually a sort of universally applied socialism: why should the wealth and resources of mother earth belong to a few, in casu, a few countries? The idea is old. Pufendorf [30] discussed it in connection with fisheries, p. 602: "Car quoi que pour l'ordinaire il y aît une plus grande quantité de Poissons dans la Mer, que dans les Rivières ou dans les Lacs; cependant, s'il étoit permis à tous les Peuples de venir pêcher sur les Côtes d'un Païs, cela diminueroit un peu la pêche et le profit des Habitans; d'autant plus qu'il y a certaines sortes de Poissons, ou de choses précieuses, comme les Perles, le Corail, l'Ambre, qui ne croissent qu'en seul endroit de la Mer, quelquefois même d'assez petite étendue. Pourquoi est-ce donc que les habitans d'un

Côte ne pourroient pas se prévaloir à l'exclusion des autres, de la fécondité ou des rares productions de la Mer voisine? Certainement on n'a pas plus de raison de s'en fâcher ou d'envier un tel avantage à ceux qui en veulent jouir seuls, que de se plaindre de ce que tout ne croît pas dans tout Païs". Whilst Pufendorf was opposed to the idea, we find in Sarpi [a] quoted by Fulton [31], p. 547, an advocate of this thought applied to the territorial sea: " .... the extent of territorial seas should not be fixed everywhere in an absolute manner but should be made proportional to the requirements of the adjoining state, without violating the just rights of other peoples. Thus a country or city which possessed large and fertile territories that provided adequate subsistance for the inhabitants, would have little need of the fisheries in the neighbouring sea, while one with small territories that drew a large part of its subsistance from the sea ought to have a much greater extent of sea for its exclusive use".

A modern supporter of the idea we find in Azcárraga [32]. First of all he draws attention, p. 93, to the ideas of Florio [b], who suggested a draft Convention in which the extension of territorial waters would depend on whether they formed part of the oceans, open seas or closed seas. Azcárraga himself, p. 97, in a footnote, suggests to fix the extent of the "aguas jurisdiccionales" (territorial waters), for each country separately, in such a way that the extent shall be directly proportional to the density of population (number per Km²) and to the length of the coast line (in Km), and inversely proportional to the surface of the national territory (in Km²). But all these writers referred to territorial waters or contiguous zones, which are artificial boundaries.

The continental shelf is a "fait accompli", a natural feature of the earth, which we have to take or to leave as it is.

If we are going to introduce a "legal edge" of the 100-fathom line, we do that in order to obtain a uniform and easily discernable limit, realizing that it is at the same time the outermost limit for exploitation possibilities for a long time to come.

But we cannot accept a minimum width of 20 miles like Govare [29] c.s. do, because it would give countries, where the 100-fathom

---

[a] Dominco del Mar Adriatico e sue Raggione per il jus belli della serenissima Republica de Venetia, Venezia 1686.
[b] Il Mare territoriale e la sua delimitazione, Milano, 1947.

isobath lays at a smaller distance off the coast, an area which would be impossible to be used for exploitation of mineral resources [a]. Neither can we agree, with Azcárraga, who suggests to grant to States rights ("control y soberanía") even if the shelf extends beyond his general suggestion of a margin for "jurisdictional waters" of 20 miles, to the isobath of 200 metres, except if this should be considered to place an excessive area under the jurisdiction of the coastal State, because of the extreme width of the shelf, in which case he suggests to limit these rights to the isobath of 100 metres (about 50 fathoms). This would mean that those countries would be deprived of a part of the shelf between the 100- and 200 metre isobaths which might be exploitable. His idea might only be considered, if the oil-technicians would tell us that building installations beyond the 100-metre line is out of the question for a long time to come. But even then, we would distanciate ourselves more from the natural features of the earth than if the 200-metre line is accepted.

All this, of course, is only applicable if we are going to grant rights to coastal States at all and only for the purpose of the exploitation of mineral resources, leaving the discussion of the marine resources to the Chapter dealing with Fisheries.

We may wind up this section by saying that the difference in width of the continental shelf was not the reason why the International Law Commission decided to state in its Report [14], p. 22: "that it would be unjust to countries having no continental shelf, if the granting of the right in question were made dependent on the existence of such a shelf". First of all because the expression "submarine areas" used in the Report was suggested by Hudson [25], p. 7, on the consideration, that "the principle of shallow waters was broader than that of the continental shelf". This suggestion, being an answer to the question, put by François in his Report [10], p. 40: "Should recognition of special rights as regards the working of the marine subsoil and the protection of marine resources be linked with the presence of a continental shelf?". Hudson explained, p. 8, that "a continental shelf was

---

[a] In the same sense François [26], p. 68: "for there is no need to attribute control and jurisdiction over the subsoil at a depth of over 200 m., where working is impossible".

not always present; in the Persian Gulf, for example". Hudson's proposal was accepted (p. 17).

In the second place, our statement is proved by the explanation given by François [26] in his Second Report, p. 67, where he repeats the words from the Commission Report, quoted above, adding after "shelf" the words: "in the geological sense" (that is what we called the continental shelf stricto sensu) and, continuing: "thus excluding in certain cases the shallow water offshore which none the less allows the working of the subsoil. If the term continental shelf were taken to include the shallow water which the International Law Commission has in mind, there would be no objection to the use of the term continental shelf without qualification".

This explanation excludes any allusion to an absence of a continental shelf in the sense of an absence of shallow waters, namely if the littoral sea-bed is so steep that practically no shelf exists. The misunderstandings about this point were, however, not yet cleared away in the bosom of the Commission.

A year later in its 113th meeting [33], (p. 13 et seq.) the Commission discussed the first article of the conclusions of François in his Second Report [26], p. 69, reading:

"1. The continental shelf is constituted legally by the bed and the subsoil of the submarine regions situated off the coast where the depth of the water does not exceed 200 metres".

Yepes referring to Chile, asked whether a State in that position would have a right to compensation because it did not possess a continental shelf. When the discussion of the proposed articles of François was finished in the 117th meeting [34], p. 7, Yepes proposed to complete the draft with a new article reading: "The rights of control and jurisdiction referred to in the present Chapter belong, up to a distance of twenty miles beyond the coast, to all coastal States, even if they do not possess a continental shelf in the geological sense". He referred to a (provisional) formula adopted the previous year: "The area of such control and jurisdiction will need definition but it needs not depend on the existance of a continental shelf". He thought that meant that the Commission had undertaken to "do something for States that did not possess a continental shelf in the geological sense", and explained that (p. 8) he had Chile and Peru in mind.

The confusion is obvious. The Commission adopted this formula for the sake of countries, for instance on the Persian Gulf, as we have explained. Therefore the Commission inserted in the formula the words: "Where the depth of the waters permitted exploitation" in its Report [14]. This formula, was not meant for countries like Chile, where offshore drilling is impossible.

A repetition of many arguments of the previous year followed and Brierly (p. 11) explained that the preceding year, ..." it had by no means been his intention to suggest that unfavourable geological conditions, imposed on States by nature, could be remedied by legal enactments. It was completely useless to grant rights of control and jurisdiction to Chile, for instance, which had neither a continental shelf nor shallow waters off its coast".

Eventually by a 6–5 majority the following text was adopted, (S.R. 117 [34], p. 15):

"The rights of control and jurisdiction referred to in the present Chapter belong, up to a distance of 20 miles beyond territorial waters, to all the coastal States which do not possess a continental shelf as defined in article 1".

The reason why some members were anxious to give rights to States having even no shallow waters, was explained by Cordova, who said (p. 16), that a right should be given to these countries to exploit the sea-bed (he meant the subsoil) from the mainland.

It is difficult to understand that this argument was allowed to play a role in discussions about the continental shelf, because such exploitation is completely outside the problem. Working from the mainland with mines or directional drilling does not interfere with navigation, nor with fisheries. It is not new, Chile, Nova-Scotia and England have mines under the sea, and this fact has never been a point of discussion or conflicts.

However, a subcommittee was set up to study the matter, (p. 19).

In the 123rd meeting [35], p. 3, François reported on the results. The subcommittee had unanimously approved to change the words "where the depth of the water does not exceed 200 metres" into "where the depth is such as to permit the exploitation of the natural resources of the sea-bed and subsoil".

The subcommittee's draft was adopted (p. 4). We will come back on this decision later.

Finally the divergence in width excludes any attempt to delimit the continental shelf in terms of distance off the coast. In the Draft Articles on the continental shelf in the Report of the International Law Commission, covering its third session [36], p. 55, under (7) we read: "The Commission considered the possibility of fixing both minimum and maximum limits for the continental shelf in terms of distance from the coast. It could find no practical need for either ....".

SECTION 6. HISTORY AND RESOURCES OF THE CONTINENTAL SHELF

About the history we can be very brief. For the purpose of this study mainly the geomorphological facts are important. Geologists will tell us about the changes which have taken place in the past and which may take place in the future. Their clock, however, is considerably slower than that of a lawyer. Changes in depth, for instance, are generally so slow, that they should not worry us nor future generations for centuries to come. Another reason, that the history of the shelf can be dealt with in a brief manner, is the fact that very little is known about it. Carsey [18] says: "Much additional information will have to be accumulated and studied before the shelf origin is fully understood", and Kuenen [5], p. 169: "It is obvious that the problem of the shelf is still far from being definitely solved".

However, there are some points relating to the history and origin of the shelf on which geologists are fairly unanimous. The first is, that the shelf belongs to the continents. Krümmel [1], p. 104 writes: "The shelf belongs to the body of the continents proper, which find their real limits from the deep sea on its outer margin, the continental-edge, mentioned above". Umbgrove [2], p. 97,: "Almost everywhere the real edge of the continent is situated below sea-level". Pratt [37], p. 658: "The continental shelf may be looked upon as part of a larger earth feature, it is the outer or seaward portion of a great shelving plain which intervenes in the region of the margins of the continents between the continental heights and the oceanic depths". Murray, in the Science of the

Sea [38], p. 248: " .... the continental shelf .... is regarded as belonging to the continental areas". The Geological Nomenclator [39], p. 7, gives for Shelf or Continental Shelf, in the Dutch column (translated): "The submarine continuation of the continental area up to about the 100-fathom line". This vision on the nature of the continental shelf has been used by lawyers to apply the doctrine of contiguity or continuity or the principle of propinquity, which will be discussed later.

Already in 1803 de Rayneval [40], p. 161, argued on geological grounds: " .... le fond de la mer, le long des côtes, peut être considéré comme ayant fait partie du continent, et qu'il est pour cela considéré comme en faisant encore partie" [a]. In the Proclamation of President Truman we read in the fourth paragraph: " .... since the continental shelf may be regarded as an extension of the land-mass of the coastal nation and thus naturally appurtenant to it .... " and in the Mexican and Argentine Declarations we find words of the same tenor as in the Peruvian Decree of 1 August, 1947,: " .... the continental shelf forms a single morphological and geological unity with the continent". Yepes, who said in the 67th meeting of the International Law Commission [25], p. 4: "The first and most important was the rule of continuity, according to which the continental shelf was only the submarine continuation of the territory above water". And finally Smith [41], p. 8: "The shelf is nothing more than an extension of the continent into the sea. It may be argued that this fact alone gives a State paramount right to assert its claim to those resources. It would seem that the principle of propinquity is another way of stating that the geography of the location gives a State a paramount right to such resources".

In our opinion, it is wrong to apply uncritically a statement valid in one field of science to another one. In geology the fact that the shelf is covered by the sea does not influence the unity of the continent and the shelf; in law it does, because the legal status of the sea is quite different from the land. Therefore François said in his first Report [10], p. 39: "This argument would appear to be of dubious value from the legal point of view".

Another item of the history, which may be of some interest

---

[a] Liv. II, Chap. IX, paragraph 10.

to lawyers, is the "age" of the continental shelf. The fact, that in some governmental documents and in publications and discussions the continental shelf is referred to as "submerged lands", suggests that this term is used to strengthen the idea that it once belonged to the riparian State and that there is nothing strange in the fact that a State claims control and jurisdiction over the resources in the shelf, or even sovereignty over the shelf. As example we mention the Press release accompanying the Proclamation of President Truman of 28 September, 1945 [6]: "Valuable deposits of minerals other than oil may also be expected to be found in these submerged areas"; and the Petroleum Act of the Philippines of 1949 (see Chapter IV) art. III: "All natural deposits .... whether found in or under the surface of .... lakes, or other submerged lands within the territorial waters or on the continental shelf ...." [a]

Geologists, in so far as they believe the shelf, or at least part of it, to have been once dry land, do agree that this has been during the last glacial period. The flooding of the shelf began, if we may choose from a vast diverging variety of figures given by scientists who indulged in this indiscreet age question, between 10,000 and 12,000 years ago, at the end of the Pleistocene [b].

Human beings have been witnesses of these glacial and interglacial periods, as has been proved, for instance, by pictures they made of mammouths, an animal characteristic of those ages. Krümmel [1], p. 109, speaking of the North Sea shelf, says: "After the regression of the inland-ice the southern North Sea land was not only colonized by big now extinct diluvial mammals, but also by palaeolithic men, who hunted these animals. Steam

---

[a] *In the Preamble of the Brazilian Decree No. 28, 840 (see Chapter IV) we read:* "*Whereas the continental shelf contiguous to continents and islands and extending beneath the high seas is in reality submerged territory* ....".

[b] This applies only for the part of the shelf, which can be explained by post glacial rise of the sea-level.

The rest has to be explained by subsidence of a much more remote age, see Umbgrove [4], p. 249, quoted above, and Shepard [27], p. 7–8, who grounds his theory on the submarine canyons having an extraordinary resemblance to adjacent stream-cut canyons on land; the fact that offshore drilling, has shown that „shallow water, or even terrestrial deposits, extend for thousands of feet below present sea-level", ...
that: "Rounded cobbles, indicative of wave action in shallow water, are found widespread on the banks and sea mounts outside the continental slopes", and that:

"Fossil evidence reported by R. S. Dietz, Maurice Ewing, and most recently by Roger Revelle indicates, that these banks were covered only with shallow water at a time that probably preceded the ice age by many millions of years ...." etc.

trawlers often catch in their trawls bones of mammoths, rhinoceroses, bisons ... from the bottom of the Dogger Bank ..."

Using the words "submerged lands" may have some psychological value, but the fact that these lands were drowned could hardly be adduced for the construction of any right or title. "Operations of nature" is one of the classical modes of losing State territory. Oppenheim-Lauterpacht [42], p. 529–530, writes: ".... a State may be diminished through the disappearance of land and other operations of nature .... if an island near the shore disappears through volcanic action, the extent of the maritime territorial belt of the respective littoral State is thereafter to be measured from the low water mark of the shore of the continent, instead of from the shore of the former island". Visser [43], p. 90, came to the same conclusion: "Often parts of the sea along the coast were originally land and were only in the course of time covered by water .... As soon as a piece of the earth is covered by water and hence the space of the sea is enlarged, it becomes part of the high seas and is subject to its legal status. The former situation is then irrelevant".

Nor does the possibility that these lands were once inhabited add any strength to an attempt to derive a title on these grounds. An attempt to apply the principle of "prescription" would likewise fail, because there was "dereliction" by the old inhabitants and therefore there was not "continuous possession".

We do not say that the littoral State could not claim any right or title here, we only say that it certainly cannot use this argument.

It is remarkable that also a comparison with the opposite principle, namely accretion, as a mode to acquire land, has been suggested as an argument in favour of the recognition of the continental shelf theory. Feith, in presenting his Report to the Conference of the International Law Association in Brussels in 1948 [44], said, p. 174:

"It is a customary rule of the law of nations that any enlargement of territory created through new formations takes place ipso facto by the accretion without the State concerned taking any special step for the purpose of extending its sovereignty". He then quotes Oppenheim [42] (op. cit. p. 516): "But every

State may construct such artificial formations as far into the
sea beyond the low-water mark as it likes, and thereby gain
considerably in land and also in territory since the maritime
belt .... is now to be measured from the extended shore".
Feith then continues to say that from an International Law
stand-point there are no objections "to impolder .... the whole
of the adjacent continental shelf" by the coastal State. And
finally: "The continental shelf theory appears to dispense with
a great deal of work in connection with the making of dykes ....
by simply recognizing, once and for all, that the coastal State
can without taking any special steps assert its rights to the
continental shelf bordering on such coastal State". But what
he mentions is an "artificial formation". It would have been
nearer to the ideas of some geologists, if he had given as an
example the natural accretion, produced through operation of
nature: alluvion. Oppenheim [42], op. cit. p. 516: ".... if the
alluvion takes place on the shore, the extent of the territorial
maritime belt is now to be measured, from the extended shore"
or similar to alluvion: a delta "is to be considered an accretion
to the territory of the State to which the mouth of the river
belongs, although the delta may be formed outside the territorial
maritime belt". Keeping in mind, as Feith did, that it is only
meant as a comparison, because in our case, there *is no accession
of land*, we are thinking here of certain theories concerning the
history and origin of the shelves. Kuenen [20], p. 76, mentions
a few theories: "Others have put forward the view that the
terrace is essentially a wave built embankment, which has been
gradually prograded towards the deep sea by the dumping
of the continental detritus on the continental slope" and "more
recently subsidence and concomitant sedimentation have been
suggested as the main processes in operation (Ewing et al.,
Umbgrove)" or in his book [5] op. cit., p. 158: "Some authors
have described the terrace to outbuilding from the continents
in the manner of a delta front". As the continental detritus and
part of the sediments are of continental (terrigenous) origin,
either having been cut out of the mass of the continental blocks
by wave and current actions or carried into the sea by rivers,
or wind, some authors thought that this would entitle the coastal
State to claim its material (Ruelas [45], p. 130).

Thought associations lead to analogies, but we have to be very careful with analogies in law! We suggest to dismiss these thoughts on the arguments that firstly the accretion theory, where alluvion or delta's are concerned are built on the premise, that land emerges from the sea, which is not the case here, and secondly that these ideas start from theories of geologists, about the history of the shelves, which are controversial and not generally accepted. After an enumeration of 5 different theories, a few of which we mentioned, Kuenen [20] remarks in 1950, p. 76: "Data upon which to choose between these or other possibilities are as yet very meagre".

Even more far-fetched seem to us the legal "associations", which are mentioned in the Memorandum of the Regime of the High Seas [22], p. 67, where the abrasion (marine erosion) theory of some geologists makes the author think of "restitutio in integrum". We have met this principle in cases between States concerning the measure of damages, but never in a case between a State and the Ocean. Apart from this, the abrasion theory again, is a very controversial one. The same principle is suggested on p. 68 and 70 in the case of the transgression-theory, which we have dealth with speaking about "submerged lands", and we are afraid that this construction would not help us out either, as long as scientists cannot produce a generally accepted and scientifically proved history of the continental shelf.

A question connected with the history is the composition of the shelf and the probabilities of the occurrence of oil.

Here the geologists have given us very promising predictions, which have partly already been shown in practice to have been justified. Lees [46], p. 108: "I believe that the continental shelves are the seaward extension of the continents and if the adjacent land area has favorable structure and stratigraphy similar conditions may be ascribed to the downwarped margin", and p. 109: "I am sure that in some sectors important oil reserves may be found, but I would stress that the value of such sectors depends on their position in the tectonic scheme and not on the sole fact of being a shelf".

Kuenen [5], p. 168: "The chances of finding great reserves of petroleum in the continental terraces are highly favorable".

Pratt [37], working on his hypothesis, already mentioned above, of the continental shelf forming part of a great shelving plane extending inland to 600 ft high and seaward to 600 ft (100 fathoms) deep gives us an optimistic outlook: p. 658: "Beneath the inland half of the plane are situated the natural reservoirs from which has come the bulk of all the petroleum the world has consumed in the past together with those which contain an even larger proportion of our proved petroleum reserves. Quoting Illing (Oil Weekly, July 15, 1946, p. 34 et seq.): ".... the mother rocks from which the oil is produced are formed in environments which occur quite commonly and particularly in a cycle of marine deposition on the continental shelf". Pratt then continues: "the organic life of saline waters, which Van der Gracht [a] and others generally believe to be the source material of petroleum is concentrated in that relative small fraction of these waters, which overlaps the continental shelves". Pratt goes on, p. 663: ".... it is believed that plankton itself may be an important source material for petroleum". He then explains why the waters above the shelf are replete with marine life, but we will deal with this aspect in the Chapter on Fisheries. He ends, p. 670, in saying: "That normal earth processes make probable large volumes of petroleum beneath the continental shelf, is a comforting reflection to geologists who feel the responsibility of assuring the nation adequate supplies of liquid fuels" [b].

The Proclamation of President Truman refers to these and similar opinions in the second paragraph of the Preamble:

---

[a] W. A. J. M. Van Waterschoot van der Gracht, The Stratigraphical Distribution of Petroleum, The Science of Petroleum, Vol. I, Oxford, Univ. Press, London, 1938, (p. 58–62).

[b] Weaver [47], p. 398, mentions oil-seeps, which are indicated on a map of the Gulf of Mexico (Fig. 109 on p. 394) as .... "Oil indications other than wells. An area near the edge of the continental shelf, off Louisiana, has had many reports of floating oil, and even of 'oil bubbling on the surface .... as coming up in three jets'. The area seems to be centered on Latitude 27° 30' North, Longitude 91° 30' West, about 25 miles beyond the edge of the shelf. This well authenticated seepage seems added argument to the probability of oil fields across the continental shelf in this area. Beyond the edge of the continental shelf, off the sharp northwest corner of the Yucatan Peninsula, numerous showings of oil on the water have been reported, which, if truly from a submarine source, could be considered significant both as to the presence of oil, and as to thickness of sedimentary section, because some of the reports are from great depth of water. The steepness of the continental slope in the vicinity of these reported seepages suggests a faulted zone".

"Whereas its (the Government of the United States) competent experts are of the opinion that such resources underlie many parts of the continental shelf off the coasts of the United States of America, and that with modern technological progress their utilization is already practicable or will become so at an early date ....". In his letter of transmittal the Secretary of the Interior, Harold L. Ickes, in his Annual Report, 1945 [48], p. X, writes: "Experts in the geology of oil lands would not be surprised if we found 22 billion barrels of oil – more than we are sure that we have on the continent – beneath one small part of the shelf that reaches into the Gulf of Mexico. Geologists also think that the shelf will yield rutile, sulphur, ilemenite, chromite, monazite and other heavy minerals". According to the Report of the International Law Association, Brussels [44], p. 172, also uranium is expected to be found.

The best proof of the pudding, however, is in the eating. Weaver [47], p. 395, says: ".... as of June 1, 1950, 22 fields west of the Mississippi River are producing within the continental shelf".

Section 7. Summary and Provisional Conclusions

We have not discussed rights of States on the resources of the continental shelf, or on the shelf itself, nor the waters above the shelf.

We only tried to throw some light on the conception "continental shelf".

We believe that the term, continental shelf, represents a certain conception amongst lawyers. For the same reason the International Law Commission clings to the word in Art. 1 of the Draft Articles on the continental shelf and related subjects, an Annex to the Report covering its Third Session [36], p. 54 et seq. Point 3 of the explanatory notes, p. 54: "because it (the term "continental shelf") is in current use". The Commission rejects the term "submarine areas", because this term used alone would give no indication of the nature of the submarine areas in question [a].

---

[a] Alfaro, in the 68th meeting of the International Law Commission [49], p. 4, suggested the term "the submarine platform", "an expression used by the earliest writers, e.g. Bustamante. That would obviate the difficulty of shallow water areas". We feel that this is a translation of the Spanish term "plataforma submarina", which means "continental shelf". It would therefore not help us out.

If we use the term continental shelf, do we use it in the geological sense or in a legal sense? Yepes, in the 68th meeting of the International Law Commission [49], p. 10–11, said: ".... a definition (of the continental shelf) could be given only by geologists or geographers" and as we have seen, in the Report to the Bar Association [15], p. 10, it is suggested to use it in the geographical-geological sense. We have two serious objections against this suggestion.

Firstly Hudson declared in the 67th meeting of the International Law Commission [25], p. 8: "that if it was desired to use the expression "continental shelf" he would ask for shallow waters to be assimilated to it". And indeed accepting the term in the geological sense would exclude such areas where the depth would admit exploitation of the subsoil but which would not be comprised by scientists or at least some geologists under the name continental shelf (Persian Gulf).

For that reason we agree with the International Law Commission that the geological concept should be discarded. In explanatory note 1, the Commission reasons that: "The varied use of the term by scientists is in itself an obstacle to the adoption of the geological concept as a basis for legal regulation of the problem". This reason is understandable when we remember the distinction between inner shelves and outer shelves, made by some scientists and not by others. Explanatory note 2 gives then what it is called *another* reason, but which is in our opinion very akin to the first one, and obviously refers to areas like the Persian Gulf. It reads: "The mere fact that the existence of a continental shelf in the geological sense might be questioned, in respect of submarine areas where the depth of the sea would nevertheless permit exploitation of the subsoil in the same way as if there were a continental shelf, could not justify the application of a discriminatory legal system to these 'shallow waters' ".

Our second objection against the use of the term in the geological sense is, that it would mean an area with varying depth-limits. Moreover as Bourcart has pointed out, the edge is sometimes difficult to find. We think, that this would cause many difficulties, when applied in law. It would lead to endless conflicts, both parties taking expert advice from geologists, who would probably not agree either. Law demands clear concepts and easily discernable limits.

So we are going to use the term in a legal sense. That means that we have to find a new definition. The International Law Commission, in doing so, gives the following definition in article 1 of the draft: ,,As here used, the term "continental shelf" refers to "the sea-bed and subsoil of the submarine areas contiguous to the coast, but outside the areas of territorial waters where the depth of the super-jacent waters admits of the exploitation of the natural resources of the sea-bed and subsoil". This definition differs from the geological conception first of all in leaving out the sea-bed and subsoil in the area of the territorial waters. This means, as we will discuss later, that the Commission is of the opinion that a coastal State has ipso iure at least the rights, we are discussing here, on the sea-bed and subsoil of its territorial waters. The Commission goes further in explanatory note 9: "Submarine areas beneath territorial waters are, like the waters above them, subject to the sovereignty of the coastal State".

Secondly it differs from the geological conception by introducing the criterion of "the technical possibility of exploitation of the natural resources". It does not expressis verbis say what sort of natural resources, but obviously the adjective "mineral" should be thought of in view of the explanation given in point 8: "In the opinion of the Commission fishing activities and the conservation of the resources of the sea should be dealt with seperately from the continental shelf". We will discuss this opinion of the Commission in the Chapter on Fisheries.

We have already stressed the necessity to delimit claims, which States may put forward, and have put forward on resources in the shelf or on the shelf itself. We gave as our opinion that in view of the vast differences in width, a delimitation relating to the distance from the coast would not be practicable. The Commission also discards this criterion in point 7, a maximum limit (for instance suggested by Azcárraga [32]) as well as a minimum limit (as suggested by Govare [29] c.s.).

The other delimitation we mentioned, was the depth, in connection with the condition of the technical possibility for exploitation. These two are very much akin. We have already said, that the condition of the technical possibility for exploitation is a reasonable one, because it corresponds with the aim of the majority of proclamations and decrees. It points directly to the origin of our problem and the reason why legal regulation of the problem is desirable. Moreover it corrects the difficulties of delimitation related to differences in depth of the edge and irregularities in depth of the edge and the shelf itself.

On the other hand this condition diverges rather from the conception "continental shelf" even in the wider sense, including the inner shelves. We pointed out that for the time being the technical possibilities are far below the average depth of the edge of 130 metres, let alone the depth of 200 metres, and also that the nature of the edge and that of the slope make working there quite impossible for a long time to come.

The Commission makes this condition into the main criterion and ties the two criteria together in saying: "where the *depth* of the super-jacent waters *admits* of the *exploitation* of the natural resources".

In note 5 it explains further: "It follows that areas in which exploitation is not technically possible by reason of the depth of the waters are excluded from the continental shelf here referred to".

In our opinion the words "where the depth admits of the exploitation" should be interpreted in an objective way. If the coastal State has not available the technical devices to drill at a certain depth, but other States have, that depth does admit of the exploitation. Even if the coastal State does not wish either to exploit or to explore the subsoil, it would nevertheless have control and jurisdiction over the resources of the adjacent shelf up to the depth, which would admit of the exploitation, if this was performed with the worlds' best devices.

Obviously the choice has not been an easy one. We refer to the findings of the subcommittee, mentioned above, and the reason why the criterion of the possibility of exploitation was adopted instead of a limitation in depth. We will now turn to the commentary note 6, given by the Commission [a].

In point 6 of the explanatory notes the choice is further

---

[a] See for the drafting of this comment the Summary Record of the 130th meeting [50], p. 11 et seq., and Summary Record of the 134th meeting [51], p. 7.

elucidated. Considering the depth criterion it says: "It seems likely that a limit fixed at a point where the sea covering the continental shelf reaches a depth of 200 metres would at present be sufficient for all practical needs". This expresses exactly what we have been trying to explain in the foregoing pages. We may add a reason given in the Memorandum on the Regime of the High Seas [22], p. 51: "a definition on that basis .... possesses the characteristics of uniformity, fixity and certitude required for legal transactions". But then the explanatory note goes on: "The Commission felt, however, that such a limit would have the disadvantage of instability. Technical developments in the near future might make it possible to exploit resources of the sea-bed at a depth of over 200 metres". The Commission prudently says "might" and is correct in saying so. Nevertheless we feel that this "might" points to a time, not in the near, but in a rather remote future and we are, as we will further explain in the Chapter about Offshore drilling technique, inclined to be doubtful whether drilling at this depth will ever be realized, especially in view of the nature of the sea-bottom at that depth.

Therefore we feel that the Commission is sacrificing a perfectly clear and easily discernable limit, marked on all sea-charts (leaving for a moment the difference between markings in metres and fathoms) for a rather vague conception of "where the depth admits of the exploitation of the natural resources", for a reason which contains a low factor of probability.

We call a delimitation based on the possibility of exploitation vague, because it runs behind the facts. When does this possibility exist? When a new device comes off the drawing desk of a drilling engineer?

We should think that this possibility can only be proved in practice. Up till now, the depth admitting of the exploitation is 30 metres.

But the Commission is going to allow oil-geologists to explore outside the 30 metre-isobath and cause damage to fisheries with their seismic exploration methods. Let us imagine that these geologists predict good results at a spot where the depth is 40 metres. An oil-company has a new device in stock and starts building an installation on that spot.

Unfortunately, after some months of work it turns out to be a dry hole, or better still, the device is a failure.

Exploitation is proved to be impossible. In other words, the coastal State had no right to build the installation, which in the meanwhile has been an "illegal" obstacle in the high seas, hampering navigation.

We fear that adoption of the Commission's proposal has a greater disadvantage of instability than the very generously fixed depth limit of 200 metres would have.

Then the explanatory note touches a point we have also dealt with, namely the possibility of directional drilling: "Moreover, the continental shelf might well include submarine areas lying at a depth of over 200 metres, but capable of being exploited by means of installations erected in neighbouring areas, where the depth does not exceed this limit."

We have explained that even this possibility seems to us to be a rather remote one.

Finally the Commission point out that it is not intended in any way to restrict exploitation of the subsoil of the sea by means of tunnels driven from the mainland. This is natural, because this way of exploiting the resources of the subsoil is quite outside our problem. We would add, directional drilling from the mainland or islands and tunneling from islands.

We said in the Introduction, that the problem is the conflict of interests, namely between exploitation of mineral resources on one side, and navigation of the other side, fisheries being in between. By tunneling or directional drilling from the mainland or an island, this conflict does not arise. Other conflicts may arise as tunneling or drilling under a neighbour's territory, but these conflicts are not characteristic of our problem.

In point 10 the Commission explains, that the continental shelf does not include the waters covering it. We will discuss this point in the Chapter on Fisheries.

In point 4 it is said, that the word "continental" in the term "continental shelf" as here used, does not refer exclusively to continents. It may apply also to islands to which such submarine areas are contiguous.

On the points, so far discussed, we suggest provisionally the following:

The term "continental shelf" being of geographical, oceano-graphical or geological origin, should not be used in law with quite a different meaning.

If we could adopt the name "shelf" and leave the adjective "continental", it would be a conception, which would mean the same in geography, oceanography and geology as it will mean in law.

It would comprise outer shelves, inner shelves and insular shelves. It seems to us, that using this term would make lengthy definitions quite unnecessary.

We may remind the fact that the once popular initials U.N.O. were changed into U.N. and this change turned out to be ac-ceptable and was as a matter of fact generally accepted without too much delay.

Of course even then delimitation is necessary. And again we suggest to divert as little as possible from the scientific delimit-ation. We are inclined to cling to the original suggestion of the 200 metre- or 100 fathom-line, in spite of our admiration for the ingenious criterion proposed by the International Law Com-mission, for the reasons given above, and other reasons to be explained in following Chapters.

The shelf would, so limited, comprise all areas which would be-come dry if the sea-level would lower 200 metres or 100 fathoms.

Furthermore we suggest in accordance with the relevant discussions above to comprise basins on the shelf, deeper than 100 fathoms but not connected with the ocean under the area of the shelf.

For canyons at the edge we think that the 100 fathomline should be followed even if it would mean a considerable dent in the shelf-contour.

And finally we suggest to adopt the 100 fathom-line and not the 200 metre-line.

Chapter II

FISHERIES

Introduction

Already in 1916, the continental shelf was mentioned in connection with fisheries. In the Report to the League of Nations, 1927 [52], p. 63, we are informed that:

"At the National Fishery Congress held at Madrid in 1916, Odón de Buen [a], who is now Director-General of Fisheries in Spain, urged the necessity of extending the territorial sea to include the whole of the continental shelf".

The first introduction of the continental shelf in International law discussions was based on the theory that the waters above the shelf were of the utmost importance for the fisheries.

Although the mineral resources have given a new impetus to the continental shelf-theory, the old ties with the marine resources are not completely severed. The American Proclamation concerning the continental shelf was issued at the same time as the Proclamation with respect to coastal fisheries, and other declarations and decrees relating to the continental shelf contain at the same time provisions concerning fisheries. In the latter in some instances fisheries are directly connected with the continental shelf theory.

Section I. Whether there exists any relation between the continental shelf and fisheries

This is the first question, which we will attempt to answer. Using the term "continental shelf" we take this term in the wide sense, including inner shelves.

Where do we find edible fish in quantities, which makes it

---

[a] The name, according to Azcárraga [32], should be: Odón de Buen y del Cos, not: "de Buren", as printed in the Report.

worth while sending out a fishing fleet? To put it more in scientific wording, how is the geographical distribution of the different species of edible fish, and where and when is the highest density of these species to be found?

Generally speaking the greatest concentration of fish will be found in areas where food is abundant. Without going into detail of the diet of the many species of edible fish, we can say that most of them either feed on animal- or zooplankton [a], like the herring, the mackerel and the sardine do, or on larger animals. The latter in their turn feed on zooplankton and this finally feeds on the floating flora, the phytoplankton. The latter therefore occupies so to speak "the base of the food chain", because all animals depend on the plant, eating it either directly or indirectly.

To answer our question it seems necessary to find out where these plants are available in abundance, and this depends on sunlight and on the food situation. As Lebour (Science of the Sea [38], p. 151) puts it: "The plant feeds by utilizing the salts and gases in the sea, breaking up carbondioxide in the presence of sunlight by means of the colouring matter, chlorophyll or some closely related substance contained in special bodies, the chromatophores, and so building up the complex protoplasm of which it is composed". This is called the photosynthetic process. "Since the phytoplankton is the main pasturage of the sea" (Sverdrup [54], p. 771) "it is heavily drawn upon by the countless numbers of herbivorous plankton animals living within or periodically invading the euphotic zone". This zone (op. cit. p. 774) "which is abundantly supplied with light sufficient for the photosynthetic processes of plants .... extends from the surface to 80 or more meters". He further distinguishes the "disphotic zone, which is only dimly lighted .... from 80 to 200 or more meters. No effective plant production can take place in this zone ....", and "the aphotic zone, the lightless region below the disphotic zone. In the deep sea it is a very thick layer in which no plants are produced and the animal life consists only of carnivores and detritus feeders".

---

[a] Planktonic organisms, following the definition of Johnstone [53], p. 3, are such as possess very limited powers of locomotion and are accordingly carried about passively, over large sea-areas, by resultant tidal streams, currents or wind drifts. In general they are microscopic in size.

Therefore the first condition necessary for plant life is sunlight, which is available in the surface-layers to a depth of 80 metres, but we may say that for a prolific phytoplankton production the top-layer up to a depth of 50 metres is the most important one.

The second factor is, whether sufficient "plant nutrients" are available. As we have seen, the plant feeds on mineral salts, diluted in the seawater. Important salts in this connection are nitrates and phosphates. About these salts we read in Sverdrup (op. cit. p. 246): "If the utilization of salts by the plants in the upper layers exceeds the supply, the concentration of these salts will decrease", and on p. 783: "The mineralized nutrient elements are not uniformly distributed in the sea, however, and they undergo cycles during which there are periods of delay between the available mineralized state and the unavailable organically bound state of the nutrient". In other words the following cycle occurs: in a certain area where nutrient salts are abundantly available the phytoplankton will reproduce rapidly and diminish the stock of nutrients. When "consumed" the salts are organically bound and become only mineralized and soluble again, when the plantcells, or the animals feeding on them, die. As the bodies sink down, the salts after decomposition of the organisms, will dissolve in deeper layers and become available as plant-nutrients only if taken back to the surface-layers through upwelling. "This leads to fluctuations in the intensity of plant production both in space and in time" .... "In the sea the mineral nutrients .... have accumulated from the land over eons of time and are stored mainly in the deeper dark-water layers where they are not accessible to autotrophic plants and from whence they must be transported by water movements to replenish the supply periodically exhausted in the productive lighted zone". The transport of these nutrient salts from the darkwater layers to the "euphotic zone" in which only the plants can thrive, takes place mainly by vertical circulation of the water.

Sverdrup, op. cit. p. 785: "Circulation of the water by upwelling, turbulence, diffusion or convection is the physical agency by which the return is accomplished ....". Turbulence and diffusion being of less importance will not be discussed here and we will limit ourselves to a short explanation of the influence of upwelling and convection.

"The *upwelling* of subsurface water (Sverdrup loc. cit.), in which nutrient salts are brought back to the lighted layers is reflected in large diatom [a] production along the California coast ...., along the western South American coast through the mechanism of the Peruvian coastal current .... [b], in regions of divergence [c] along the equatorial countercurrents, and along the west coast of Africa .... ". About the importance of upwelling water, and where it occurs, a question which interests us most, we find some further data in Brongersma [55].

Apart from the more important places mentioned above, Brongersma [55] gives the following, p. 20 and 22: "A narrow strip along the Atlantic coast of the Iberian peninsula; part of the coasts of Venezuela and Colombia, coasts of Brazil near Rio de Janeiro and a limited region near Bahia; the coasts of Somaliland south of Cape Guardafui, surroundings of Aden and Perim and further eastwards along the Arabian coast to Cape El Hadd, southern coast of Ceylon; northern part of Chile, Gulf of Panama" (but not all of them are all the year round).

Brongersma points out p. 75: "A high production occurs just as well in regions of ascending water lying far off the coast, for instance in the divergences in the vicinity of the equator, near the ridge between Cape Guardafui and the Island of Socotra etc. .... However, as littoral conditions greatly favour the production the latter will be far greatest in regions of upwelling lying near coasts".

Convection currents, appearing if the density of the surface water is increased beyond that of the underlaying strata, in regions where there are marked seasonal changes in temperature between winter and summer, are also important for the renewal of nutrients in the surface layers. Sverdrup (op. cit. p. 790): "The phenomenon is, of course, dependent upon the surface waters

---

[a] Diatoms is a collective name for a large family of phytoplankton. Johnstone [53], p. 6: "They have extraordinary importance in the general schemes of marine life".

[b] A more exact statement is made on p. 702: "The upwelling is caused by the southerly and south-south eastward winds, which prevail along the coasts of Chile and Peru, and carry the warm and light surface waters away from the coast, resulting in cold water being drawn from moderate depths toward the surface". This happens mainly in regions situated between latitudes 3° S and 33° S.

[c] Regions of divergence (loc. cit. p. 787) "are defined as those regions where surface water masses of the ocean flow away from each other or away from the coast so that water from the deep must rise, as a feature of upwelling, to replace them".

cooling to a point where their density becomes sufficient to cause them to sink and be replaced by upward movement of lighter and incidentally nutrient-rich waters from below ..." and (This) "is a seasonal rather than a continuous process". This process plays a role for instance in the Antarctic regions (see Brongersma, loc. cit. p. 74–75).

Finally an important "stimulation of phytoplankton growth, commonly observed to be coincident with the mixing of two bodies of water with different characteristics may also result from an inoculation of a sparse population living under poor conditions into more favourable waters, which have lacked spores suitable to take advantage of the good physical-chemical environment". (Sverdrup, op. cit. p. 795). A good example is the mixing of the cold Labrador current and the warm Gulfstream waters on and near the New Foundland Banks, which are particularly rich in marine life. Here the influence of horizontal currents is shown to be favourable to the phytoplankton. The opposite may happen when the plant-cells, for instance neritic diatoms (belonging to the coastal sea up to 200 m. depth), have been carried far seaward and degenerate through lack of nutrients. (Sverdrup, op. cit. p. 792).

With the factors temperature and salinity we will deal only in a very brief manner. Although they are important as regulators in respect to growth and reproduction of different species of phytoplankton, the latter are all influenced in a different way. And since the zoo-plankton probably are little particular about the available "vegetables", their development will be pretty independent on the factors mentioned. All these factors have to be seen in their combined effect, and according to Liebig's law of the minimum, production is limited by the factor occurring temporarily in minimal quantity, the so-called "limiting factor".

Two facts may be added to this survey of influences on plant production. Sverdrup, op. cit. p. 794: "The phenomenal outbursts of diatoms in arctic regions coincident with the springmelting of the ice have been explained as probably being associated with the rapid germination of spores (which are very resistant to extreme temperatures) that have been locked in the ice and are released through melting. It is known that these regions simulate biological conditions of coastal areas and the population is large-ly neritic".

So far those factors have been considered which are instrumental in a prolific production of phytoplankton.

In this connection a source of mineral salts not mentioned before, should be added, viz. the solubles supplied by the big rivers.

As to the geographical distribution of phytoplankton, we can now draw our conclusions: Phytoplankton is abundant in part of the Arctic and Antarctic waters, along many coasts especially where upwelling waters or big rivers supply the necessary nutrients.

Far off shore the oceans generally speaking, are lacking phytoplankton in great quantities, except for those places where diverging currents or submarine ridges cause upwelling, and other necessary factors are available.

The production in the temperate and higher latitudes is influenced by seasons. Sverdrup remarks, op. cit. p. 783: "Even the most casual observer is impressed with the browngreen colour that is imparted to the waters of coastal areas and over banks during these periods of diatom 'bloom'. The production may go on at varying intensities during the summer and frequently becomes augmented again in the autumn. But during the winter months, and sometimes at other seasons as well, there is a dearth of plants and the waters again become deeper blue in colour. Similar conspicuous changes in water colour occur when one leaves the rich coastal waters and proceeds to the barren open sea. It has been said that 'blue is the desert colour of the sea', and this is indeed true of vast ocean stretches where phyto- and zoo-plankton are at a minimum and the water is not otherwise discoloured and turbid with inanimate dissolved or suspended material" and, p. 942: "As to production in lower latitudes, we can say that in general the open sea is relatively sterile". And finally phytoplankton is limited to the upperlayers of the waters.

Although, generally speaking, phytoplankton is more abundant near the coasts, it does not seem that the existence of a continental shelf is a conditio sine qua non. The plant production is rather independent from the existence of a shelf, and upwelling is a far more decisive factor. During the "Meteor" expedition, on the basis of extensive investigations, "Hentschel demonstrated a relatively much heavier phytoplankton production along the west African coast, (Sverdrup, p. 786), (and we add: with a relatively narrow shelf), where marked upwelling is known to take place, than along the east coast of South America (we add: where a relatively wide shelf exists), except at the south-eastern tip where the influence of the Falkland current is operative".

Indeed, if we look at the map in Chapter I, we see that several places, mentioned by Sverdrup and Brongersma, are coasts with a narrow and even a very narrow continental shelf, like the west coast of Africa, the coast of Portugal and the south coast of Arabia, the coast of California and the coasts of Peru and northern Chile. In the Peru Current Gunther found in 1936 heavy phytoplankton production near the coast mainly, but also in some sections in almost undiminished intensity to a distance of 320 kilometres seaward (Sverdrup, p. 942).

Where the rivers are concerned, they do not debouch only on coasts with broad shelves, but also on coasts with narrow shelves, as the Congo river for instance.

In the northern part of the Arctic zone there is no question of a shelf nor is there any shelf of any significance in the Antarctic seas, where enormous numbers of organisms are found, most of the few shelves being constantly frozen.

Although we demonstrated that the distribution of zoo-plankton, feeding on the phytoplankton, will be very much dependent on the production of the latter, we have not yet dealt with the, rather obvious, fact that the distribution of edible fish is, in its turn, dependent on the places where zoo-plankton thrives.

Surface Fisheries.

A part of the zoo-plankton consists of planktonic eggs and larvae of bottom- as well as swimming animals, collectively known as the temporary plankton. This is especially abundant in the neritic waters (coastal waters to a depth of 200 metres) (Sverdrup, op. cit. p. 814). The rest, called permanent or holoplankton, inhabits the neritic waters as well as the oceanic waters. But roughly we may say that zoo-plankton is generally abundant where their food, the phytoplankton, is found, which means that largely the same applies as has been said about the geographical distribution of phytoplankton. "The direct corollary (to the abundance of plant production in the coastal waters) is the abundant animal life in the coastal regions as opposed to the more sparse population of the waters far from shore" (Sverdrup, op. cit. p. 898), and, as he says, (p. 279): "Therefore per unit area of the sea, the neritic province is far more productive than the oceanic province and is consequently the region of greatest importance to marine life in general. Here fish of greatest economic importance are taken, not only because of greater availability, but also because it is their natural habitat". Brongersma [55] describes the same relation between abundance of plants and animals, op. cit. p. 76: "As animals live directly or indirectly on phytoplankton, the number of animals is also very great in regions of upwelling. In the Peru coastal Current the abundance of invertebrates, fishes and birds (feeding on the fish) is astonishing. In the upwelling water along the coast of S.W. Africa invertebrates and fishes are also very abundant .."

Fishes feeding directly on zoo-plankton, like herring, sardine, mackerel and anchovy, will be found where the most dense concentration of zoo-plankton is available. Sverdrup, op. cit. p. 907, mentions that "among the pelagic fishes the best material illustrative of correlation with plankton comes from studies of the herring .... Hardy, Lucas, Henderson and Frazer .... in correlating the numbers of Calanus copepods of the plankton with the amount of herring caught by fishermen in the same areas .... have shown in most instances that the greatest number of adult herring are caught in Calanus-rich waters".

Fishes preying on larger animals will in turn be found in places frequented by their prey. Fast swimming predators, among which are the surface fishes tuna, barracuda and salmon, feed largely on plankton-feeding fishes. "Such fishes must therefore stand as an important link between the animal plankton and the larger piscine predators unprovided with direct means of gaining sustenance from the small planktonic life" (Sverdrup op. cit. p. 896). Of the big mammals of the sea, the whalebone whales are plankton-feeders (the blue whale for instance) ".... their numbers may be correlated with the abundance of planktonic food of their preference" (Sverdrup, p. 892).

Predator mammals of the sea are the toothed whales (sperm

whales feeding on squids at great depths, and the killer whales) porpoises, dolphins, the seals, the sea-lions and walruses. They feed on fish and crustaceans (see Sverdrup, p. 897).

Another factor, apart from food, influencing distribution at certain times is reproduction. We have already mentioned the fact that eggs and larvae are especially abundant in shallow waters. That fish in most cases deposit their eggs in coastal waters and on banks can be explained by the fact that there the optimum conditions for phyto- as well as zoo-plankton on which the young fish feeds, are present. Exceptions are the anadromous salmon, which, as the name indicates, ascend rivers to spawn, the European and American eel, which spawn in the Atlantic far from land, and the sardines, which, according to the Report on a Survey of the Fishery Resources of the United States and its Possessions [56], p. 9, "spawn .... in the open sea as far out as 300 miles though generally 50 to 200 miles off shore (Pacific coast)".

"*Salinity* influences the character or type of animals that will be present in any region rather than the rate of reproduction or the total amount of animal organic material produced" (Sverdrup, op. cit. p. 841).

Temperature plays an important role on reproduction, and where and when spawning takes place and hence on migration. Generally speaking, however, it is a factor which plays a role in combination with other factors; all these factors together resulting in the distribution as described.

From all these facts we derive the same conclusion as before, namely that coastal waters are the most productive, but independent of the existence of a continental shelf. In other words, there is no reason to tie production of fish to the existence of a shelf.

Bottom-Fisheries.

As to the bottom fish like plaice, flounders, halibut, cods and rays, we can be very brief, because here the technical devices to catch them, usually the trawl, are limited to a certain depth, which corresponds more or less with the depth of the shelves, granted that the bottom is smooth enough.

Conclusion.

The answer to the question whether there exists any relation between the continental shelf and fisheries can be worded as follows:

The continental shelf can be very productive if the conditions influencing abundance of fish are fulfilled, which is particularly so in regions of upwelling and convection. The last two factors, however, do take place also in regions where no, or only a narrow, shelf exists, along coasts as well as far off shore. In other words, it is true that the continental shelf is very often a place where an abundance of fish is found, but it is not the only place and hence it should not be made into a criterion for delimitation of rights concerning fisheries.

This conclusion finds further confirmation in the Report on a Survey of the Fishery Resources of the United States and its Possessions [56], (p. 1): "Unlike conditions on the North Atlantic coast, food-rich water in the Pacific is not confined to the continental shelf, extends many miles to sea over deep water, supports large populations of many kinds of pelagic fishes. California catch 1942: 1,173 million pounds", and (p. 9-10): "Sardines spawn ... in the open sea as far out as 300 miles, though generally 50 to 200 miles off shore (Pacific Coast). By the time the fish are 10 inches long .... practically 100 per cent of them become available to the fishery and are aggressively sought. The younger fish are found first in the fishing regions nearest the spawning areas, and later in the more remote grounds". (The major known spawning area being indicated on p. 12 between San Diego and Monterey. We add that the continental shelf between these places is not wider than 15 miles at a few places, but mostly much narrower). And p. 26: " .... Anchovy is found from the Queen Charlotte Islands, British Columbia, on the north, to Cape San Lucas, Lower California, on the south, and an unknown distance to sea — probably less than three hundred miles", and p. 39: "The tunas are a world resource, ranging over vast distances and migrating across oceans .... ".

Japan's fisheries (of tuna) extended over vast distances of the Pacific, and her production came to 460 million pounds a year ... and p. 40: "They (the tuna's) are oceanic fishes not limited to the continental shelf .... There are indications that albacore migrate across the Pacific, blue fin across the Atlantic" and p. 21: "Halibut live on banks extending from shore to about 250 fathoms deep (on the Pacific Coast)", and p. 57: "Whiting are found on sandy or pebbly bottoms from the shore line to a depth of about 300 fathoms" (this must be on the continental slope) and p. 122: "The territory of Hawaii .... the offshore fisheries provide, in normal times, about 86 per cent of the total landings in the islands. They (tunas, mackerel, scads and swordfish) are pursued in deep water as far as 100 miles offshore .... deepsea fisheries average 1933–1940: 12,662,000 annual yield in pounds" (we add that the continental shelf round Hawaii, according to the 200 metre-isobath on the chart is only a few miles wide).

Of course, the continental shelf (in the wide sense) is in many parts the most important place for fisheries and especially for bottom-fisheries. The Report mentioned above p. 2: "Large areas of the shelf rise to form submerged plateaus, called banks, on which lives a vast population of bottom-living fish. Most important among them are the haddock, rosefish, cod, flounders and pollack", and p. 54: "Haddock is the dominant species .... on Georges Bank (Gulf of Maine), cod becomes increasingly dominant off Novia-Scotia and exceeds any other bottom fish on the New Foundland Banks and beyond". The continental shelf is for practical reasons often the limit for the fisheries. The Report states on p. 19: "Since soles or flounders are caught mainly by trawling, the fishing grounds for them are limited to the area in which trawls can be used. To date little successful trawling has been done in depths greater than 100 fathoms". Here the criterion of the technical possibilities plays a role. As we have mentioned above, some bottom fish live at depths of 300 fathoms (whiting), but trawling at that depth meets with many difficulties, one being the roughness of the sea-bottom (continental slope). (Halibut living on banks on the Pacific Coast extending from shore to 250 fathoms deep are, however, "caught with long lines, made up in units called skates ....", loc. cit. p. 21). Other bottom animals like clams, crabs, lobsters and shrimps are mostly caught in relatively shallow waters.

Further evidence that fisheries are independent from the continental shelf is produced by Kask [57], p. 7 and 10: " .... the shallow water areas account for only five or six per cent of the whole Pacific Ocean area. Is the remainder of this great and deep body barren or can it be counted on to contribute its share

to man's needs? There are some indications that it might contribute more than was ever dreamed .... For instance it is no
longer speculation that the high seas tuna fishery based in Southern
California caught enough of this species in 1948 to fill nearly six
million cases of cans, and that this catch represented the greatest
cash crop of fish taken out of any ocean. And the tuna fishery is
a very new fishery .... Japan was before the war, and she still
is, the greatest fishing nation on earth. Before the war, her one
and a half million fishermen accounted for between one-fifth and
one-quarter of the world's catch. Driven by a national need for
more food and encouraged by a sympathetic Government which
was preparing to wage war the Japanese fishermen began to
delve into the unexplored depths of the ocean for their catches,
as well as spreading over the world's known fishing grounds. Their
experiments were only in their initial stages when the war cut
them short but not before Japanese scientists had preserved in
their laboratories specimens of good edible fish taken far from
land and from ocean depths up to 500 fathoms .... At times
during cruises in deep water 'false' or 'phantom' bottoms are
recorded by depth measuring instruments; that is, impulses are
returned from objects in intermediate layers of water in areas
where it is known that the actual bottom is at a greater depth.
These 'phantom' bottoms are reported at times to be many miles
in extent and the position of these dense intermediate layers shifts
both vertically and horizontally. This has led to much speculation as to what could make up this dense layer. Last year one
ingenious Norwegian investigator sent a movie camera down to
the depth indicated. When his film was developed he proved
that in one case at least the "phantom" bottom was a school of
cod-fish".

Shapiro [58] gives us the following information on Japanese and
American tuna fisheries, p. 1.:
"The long-line gear is essentially a method by which hooks
are lowered to fishing depths of about 100 to 350 feet below surface-
level .... Among the more important of the areas developed
by the Japanese long-line fishermen are the winter albacore
grounds located near Midway Island in the mid-Pacific, the tuna
grounds of the Ryukyu Archipelago, and the yellowfin tuna

grounds in the southwest Pacific Ocean", p. 3: "American expansion in the tuna industry is continuing. This is evident from the fact that, in their search for new fishing grounds, many of the larger fishing companies have under serious consideration the possibility of extending their operations westward into the mid-Pacific regions" and p. 19: "On the basis of their exploratory work, Japanese investigators claim that the tunas caught by the long-line gear are primarily on a feeding migration and it is best to set the gear across the path of movement of the fish or near an obstacle in their path. .... areas where the tunas are available are more easily located by looking for barriers to movement. Barriers to movement may be an island, a reef, a submerged bank, or a difference in water temperature between two water masses .... The exploratory vessels look for localities: .... 3. Where areas of convergence or divergence between two adjacent currents or masses of water are present ....".

Finally in Report No. 104, Natural Resources Section, S.CA.P. 1948 [59], giving detailed information of Japanese tuna fisheries, we find in figure 10, on p. 27, the fishing grounds for the albacore stretching from Japan as far as about 177° west longitude in a broad ocean area between 42° and 26° north latitude. A same long stretch goes south roughly between 140° and 160° east longitude down to the equator. The last area forms approximately also the fishing ground for the yellowfin tuna, according to figure 12 on p. 31.

SECTION 2. OPINIONS EXPRESSED AT CONFERENCES

Taking our conclusion as a basis, we can now review critically what has been said about the relation of the continental shelf and fisheries and what has been written in the proclamations and decrees.

For many years the opinion has been expressed that the territorial waters should be extended for the sake of fishery rights. As has been mentioned before, at the National Fishery Congress at Madrid in 1916, Odón de Buen urged the necessity of extending the territorial sea to include the whole of the continental shelf. We have been criticising the opinion that, if the right of jurisdiction and control of mineral resources depended on the presence of a continental shelf, an injustice would be committed towards countries without a continental shelf, such as Chile. We founded our criticism on the presumption, that those countries would not benefit by such rights, because it would, technically, be impossible to work the resources at a depth more than 200 metres and on the ground, that we have to accept the geographical assets or absence of assets as they are.

In the case of fisheries, however, we believe that the idea of de Buen would lead to discrimination against those countries which have no continental shelf, because we have seen above, that fisheries are not limited to coasts where a continental shelf exists. This extension of territorial waters for fishery purposes is based on the wish to exclude fishermen from other countries, or on the wish to apply measures of conservation to a wider area, or on both. If such wishes were to be fulfilled, why should countries, which have a perfectly good reason for their demands be excluded, for the irrelevant reason that they do not possess a continental shelf? François, therefore, very rightly put his question in his First Report to the International Law Commission [10]: "whether the grant of special rights in respect to .... marine resources should be made conditional upon the existence of a continental shelf". (We left the "mineral resources" out, because that part of the question has been discussed in Chapter I). François' statement, which could induce an affirmative answer to his question, namely (loc. cit.): "A depth of 200 metres .... is also the depth which marks the extreme limit of the optimum biological condition both for plants and animal-species used for human consumption", seems not to be in agreement with our conclusions, derived from the facts, given in the beginning of this Chapter. We are inclined to give a negative answer to this question.

How was the continental shelf introduced in legal discussion? In the observations by de Magelhaes [a] on p. 63 of the Report to the Council of the League of Nations, 1927 [52], we read: "The maritime species of fish, particularly the edible varieties, are not uniformly distributed throughout the whole sea. Whether they

---

[a] Dr. Barbosa de Magelhaes, Prof. at law, Univ. of Lisbon.

are sedentary or migratory, their biological characteristics demand special conditions, which, generally speaking, are only to be found together in proximity to the coast, or at a relatively short distance therefrom .... What fishermen knew, however, because they had occasion to notice it every day, was that those edible varieties of fish, which were to be found quite near the coasts, did not inhabit the high seas". He then describes the continental shelf and continues: "The other much vaster region, which extends beyond this step (edge), is the abysmal region, the rare species of fish found in this region are generally inedible. On the other hand, those which inhabit the continental shelf are for the most part edible".

"As has already been observed, the width of the continental shelf is very variable .... The following conclusion may be drawn from these facts:

a) Edible varieties of fish are crowded together, so to speak near the coasts of the countries, whose continental shelf is narrow, whereas they are scattered about the high seas off coasts whose shelf is more extensive;

b) Consequently, fishing is more sought after on narrow shelves, because it is more fruitful there than elsewhere".

After having explained the necessity to extend the territorial waters in order to secure "exclusive enjoyment of fishing rights" for the coastal population, as well as "to take the necessary steps for the protection of fish", he quotes from a Memorandum of 1921 by the Portuguese Admiral Almeida d'Eça, who, discussing the necessity of protection, remarks that the coastal State cannot enforce its regulations outside the territorial waters: "The outer limit of the territorial waters, as now recognized, does not, however, coincide with the greatest depth at which edible species of fish are to be found. For the edible fish the barrier is the drop from the continental shelf; they are not to be found beyond this line ....".

It seems that both the opinions are based on a premise which is not in accordance with the facts. It was erroneous to presume that no edible fish inhabits the seas beyond the edge of the continental shelf. It is perfectly true that the waters above the shelf and the sea-bed of the shelf are often rich fishing grounds, but to reverse this fact and to say that the shelf is a conditio sine qua non for fish or that no edible fish can be found outside the shelf waters is quite incorrect [a].

The "Report on the exploitation of the products of the sea" (Buenos Aires, Dec. 8, 1925), by the Rapporteur Suárez [b], loc. cit. p. 121, starts with the proposal: "Should not a special technical conference be convened to draw up immediately, without regard to the extension or maintenance of maritime jurisdiction extending to the 3-mile limit, uniform regulations for the exploitation of the industries of the sea, whose wealth constitutes a food reserve for humanity, over the whole extent of the ocean bed forming part of the continental shelf, i.e. the region along the coast where the depth does not exceed 200 meters?" The rest of the Report, however, is more carefully worded: "The exploitation of the products of the sea requires regulations the most urgently in the waters nearest the coast, because it is in these regions, and particularly on the shoals, that the species most useful to man have their habitat. In the open sea away from the continental shelf, where the depth exceeds 200 meters, only a few species useful to man are to be found in the upper levels of the sea. Apart from the waters in the immediate vicinity of the beach, it may be said to

---

[a] *It is remarkable that the Portuguese Decree regulating fishing by steam vessels, of the 9th November 1910[153], p. 19, does not contain these erroneous ideas but is on the contrary founded on principles which are more in accordance with the biological facts. We quote the following passages:*

"Whereas deep trawling by steam vessels at depths of under 100 fathoms within the limits of the continental shelf is extremely harmful to fisheries, because this method destroys the feeding grounds on the sea bed and therewith the young fry feeding, sheltering and developing there, a process rapidly leading to the destruction of the marine life along the coasts affected since, as a consequence, it becomes impossible to replace the stocks of fish at depths of over 100 fathoms, the habitat of the fully developed fish, so that an important source of wealth is destroyed;

Whereas this has occurred on all coasts where such a system has been used, even along the vast continuous continental shelf which runs from the Bay of Biscay northwest along the coasts of France, Belgium, Holland and Germany, as far as the Norwegian coast where it turns southwest and extends to within fifty miles of the west coast of Ireland, which means that many steam fishing vessels belonging to those States are coming further to deplete the resources of our narrow continental shelf area, as has already occurred in the case of Morocco ....".

Then follow the restrictive provisions. Art. 2 reads: "Fishing by this method (trawling) may only be carried out beyond the bathymetric line of 100 fathoms, and never at a distance of less than three miles from the coast".

[b] Dr. José León Suárez, Dean of the Faculty of Political Science at the University of Buenos Aires.

The Committee of Experts for the progressive codification of International Law asked Suárez to examine the question of the exploitation of the products of the sea. His Report [60] was sent as an Annex with the Questionnaire No. 7.

be natural law that the intensity and variety of marine life is inversely proportional to the depth of the water" [a].

But even this statement is not yet quite reflecting the actual situation.

We cannot pass over these opinions as being just private ones, because Governments have acted on these and similar ideas. In the "Observations" of the Portuguese reply on the questionnaire relating to the "Exploitation of the products of the sea" (loc. cit. p. 192 et seq.) the Portuguese Government refers to the 6 International Fishery Congresses which had already been held and all of which came to the conclusion that, as a measure for the protection of maritime species, it was essential that the zone of the territorial waters be extended, for purposes of fishery, to 12 or 15 miles. The reply goes on: "This is practically the only step required, and we base our opinion on the following facts:

1. ....   2. ....

3. Except in special cases, such as that of the cetaceans, to which we make special reference, and that of the North Sea, which is entirely a continental shelf, the point of interest for each State is its own continental shelf, where the edible species dwell.

4. If the zone of the territorial waters were, for purposes of fishery increased to 12 or 15 miles, a step which has already been taken under the legislation of various countries for other, albeit less important purposes — each State would have its continental shelf included in its own territorial waters and would consequently be able in that area to adopt such rules as it might hold to be most desirable for the preservation of species" [b].

---

[a] Feith, in his Report to the Conference of the International Law Association at Brussels in 1948 [44], p. 204, quotes Suárez proposing to acknowledge the right of coastal States to the continental shelf: "There is no stable permanent and convenient solution except to adopt the rule of the continental shelf with some modifications, according to circumstances", but his conclusion that Suárez was the first to formulate the "continental shelf-theory" is incorrect. Odón de Buen and Almeida d'Eça did so before him, and all of them only in relation to the marine resources.

[b] At the Codification Conference at the Hague in 1930 [61], Portugal did not go as far in its wishes. Its delegate M. de Magelhaes said, p. 134–135: "I now pass to the special reasons which led the Portuguese delegation to claim this breadth. The coast line of Portugal is very long in comparison with the area of her territory. Her continental plateau has a special configuration which has a far reaching effect upon fishing. This plateau is excessively narrow. It is one of the narrowest in the world. It is only there that fishing can be carried on, further out, the sea falls to great depths, abysmal depths inhabited only by rare species which, generally speaking, are not edible. ....
Thus, on the one hand, Portugal has a very narrow continental plateau, on the other,

Portugal, having a relatively narrow shelf, has obviously not realized that other countries have much wider shelves. But about their idea to include the continental shelf in the territorial waters we can say that the same incomplete conception has been the base of this reply, which, if it had been followed up, would have lead to an unjustifiable discrimination towards countries with an extremely narrow shelf and an extension of sovereignty out of proportion of the interests involved to those which have an extremely wide continental shelf. We repeat, that the discrimination is here accepted as an argument, because it is technically possible to obtain the marine resources beyond the shelf, whereas this possibility does not exist in the case of mineral resources.

SECTION 3. PROCLAMATIONS, DECLARATIONS AND DECREES

The Proclamation by the President of the United States with respect to coastal fisheries in certain areas of the high seas, of September 28, 1945 [62] [a], reads as follows:

"Whereas for some years the Government of the United States of America has viewed with concern the inadequacy of present arrangements for the protection and perpetuation of the fishery resources contiguous to its coasts, and in view of the potentially disturbing effect of this situation, has carefully studied the possibility of improving the jurisdictional basis for conservation measures and international coöperation in this field; and

Whereas such fishery resources have a special importance to coastal communities as a source of livelihood and to the nation as a food and industrial resource; and

Whereas the progressive development of new methods and techniques contributes to intensified fishing over wide sea areas and in certain cases seriously threatens fisheries with depletion; and

Whereas there is an urgent need to protect coastal fishery resources from destructive exploitation, having due regard to conditions peculiar to each region and situation and to the special rights and equities of the coastal State and of any other State which may have established a legitimate interest therein;

Now, therefore, I, Harry S. Truman, President of the United

---

this very fact induces foreign fishermen to go there, so that the Portuguese are in a very critical position unless they can have a monopoly of fishing within a belt of at least six miles .... More than 50,000 men are employed in fishing in Portugal, which represents a very high percentage of the total population of the country".

Here again the premises are not in agreement with the facts we mentioned.

[a] See also [153], p. 112.

States of America, do hereby proclaim the following policy of the United States of America with respect to coastal fisheries in certain areas of the high seas:

In view of the pressing need for conservation and protection of fishery resources, the Government of the United States regards it as proper to establish conservation zones in those areas of the high seas contiguous to the coasts of the United States wherein fishing activities have been or in the future may be developed and maintained on a substantial scale. Where such activities have been or shall hereafter be developed and maintained by its nationals alone, the United States regards it as proper to establish explicitly bounded conservation zones in which fishing activities shall be subject to the regulation and control of the United States. Where such activities have been or shall hereafter be legitimately developed and maintained jointly by nationals of the United States and nationals of other States, explicitly bounded conservation zones may be established under agreements between the United States and such other States; and all fishing activities in such zones shall be subject to regulation and control as provided in such agreements. The right of any State to establish conservation zones off its shores in accordance with the above principles is conceded, provided that corresponding recognition is given to any fishing interests of nationals of the United States which may exist in such areas. The character as high seas of the areas in which such conservation zones are established and the right to their free and unimpeded navigation are in no way thus affected".

In a memorandum handed in the State Department to an official of the Netherlands Embassy in May, 1945, the Government of the United States expresses its concern with regard to the progressive development of new methods and techniques which contribute to intensified fishing over wide sea-areas and seriously threaten fisheries in some areas with depletion. In an oral explanation to this memorandum it was said, that information had been received about plans to use floating fish canning factories like the whaling motherships, which would operate with their own fishing fleet far from the motherland and after these operations transport the products directly to the market countries. Chapman [63], p. 70–71, wrote about these motherships: "A large ship and a group of smaller fishing vessels go out as a group. The large ship acts as supply and repair vessel for the small vessels. The small vessels catch the fish and transfer the catches to the big ship for processing or refrigeration .... Canada and the United

States by mutual sacrifice, expense, and strict regulation of their fishermen, have built up the Pacific halibut banks so that they are among the richest fishing grounds in the world. If there is nothing under accepted International Law that would prevent a third nation from sending a mothership expedition to skim the cream off these halibut banks, what is the use of building up fisheries resources in this manner? .... In this way Japanese mothership operations [a] have worked in Bristol Bay; and English motherships have operated in the Greenland halibut fishery". Another factor mentioned in the memorandum was the need to protect coastal fishery resources from destructive exploitation, with due regard to conditions peculiar to each region and situation and to the special rights and equities of the coastal State and any other State which may have established a legitimate interest.

In this connection the Alaska salmon fishery was mentioned in the oral explanation and in a Press Release, dated September 28, 1945[6], p. 484: "Through painstaking conservation efforts and scientific management the U.S. had made excellent progress in maintaining the salmon at high levels. However, since the salmon spends a considerable portion of its life in the open sea, uncontrolled fishery activities on the high seas" (about 100 miles offshore on the route followed each year by the salmon), "by nationals of either the U.S. or other countries have constituted an ever present menace to the Alaskan salmon fishery (and salmon culture)".

Uncontrolled fishery, it was said, had for instance destroyed the halibut fishery on the Iceland coast, whereas this fishery is still satisfactory on those grounds, where, in co-operation with other countries, the United States have taken timely the necessary measures for protection. The memorandum continues to say

---

[a] On Japanese motherships in general, we find some data in Report No. 31, Natural resources Section, S.C.A.P., 1946 [64], as far as the Japanese salmon industry is concerned. On p. 18 it says that: "Since 1927, the Japanese have developed the operation of floating factory ships .... Until 1927 salmon were caught close to the coast as they approached the mouths of rivers on their spawning migration. Floating canneries, with their fleets of fishing vessels, changed this method by being able to fish in deeper water offshore.

.... At the peak of its floating cannery operations, the Nichiro Fisheries Company owned nine factory or motherships, ranging in size from 1,500 to 6,200 gross tons each .... The factory ships were completely equipped for canning, salting, and freezing ....".

that, in view of these factors, the Government of the United States has concluded, that the productivity of important fishery resources contiguous to its coasts may be destroyed unless a basis for the extension of protective jurisdiction beyond present limits is found and adopted.

Although the policy provides that fishery activities within conservation zones established under it "may, when conditions warrant, be limited to" the United States or to the United States and other States which collaborate in establishing the zones, the Government does not contemplate that this will effect general exclusion from these zones of all fishing enterprise other than mentioned above. (It seems to us that this cannot mean anything else than that any other fishing enterprise has to submit to the rules and regulations laid down by the United States alone or in agreements between them and certain interested countries). The rights of all States which have taken any substantial part in a fishery are thus preserved.

The Government of the United States does not intend that well established or historic fishing activities being carried on by the nationals of other States shall be disturbed. (Again in our opinion on the condition that these States enter into an agreement with the United States). Instead, the memorandum goes on, co-operation with such States in the regulation and control of the fisheries is contemplated. So far the extracts from the memorandum and the Press Release.

One of the main reasons for issuing the Proclamation seems to be the interception of salmon on their way to the territorial waters of the United States. We have to investigate this a little bit deeper. We have mentioned already the migratory habits of certain fish species. Sverdrup [54], in connection with reproduction (op. cit. p. 811): "Other species of salmon also may journey over 650 km., presumably returning to the stream in which they were born (Am. Ass. Adv. Sci. 1939)" and in the Report on the Fishery Resources of the United States [56], p. 3: "The spawning seasons of the salmon vary, but in general, last from late summer to early winter. As the fish approach sexual maturity they reassemble and swarm up the rivers of their birth until they find a place to spawn. .... All Pacific salmon die after their single spawning season. The eggs incubate, buried in the gravel, during the fall and winter. They hatch early in spring .... the young fish, now called fry, struggle up through the gravel and begin searching for food. Pink and chum salmon now proceed almost immediately to the ocean, but the other species generally stay in fresh water from 1 to 4 years before migrating to sea. After reaching the ocean, they remain there 2 to 5 years, according to species and latitude ...." and p. 8: "A year class of salmon is available to the fishery only in its last year of life, and only during the rather short period of change from ocean to stream life". Of course a certain amount of salmon is allowed to proceed to the spawning grounds to sustain production and the rest is caught.

The Report mentions the legislative control of salmon in Alaska under the jurisdiction of the Federal Government, p. 8: "Laws and regulations define fishing areas, prescribe maximum and minimum size limits on certain types of gear, prescribe opening and closing dates for the fishing season and provide for weekly closed periods; they prescribe maximum catch quotas in certain areas and in general so regulate the fishery that the maximum catch may be taken without endangering future yields"

These regulations can obviously not be enforced, at least on foreigners, outside the territorial waters, and thus not prevent misuse and depletion of this marine resource.

According to Selak [65], p. 671, there was during the 1930's and until the outbreak of World War II "concern over possible depletion of the Pacific salmon fisheries centering around the Bristol Bay [a], Alaska, largely because of the activities of Japanese fishermen." In a Note he draws attention to (U.S. Dept. of State Press Releases, Vol. XVIII (Jan. – June, 1938), p. 412–417) "The U.S. Government made a statement in this regard to the Japanese Government, stating, inter alia, that: 'The American Government must view with distinct concern the depletion of the salmon resources of Alaska. These resources have been developed and preserved primarily by steps taken by the American Government in co-operation with private interests to promote propagation and permanency of supply. But for the efforts, carried on over a period of years, and but for consistent adherence

---

[a] See for instance the article of K.B. [66], p. 102–103, about the Japanese attempt to obtain a foothold in the Alaskan salmon fisheries and the American concern about the Japanese interest in pelagic salmon fishing in the Bristol Bay area.

to a policy of conservation, the Alaska salmon fisheries unquestionably would not have reached anything like their present
state of development .... the American Government believes
that the safeguarding of these resources involves important
principles of equity and justice. It must be taken as a sound
principle of justice that an industry such as described which has
been built up by the nationals of one country cannot in fairness
be left to be destroyed by the nationals of other countries ....' ".

It boils down to the fact that these salmon are born in American
rivers, proceed to the ocean and return after some years to
their homeland to spawn. On their way home they are intercepted
by United States and foreign fishermen outside the U.S. territorial waters. But why should the coastal State have the right
to prevent this? What gives the coastal State a title on these
salmon? The fact that they were born in American territory?
It looks as if a country adhering to the *ius soli*-principle, applies
this legal concept, giving a human being the nationality of the
country where he or she is born, to animals. Surely this analogy
is unacceptable. A country cannot possibly claim property
rights on fish hatched in its rivers or territorial waters and prevent
foreign fishermen to catch this fish after it has migrated to the
high seas. This American conception reminds us to a very
similar one, held by the American Government in the Behring-
Sea arbitration.

In the negotiations "the American Government first pled
a virtual mare clausum for the whole of Behring-Sea; then that
they had jurisdiction up to 100 miles from land; and lastly that
they had special *property in* and right of protection over the
fur-seals in Behring-Sea and frequenting the islands for breeding
purposes. The Tribunal of Arbitration decided that they had
not this right of protection or *property* when such seals are found
outside the ordinary 3-mile limit." (Fulton [31], p. 695).

Even Chapman [63], p. 67, defending the United States new
policy on high sea fisheries admits that: "Fish are owned property
when they are reduced to possession; fishery resources of the
open sea, however, are owned by every one or no one. They
are the property of no nation".

On the other hand, it is true, that a complete interception
on the route followed by the "homing" salmon, would result

in a quick extinction of the race in the area concerned. This would amount to financial losses and unemployment for a part of the coastal population.

The Report mentioned above, on p. 3 states that the Pacific salmon yield around 600 million pounds of fish a year. "Close to 90 per cent of the United States production and 55 per cent of the world production of salmon comes from Alaska .... 70 per cent of Alaska tax revenues come from salmon industries". If we take into consideration that also this source of income for the intercepting foreign fishermen would dry up, if no salmon were allowed to "repatriate" in order to spawn, it would be logic to conclude that inside as well as outside the territorial waters regulations on the lines as mentioned above would be necessary to prevent this complete interception. This could be achieved by Convention between the coastal country where the salmon spawns and the nations to which the intercepting fishermen belong. Is this not exactly the system laid down in the Proclamation? "Where fishing activities have been or shall hereafter be legitimately developed and maintained jointly by nationals of the United States and nationals of other States... conservation zones may be established under agreements between the United States and such other States". Such agreements would comprise the interested States as suggested above. "May be established" would suggest, that if the other nations concerned are not willing to come to an agreement, that the planned conservation zone in the area where United States nationals and nationals of other States jointly fish, would not come into being. This was also the opinion of Hudson in the 69th meeting of the International Law Commission [67], p. 23: "Of source, if a country was unwilling to conclude such an agreement, nothing could be done about it". François answered (loc. cit.) that he thought "that the difficulty in practice would be precisely that of obtaining the consent of all the countries engaged in the fishing operations. That solution was therefore not a very effective one".

Another point is, what is meant with "legitimately developing and maintaining fishing activities?". It is perfectly normal that a State lays down rules for its own nationals to be followed up outside its territory. Establishing conservation zones for that purpose does not infringe on International Law. There would be no objection to an American fishery patrol vessel or cruiser supervising American fishermen in these zones, for instance to see that no interception of salmon takes place in a closed period. But what would happen if a foreign fisherman appeared on the scene and started fishing? Would that be considered to be "illegitimately developing fishing activities"? Does it mean that fishing activities can only be developed according to law, and to what law? Do these foreigners have to register anywhere to be licensed to fish? It reminds one of the fight over licenses demanded by the British for fishing near their coasts and the endless struggle with the Dutch in the 17th century, so well described by Fulton [31].

The character of the high seas, as understood up to now in International Law, would be affected if any interference with foreign fishermen would take place.

It may be a warning, that the Proclamation of President Truman has already been interpreted in a way which justifies the doubts about the working in relation to third parties.

Bishop [68], p. 10–11, discussing this Proclamation, especially the part referring to areas where apart from the United States other States are interested, says: "Thus, in the region of Alaska and Canada, where the United States and Canada have built up the Pacific halibut fisheries, fishermen of each nation fishing in the high seas off the other's shores as well as off their own, the two states by joint agreement may establish and enforce conservation regulations, which must be followed by *anybody* fishing in such areas, *even if he be from a third state*". (underlining by us). ".... Such conservation measures cannot achieve full success unless they are made applicable to all persons and vessels of whatsoever nationality engaged in fishing in the area".

Fenwick [69], speaking about both Proclamations of President Truman, says on p. 423: ".... it is clear that both proclamations contemplate encroachments upon traditional rights to the use of the high seas, and that in consequence international agreement must be reached upon the new claims before they can be said to constitute a rule of international law".

As to the zone where only United States fishermen are active, it is remarkable that in the Press Release [6], p. 484, the wording differs from the Proclamation, where it explains that in areas where fishing activities have been or shall hereafter be developed by nationals of the United States alone .... "zones will be set up in which the United States may regulate and control *all* fishing activities". The word "all" does not appear in the Proclamation, but where the Press Release is meant as an explanation, this wording may turn out to be significant.

As to the question whether this Proclamation is linked up with the continental shelf, the first impression will be, that the shelf is nowhere mentioned in it. The Proclamation has, however, been issued together with that concerning the natural resources of the continental shelf and in one Press Release they have been explained together. This fact gives rise to the thought that a common idea may have inspired both Proclamations. The passages in the continental shelf Proclamation: ".... the urgency of conserving and prudently utilizing its natural resources" and that in the Proclamation with respect to coastal fisheries: "the pressing need for conservation and protection of fishery resources", both passages being used as the reason why the following policy is adopted, would justify the belief of a common idea behind these Proclamations.

Both these documents, according to this train of thought, are built on the law of conservation, "le droit de conservation", advocated by Fauchille [70], who describes his idea as follows, p. 147 (par. 492 [9]): "Le système que nous adoptons est celui du droit de conservation. Que faut-il entendre exactement par cette formule? La conservation de soi-même constitue un droit fondamental des Etats: elle est même pour eux un devoir. Un Etat est dès lors autorisé à prendre toutes les mesures destinées à assurer son existence, à se défendre contre tous les actes pouvant porter atteinte aux éléments de celle-ci, c'est-à-dire à son territoire, à sa population, à sa richesse matérielle ....", and p. 148: "Il lui appartiendra enfin de protéger ses propres intérêts économiques et ceux de ses ressortissants. Et tout cela, non seulement sur son sol même, mais encore sur les eaux environnantes, qui, nous l'avons dit, sont une partie du vaste océan, comme lui libres de toute souveraineté" .... "C'est ainsi le droit de conservation de l'Etat qui explique pourquoi il doit avoir certains droits sur la partie de la mer contiguë à son sol; c'est lui qui, en fin de compte, justifie et légitime l'existence de ce qu'on appelle la 'mer territoriale'. Telle étant la raison d'être des droits de l'Etat sur la mer littorale, il va de soi qu'il doit y posséder tous les droits nécessaires à la sauve-garde des éléments de son existence, mais qu'il ne doit pouvoir y exercer que les seuls droits indispensables à cette sauve-garde et seulement dans les limites réclamées par elle: en-dehors de là, la mer côtière doit, comme la haute mer, demeurer pleinement ouverte à l'usage de tous".

If this is the common idea at the basis of both Proclamations, we should object against both, because the right of conservation as meant by Fauchille has not been adopted in International Law. His theory is unacceptable because it would allow each State to decide as to the extent of the right of conservation, driven by its own egoistic motives, which would lead with unfailing certainty to conflicts with other States.

Whether this is the common idea at the basis of both Proclamations is not certain, but it becomes plausible in the light of the theory which was suggested by the United States in the Behring-Sea case. To quote Hall [71], p. 309 note 2: "In the arguments laid before the Behring-Sea Arbitral Tribunal on behalf of the United States, it was advanced as a proposition of law that a State has a right to make enactments under which it can assume jurisdiction upon the high seas, exercisable at an indefinite distance outside territorial waters, for the purpose of safeguarding property, and of protecting itself against acts 'Threatening invasion of its interests'. The laws so passed were alleged to be 'binding upon other nations because they are defensible acts of force which a State has a right to exert'. In support of the supposed right, the practice of nations was adduced in the form of 'Hovering acts', of fishery regulations, etc.".

We believe that even if a common idea has inspired both Proclamations, the Proclamation with respect to coastal fisheries is independent from the continental shelf theory. The idea of creating zones is not new. But the zones suggested by Suárez, in his Report [52] (p. 124) concerned whales and were quite different in character. He proposed exploitation of each zone in turn and for a limited period; uniformity of methods (without going into details of a nature to hamper industrial freedom), in order to insure the full utilization of the products of the chase, which today are squandered, owing to the thirst for immediate gain at all costs; adoption of general rules regarding the ages at which whales and seals should not be killed even when found in zones and during periods not subject to prohibition.

In spite of our criticism, we have to admit that the efforts of the United States in regulating fisheries might serve as an example to other States. A certain impatience with fishermen not wanting to see the general interest is therefore understandable. Let us hope that our criticism is exaggerated and that all conservation regulations will be laid down in agreement with other States interested, and that all these States will indeed show willingness to co-operate.

Nothwithstanding the excellent scientific work done by the International Council for the Exploration of the Sea at Copenhagen the protection of fish in the seas and oceans off the European coasts is still far from satisfactory.

Mexico. Presidential Declaration, 29 October, 1945 [72], p. 185 [a].
The relevant passages follow after an explanation of the continental shelf-theory and those concerning mineral resources:

"In the years before the war the Western Hemisphere was obliged to watch permanent fishing fleets of extracontinental countries engage in the immoderate and exhaustive exploitation of this immense wealth, which, although certainly it must contribute to the welfare of the world, must obviously belong in the first instance to the country which possesses it and to the continent of which that country forms a part. By reason of the very nature of this wealth, it is indispensable that this protection should be exercised by extending the control and supervision of the State to the places or zones indicated by science for the development of breeding-grounds of the high seas, irrespective of the distance separating them from the coast.

For these reasons, the Government of the Republic claims the whole continental shelf adjacent to its coasts and all and every one of the natural riches, known or still to be discovered, which are found in it, and will proceed to supervise, utilize

---

[a] *See also* [158], *p. 13 (Slightly different translation).*

and control the zones of fishing protection which are necessary for the conservation of this source of well-being.

The foregoing declaration does not mean that the Mexican Government is attempting to disregard legitimate rights of third parties, or that the right of free navigation on the high seas is affected, inasmuch as the sole end sought is to conserve these resources for the welfare of the nation, of the continent, and of the world".

Although according to Azcárraga [32], p. 60, and Young [73], p. 851–852, proposals were sent to the Congress, for the amendment amongst others of art. 27 of the Constitution, to the effect that national ownership would be claimed of the sea-waters covering the continental shelf, to the extent fixed by International Law, no data are available that these proposals have been accepted. According to Selak [65], p. 676, the Declaration mentioned above has never been put into effect.

The zones claimed in the Declaration are independent from the continental shelf. Although these zones are called "zones of fishing-protection", the word "utilize" seems to indicate that exclusive fishing-rights were meant to be claimed in these zones.

Argentina. Declaration proclaiming sovereignty over the epicontinental sea and the continental shelf, October 9th, 1946 [74] *a*.

We quote from the Preamble: "Whereas .... The waters covering the submarine platform constitute the epicontinental seas, characterized by extraordinary biological activity, owing to the influence of the sunlight, which stimulates plant life (as exemplified in algae, mosses etc.) and the life of innumerable species of animals, both susceptible of industrial utilization;...."

We quote from the operative part of the Declaration: "Article I. It is hereby declared that the Argentine Epicontinental Sea and Continental Shelf are subject to the sovereign power of the Nation"; "Article 2. For purposes of free navigation the character of the waters situated in the Argentine Epicontinental Sea and above the Argentine Continental Shelf, remains unaffected by the present Declaration; ...."

---

a c.f. the Decree No. 14,708, 11 October 1946 [153], p. 4.

In this Declaration a principle of International Law, the freedom of the high seas is impaired on the basis of a scientific statement, which is, however, not limited in its validity to the region for which it is invoked here. The influence of sunlight, which stimulates plant life is equally active outside the so-called epicontinental seas and will produce "extraordinary biological activity" wherever the other necessary conditions for plant life, as for instance a sufficient amount of nutrients, are available.

Art. 2, although stating that the character of these waters for purposes of free navigation remains unaffected, does, however, affect the free fisheries, being the other basic right resulting from the principle of the freedom of the seas.

The Preamble, moreover, refers to the Governments of the United States of America and of Mexico having; "issued declarations asserting the *sovereignty* of each of the two countries over the respective peripheral epicontinental seas and continental shelves".

This statement seems incorrect, neither in the United States Proclamation nor in the Mexican Declaration the word "sovereignty" in relation to "epicontinental seas" is used. Either this is a case of incorrect reading or else the wording has been interpreted that way. The last explanation would indicate that the legal status of these "epicontinental seas" or zones of fishing-protection or conservation zones as they are actually called are found to be incompatible with that of the high seas, in spite of the assurances in both decuments, that the right of free navigation on the high seas, and in the American Proclamation that the character as high seas of the areas concerned are not affected. Freedom of fisheries, which is one of the typical rights inherent to the freedom of the high seas, does not exist in these zones. In that train of thought the zones resemble more the territorial waters, where freedom of navigation, but no freedom of fisheries exists and which are generally thought to be under the sovereignty of the coastal State. The reasoning is somewhat farfetched and open to criticism, but not altogether unintelligible.

How far these "epicontinental seas" extend from the coast is not mentioned. Apparently as far as the shelf stretches, because these waters are defined as "the waters covering the submarine platform". The continental shelf is taken in the geological sense.

The *Chilian* Presidential Declaration of 25th June, 1947 [72] [a], p. 188, starts with a similar assertion as to the claims of sovereignty in the United States and Mexican documents adding, without doubt rightly the Argentine Decree.

The Preamble then continues:

"3. That particularly in the case of the Republic of Chile there is a manifest advantage in issuing an *analogous* proclamation of sovereignty .... because owing to its topography and lack of mediterranean extension, the country's life is bound up with the sea and with all the present and future riches contained in the sea, to a greater degree than in the case of any other nation.

4. That an international consensus recognizes that each country has the right to consider as national territory the whole extent of the adjacent epicontinental sea and continental shelf".

This is a remarkable statement, and certainly not all international lawyers will agree with it. In the Preamble of the Argentine Declaration is a similar passage, which is, however, more carefully worded: "In the international sphere *conditional* recognition is accorded to the right of every nation to consider as national territory the entire extent of its epicontinental sea and of the adjacent continental shelf". This paragraph does not appear expressis verbis in the Preamble of the Peruvian Decree, but it did appear in the original Preamble of the Decree Law of Costa Rica of 27th July, 1948. We will deal with these two Decrees presently.

The operative part of the Declaration then continues (there is a mistake in the translation, which will be dealt with presently):

"2. The Government of Chile confirms and proclaims the national sovereignty over the seas adjacent to its coasts, whatever their depth may be, to the full extent necessary to reserve,

---

[a] *See also* [153], *p. 6 (slightly different translation. Here the date of the Declaration is given as 23 June 1947).*

protect, conserve, and utilize the natural resources and wealth of whatever nature, found on, in, or under said seas, placing under Government supervision the fishing and marine hunting industries in order to prevent this type of resources from being exploited to the prejudice of the inhabitants of Chile and diminished or destroyed to the detriment of the country and of the American Continent.

3. Demarcation of the zones of protection of maritime hunting and fishing in the continental and island seas which are under the control of the Government of Chile will be made in virtue of this declaration of sovereignty, whenever the Government considers it suitable, by ratifying, amplifying, or in any manner modifying the said demarcations in conformity with the knowledge, discoveries, studies, and interests of Chile which may be made known in the future; at present said protection and control are declared over all the sea included between the perimeter formed by the coast and a mathematical parallel projected out to sea at a distance of two hundred marine miles from the continental coasts of Chile. With respect to the Chilian islands, this demarcation will be made by marking out a sea zone contiguous to the coasts of these islands, projected parallel to these coasts for two hundred marine miles from the whole circumference.

4. The present declaration of sovereignty does not disregard similar legitimate rights of other States, on the basis of reciprocity and does not affect rights of free navigation on the high seas.

Santiago, 25 June, 1947".

This Decree will be discussed together with the next one.

Peru. Presidential Decree 1st August, 1947 [72], p. 190 [a].

The Preamble after having stressed the necessity for conservation of fishing resources goes on: ".... That the fertilizing wealth deposited by guano birds on the islands of the Peruvian coast also requires for its safeguard the protection, conservation, and regulation of the use of the fishing resources which serve to nourish the said birds; ....".

Then a similar reference is made relating to the sovereignty

---

[a] *See also* [168], *p. 16.*

over the epicontinental waters as in the Chilian Decree, including the latter as well.

Then follows the operative part of the Decree, (with the same translation mistake): . . . .

"2. The national sovereignty and jurisdiction are exercised as well over the sea adjacent to the coasts of the national territory, whatever its depth, to the extent necessary to reserve, protect, conserve, and utilize the natural resources and wealth of all types which are found in or under the said sea.

3. As a consequence of these declarations, the State reserves the right to establish the demarcation of zones of control and protection of the national wealth in the continental and island seas which are under the control of the Government of Peru, and to modify the said demarcation in accord with supervening circumstances, by reason of new discoveries or studies, or national interests which may become apparent in the future; and declares at present that it will exercise the said control and protection over the sea adjacent to the coasts of Peruvian territory in a zone lying between those coasts and an imaginary line parallel to them, drawn on the sea at a distance of two hundred (200) marine miles, measured by following the line of the geographical parallels. With respect to the national islands, this demarcation will be drawn by marking out a zone of the sea contiguous to the coasts of the said islands, up to a distance of two hundred (200) marine miles measured from every point on the circumference of the islands.

4. The present declaration does not effect the right of free navigation of ships of all nations, in conformity with international law".

In both the Chilian Declaration and the Peruvian Decree, the Preamble makes reference to the continental seas, and epicontinental seas.

Only in the Peruvian Preamble these terms are defined where it stresses the necessity of conservation of fishing resources, "and other natural wealth which is found in the epicontinental waters which cover the submarine platform, and in the continental seas adjacent to it . . . .".

In the actual declaration and decree, no reference is made to the continental shelf, in so far as fisheries are concerned.

This is natural because, although the continental shelf along the northern coast of Peru is still fairly wide (maximum about 70 miles) from 15 degrees southern latitude to the south it is narrow, whereas for Chile the shelf is extremely narrow and in certain areas hardly existent. Therefore these countries seek protection for the fisheries naturally outside the waters covering their continental shelf.

We have seen in the beginning of this Chapter that the Pacific Ocean along the coasts of these countries is extremely rich in marine life. Schweigger [75] investigated the distribution of fish along the Peruvian coast. On the maps 1A–1G (p. 9–15) he gives the results of his observations concerning the occurrence of different species of fish and other marine animals in the Pacific from about 4 degrees south latitude to nearly 17 degrees south latitude, along the Peruvian coast.

Compared with the maps 2 (A–G), (p. 18–20), these maps prove that, although the largest density of fish is met on the continental shelf, many species are found outside, at considerable distances from the shelf.

The same can be said about the places where fisheries take place, shown on the maps 7 (A–G), p. 98–101.

Schweigger concludes his discussion on distribution of fish by saying, p. 102–103, that the accumulation of fishes does not follow the configuration of the continental shelf.

Data concerning the places where fisheries take place, which would be the best proof of our thesis, that fish live also in areas far outside the continental shelf, are unfortunately not everywhere available.

It is clear that fisheries far outside the coast are more costly and need capital behind them. The west coast of Africa, very rich in marine life as we have seen, doubtlessly fished by natives near the coast, has not yet been exploited at greater distances from the coast. In Peru it is only the last five years, that a great extension of fisheries has taken place, with the help of American capital. When Schweigger wrote (1943) the fisheries were not mechanized and the small boats with their hand equipment could not go far from the coast, without the means to keep the fish fresh for a longer period (p. 253). Scully [76], p. 7–8, describes the recent development of the Chilian and Peruvian fisheries:

"From Talcahuano, Chile, to Ecuador's Galápagos Islands, this area had suddenly become the great new fishing grounds of the world .... Last year most of the United States' 189,000,000 pounds of tuna were caught in these waters, frozen, and shipped to California for packing. More than eighteen million cans of South American-packed bonito, a delectable fish new to most people in the United States, were sold here. Along with these came five thousand tons of frozen swordfish and lesser lots of other species. This new inter-American industry has made jobs in South America for some thirty thousand fishermen, cannery-workers, and boat-builders.... From the massive Andes mountains that rise within sight of the coast, dozens of rivers carry huge loads of life-giving nitrates and phosphates into the Pacific. But, instead of settling and being lost, these elements are constantly stirred to the surface by the swift, cold Humboldt Current that sweeps up from the Antarctic, hugging the coasts of Chile and Peru until it swerves westward off Ecuador to wash the Galápagos Islands .... While the current itself is only 20 to 80 miles wide along the coast, it is impossible to bound the area influenced by its life-making activities, especially after it turns westward. But twenty-three species and five hundred million pounds of fish were taken last year in big numbers".

From the F.A.O. Statistics [77], p. 39, it appears that the Chilian catch has increased from 30,572 metric tons in 1938 to 76,246 in 1949, and from an article on "Better utilization of fisheries resources in Latin America" [78], p. 5, the figures for Peru in 1939 were 4,849 metric tons, and in 1949 45,260.

If the long since uttered desirability to extend the zone for exclusive fishery rights or at least to lay down conservation measures is to be materialized at all, these countries have certainly as much reason to claim the said rights as those which are in the possession of a continental shelf.

Peru has an extra reason, because the fish form the food for the guano birds, which are an economic asset to the country. Murphy [a], p. 64, quoted by Brongersma [55], p. 76, remarks concerning the coastal water of Peru: "the existance in littoral waters of a vast abundance of marine organisms, upon which

---

[a] R. C. Murphy, "The oceanography of the Peruvian littoral", Geogr. Review, Vol. 13, p. 64–85, 1925.

are dependent in turn unsurpassed fishery resources, as well as the remarkable Peru guano industry". Sverdrup [54], p. 942: "The great production of this oceanic system (The Peru Coastal Current) is manifest in the tremendous quantities of marine birds in this area. Some idea of the magnitude of production can be had from Schott's [a] Report (1932) "that on one small island of the Chinchas group there are estimated to be some five or six million marine birds, such as cormorants, pelicans and gannets, which daily remove at least 1,000 tons of small fish from the surrounding water. The great Peruvian guano deposits on the shores of these regions may be as much as 30 metres in thickness ...."".

Chile has an extra reason because of the length of its coastline compared with the surface of its territory.

The claim of 200 miles sounds extravagant, it is true, but the basis of their claims is not unsound.

In both documents sovereignty is claimed over the seas adjacent to the coasts, whatever its depth, and according to the translation *"to the* (full) *extent necessary to reserve, protect, conserve* and *utilize* the natural resources ....".

In both documents it is then declared that at present this protection and control will be exercised in a zone between the coast and a parallel at a distance of 200 miles from the coast. Hence it follows that at present both documents claim sovereignty over a part of the high seas far beyond the territorial sea. Possibly through using this or a similar translation the opinion has been expressed that sovereignty was claimed *only as far as* necessary for the protection or conservation of the fisheries. A very limited sovereignty indeed.

François said in the 69th meeting of the International Law Commission [67], p. 22: "Chile did not claim sovereignty for all purposes, but solely for the purpose of protecting natural riches ....", and Hudson (loc. cit.): "It did not .... represent a mere proclamation of full rights of sovereignty".

We, however, believe that the translation of this passage in both documents is wrong. Paragraph 4 of the Chilian Preamble, which we have criticized above, stated that every country has

---

[a] G. Schott, "The Humboldt Current in relation to land and sea conditions on the Peruvian coast", Geography V, 17, p. 87–98, Manchester, 1932.

the right to consider *as national territory* the *whole extent* of the
adjacent epicontinental sea and continental shelf. In the Spanish
text "for the whole extent" we read "toda la extensión", meaning
"the whole area, space or compass". The translation here is
correct, because "extent" cannot mean anything else here than
area, space or compass.

In paragraph 2 of the operative part of the Chilian Declaration
and of the Peruvian Decree the same Spanish words are used:
"El Gobierno, de Chile .... proclama la soberanía nacional
sobre los mares adyacentes a sus costas .... *en toda la extensión*
necessaria para reservar ...." etc. Here the underlined words
have been translated by: "to the full extent" which means in
English to the full degree. Hence the opinion that the sovereignty
is limited. This translation is wrong and should read "to the
full area, extensiveness or space, necessary to reserve, etc." [a].
It is clear that *if* sovereignty is claimed to the full *degree* "necessary to reserve", that only a limited sovereignty would be meant.
If, however, sovereignty is claimed for the *area* "necessary to
reserve" the words "necessary to reserve" do not limit the
degree of sovereignty, but the area over which sovereignty is
claimed. In this area the sovereignty is full and only conditioned
by the last paragraph saying that the declaration does not
affect rights of free navigation on the high seas. There is no
reason to translate "toda la extensión" in par. 4 of the Preamble
differently from the same words in par. 2 of the actual Declaration. Moreover why is the statement in par. 4 of the Preamble
made if it was not meant to be followed up in the Declaration?
Brierly, in the same meeting of the International Law Commission [67], p. 21, said that "the Chilian proclamation represented
a claim to extend the country's territorial waters to a very
large area".

---

[a] The word "extent" according to the Shorter Oxford Dictionary means "space
or degree to which anything is extended". If we take the meaning of "degree" the
sovereingnty is conditioned. If we take the meaning "space" the sovereignty is full
in that space. 2. International Law Quarterly ,1948[79], p. 135, gives a better translation: "within those limits necessary in order to reserve etc.", in the Chilian Declaration, and p. 137: "in the extension necessary to reserve etc.", in the Peruvian Decree.
Selak [65], p. 673, uses the incorrect translation "to the extent" and Young [73], p. 853:
"to whatever extent" which is better, but does not exclude the wrong interpretation.
He states, however, that full sovereignty was asserted over the epicontinental sea
in both documents. *The Int. Law Quarterly-translation also in* [153] *p. 6 and 17.*

Another point we cannot agree with, even if the translation had been the right one, quod non, is that only conservation rights are claimed. They claim the sovereignty and jurisdiction to *reserve*, protect, conserve and *utilize*, the natural resources (in the Chilian Declaration "aprovechar", in the Peruvian Decree "utilizar"). The wording, it is true, is somewhat queer, because one could claim the whole area of sea necessary to protect or conserve the natural resources: in order to protect them properly one needs a minimum of space necessary to make the protection effective. The area of sea necessary to utilize the resources i.e. the fish, seems to be rather an elastic one. At any rate for the protection *and control*, at present an area of 200 nautical miles parallel to the coast is deemed to be necessary and in that area Chile and Peru claim the *full* sovereignty. This was necessary, because otherwise conservation regulations could only be enforced against their own nationals and not against fishermen of another nationality. Moreover because the utilization of the fish is one of the reasons for claiming sovereignty, an exclusive fishing right is meant. We agree with François [80] (p. 929), that the Chilian Declaration (and the Peruvian Decree) are not in agreement with the Law of Nations. There may be just reasons for these countries, and in particular for Chile, with its long coastline compared with the surface of its territory, to claim a wider margin than the 3-mile limit, for exclusive fishery rights, as we believe that Norway has a just reason for such a wider claim, because of particular local circumstances, but a 200-mile margin is, according to our opinion, extravagant.

As to conservation regulations, these should, as we will explain later, be agreed upon with other States, which have fishery interests in those regions and if no agreement is possible, an international body should be empowered to lay down such regulations in that area.

Concerning conservation we therefore disagree with the method used.

Costa Rica first issued a Decree Law of 27 July, 1948 [72], which was very similar to that of Chile. But this Decree was replaced by Decree Law No. 803 of 2 November, 1949 [153], p. 9.

In Art. 2 of this Decree (compare with Art. 2 of the Chilian

Decree) the word "sovereignty" is replaced by the words "rights and interests" and the wording of the original Art. 3: "The demarcation of the zones of protection .... shall be made in accordance with this declaration of sovereignty" .... has been changed in the new article 3 by leaving the word "sovereignty" out.

The effect of this Decree is, however, exactly the same as the previous one.

Honduras: a. Congressional Decree No. 102, amending the Political Constitution, 7 March, 1950, "La Gaceta: Diario Oficial de la República de Honduras", Vol. 75, No. 14.055 (16 March, 1950), p. 2. Translation by the Secretariat of the United Nations: [153], p. 11.

The articles 4 and 153 of the Political Constitution are amended and shall read as follows: Art. 4:

"The limits of Honduras and its territorial division shall be determined by law. The submarine platform or continental and insular shelf, and the waters which cover it, in both the Atlantic and Pacific Oceans, at whatever depth it may be found and whatever its extent may be, forms a part of the national territory".

And Article 153 (quoted as far as necessary):

"The following belong to the State: Full, inalienable, and imprescriptible dominion of the waters of the territorial seas to the extent of twelve kilometers measured from the low-water mark, ... also the dominion, likewise full, inalienable, and imprescriptible, over all the resources which exist or may exist in its submarine platform or continental and insular shelf, in its lower strata, and in the area of the sea included within vertical planes constructed on its boundaries".

Marine resources are thus claimed, but it is not clear whether the boundaries mentioned in Art. 153 refer to the territorial seas or to the limits or "boundaries" mentioned in Art. 4 (the word "boundaries" appears in the translation of Appendix No. 8 to Annex to Letter, dated 19th September, 1951, from the Agent for the Government of the United Kingdom to the Registrar of the International Court of Justice [81]).

b. Congressional Decree No. 103, amending the Agrarian Law, 7 March, 1950, "La Gaceta: Diario Oficial de la República de Honduras", Vol. 75, No. 14.055 (16 March, 1950), p. 2. Translation by the Secretariat of the United Nations [153], p. 12.

Art. 1. "The first article of the Agrarian Law is amended and shall read as follows:
.... The following belong to Honduras .... (3)", cf. Art. 4, quoted above.

c. Legislative Decree No. 104, dated 7 March, 1950 (quoted from Appendix No. 8[81]) [a]: "The National Congress decrees: Art. 1 — Articles 619 and 621 of the Civil Code are hereby amended to read as follows: [b] .... Art. 621: "The adjacent sea to a distance of twelve kilometres, measured from the limit of the lowest tide, constitutes territorial waters under national ownership: however, the sovereignty of the State extends to the submarine platform, or continental and insular shelf and the waters which cover it, whatever be its depth and however far it extends, without affecting the right of free navigation in conformity with International Law".

d. Congressional Decree No. 25 (approving Presidential Decree No. 96 of 28 January, 1950), 17 January, 1951, La Gaceta, Vol. 76, No. 14,306 (22 January, 1951), p. 1. Translation by the Secretariat of the United Nations, [153], p. 302–303.

.... "Whereas legal doctrine has acknowledged and international law has declared that the said shelf belongs lawfully to the riparian States, which are entitled to proclaim their sovereignty over it and over the waters covering it, as is shown by the statements of .... (then follow the documents from the United States, Mexico, etc. mentioned above) ....

Now therefore the President, in the Council of Ministers, decrees as follows:

Article 1. It is hereby declared that the sovereignty of Honduras extends to the continental shelf of the national territory, both of the mainland and of the islands, and to the waters covering it, at whatever depth it lies and whatever its extent, and that the nation has full, inalienable and impre-

---

[a] *See also* [153], *p. 301 (slightly different translation; The Decree is here called: "Congressional Decree No. 104").*

[b] *Art. 619: "*.... *Ownership of all natural wealth, existing or that may exist, in its submarine platform or continental and insular shelf, in its lower strata and in the sea space included within the vertical planes rising from its limits, shall be vested in the State"* .....

scriptible domain over all wealth which exists or may exist in it, in its lower strata or in the area of water bounded by the vertical plane passing through its borders.

Article 2. The zone of protection of hunting, fishing and exploitation of the mainland and island waters falling by virtue of this Decree within the State's jurisdiction shall be delimited in accordance with this declaration of sovereignty whenever the Government shall see fit, and such delimitation shall be ratified, extended or amended as the national interest may require.

Article 3. The protection and supervision of the State is hereby declared to extend in the Atlantic Ocean over all waters lying within the perimeter formed by the coast of the mainland of Honduras and a mathematical parallel drawn at sea 200 sea-miles therefrom. With regard to the islands of Honduras in the Atlantic, such delimitation shall enclose the zone of sea contiguous to their coasts and extending for two hundred sea-miles from every point thereon.

Article 4. Subject to reciprocity, this declaration does not deny similar lawful rights of other States, nor affect the freedom of navigation recognized in international law, nor derogate from the rights of sovereignty and domain held by the State of Honduras over its territorial waters".

The doubt expressed above about the extent of the claim concerning marine resources is taken away by this Decree. Obviously exclusive fishery rights are claimed.

We dot not know how the eastern borderline of this zone is drawn from the point where the border between Honduras and Guatemala meets the coast, but it seems that the area runs right across the waters adjacent to British Honduras.

Panama. b. Decree No. 449, for the regulation of shark fishing by foreign vessels in the waters under the jurisdiction of the Republic, 17 December, 1946. "Gaceta Oficial", Vol. 43, No. 10,181 (24 December 1946), p. 2. Translation by the Secretariat of the United Nations [153], p. 16.

Article 3. "For the purposes of fisheries in general, national jurisdiction over the territorial waters of the Republic extends to all the space above the sea-bed of the submarine continental shelf. For this reason the product of any fishing within the limits indicated is considered a national product, and is therefore subject to the provisions of the present Decree".

Here again exclusive fishing rights are claimed.

El Salvador. Political Constitution, 7 September 1950, Diario Oficial, Vol. 149, No. 196 (8 September, 1950), p. 3105 [153], p. 300.

. . . .

Article 7. "The territory of the Republic within its present boundaries is irreducible. It includes the adjacent seas to a distance of two hundred sea-miles from low-water line and the corresponding air space, subsoil and continental shelf.

The provisions of the foregoing paragraph shall not affect the freedom of navigation in accordance with the principles recognized under International Law.

The Gulf of Fonseca is a historic bay subject to a special régime".

. . . .

Ecuador. Decree of the Congress of the Republic of Ecuador, dated 21st February, 1951, relating to territorial waters (quoted from Appendix No. 9 [81]).

"The Congress of the Republic of Ecuador considering.

Whereas it is urgent to determine in an exact form the jurisdiction of Ecuador over the territorial waters;

Whereas the American Community of Nations adopted the Resolution on Territorial Waters recorded at the 1st and 2nd meetings of Ministers for Foreign Affairs, held in Panama and Havana in the years 1939 and 1940 respectively, at which it was recommended that 'The American States should adopt in their particular legislation the principles and rules contained in such declarations' and

Whereas as a consequence of military progress the nations are enlarging the limits of their jurisdiction over territorial waters.

Decrees.

Article 1. The continental shelf or zocle adjacent to the Ecuadorian coasts and all and every natural resources found thereon belong to the State, which will control the exploitation of such resources and the protection of the corresponding fishing areas.

Article 2. The Ecuadorian continental shelf is considered to comprise the submerged land, contiguous to continental territory, which is covered by not more than 200 metres of water.

Article 3. National territorial waters comprise a minimum distance of 12 nautical miles measured from the outermost promontories of the Ecuadorian Pacific Coast as well as the inner waters of the gulfs, bays, straits and canals comprised within a line drawn between such promontories.

Also considered as the territorial sea are those waters comprised within a perimeter of 12 nautical miles measured from the outermost promontories of the farthest islands of the Colón Archipalego, the stipulations of Art. 1 of this law being applicable in this case.

Article 4. Should in accordance with the terms of any International Conventions or Treaties on this subject such as the Treaty of Mutual Assistance, the maritime areas agreed upon for policing and protection be greater than those laid down in this law, the terms of such Treaties will prevail and will be enforced as part of this Decree within the extent and range of such Treaties".

Article 1. seems to claim only protective rights for fishery in the waters covering the shelf, although the Decree of the President of the Republic of Ecuador, dated 22nd February, 1951, related to the Law on sea fishing and hunting (Appendix No. 9[81]) states in Art. 2: "For purposes of sea fishing and hunting in general the territorial waters of the Republic will be considered to comprise 12 nautical miles ...." etc. cf. Art. 3 above [a].

Iceland. Law No. 44, concerning the scientific conservation of the continental shelf fisheries, 5 April, 1948. "Stjórnartidtindi" 1948, A. 4, p. 147. Translation by Secretariat of the United Nations [153] p. 12.

Article 1. "The Ministry of Fisheries shall issue regulations establishing explicitly bounded conservation zones within the limits of the continental shelf of Iceland, wherein all fisheries shall be subject to Icelandic rules and control; provided that the conservation measures now in effect shall in no way be reduced. The Ministry shall further issue the necessary regulations for the protection of the fishing grounds within the said zones....

The regulations shall be revised in the light of scientific research.

Article 2. The regulations promulgated under Article 1. of the present law shall be enforced only to the extent compatible with agreements with other countries to which Iceland is or may become a party".

---

[a] *Brazil. Decree No. 28,840 integrating into national territory the adjoining part of the continental shelf, 8 Nov. 1950 Diario Oficial, Vol. 89 No. 264 (18 November 1950) p. 16,617. Translation by the Secretariat of the United Nations [153], p. 299. .... Article 3. "The rules governing navigation in the waters covering the aforesaid continental shelf shall continue in force without prejudice to any further rules which way be made, especially as regards fishing in that area".*

During the Conference of the International Law Association at Copenhagen, 27 August — 2 September, 1950, Mr. P. Fischer of Denmark, speaking of the Icelandic Act said that in accordance with this Act, the Icelandic Government on the 22nd April, 1950, issued regulations regarding the protection of fish at the North Coast of Iceland, prohibiting to use trawls and seines within 4 nautical miles from shore; only Icelandic fishermen were allowed to fish in this area. These new Icelandic provisions indicate an extension of the Icelandic territorial waters in respect of fishing as these waters in recent times have been limited to only 3 nautical miles.

It seems that particularly in countries of South and Central America an epidemic has broken out, characterized by an insatiable thirst for salt water and a great gusto for fish. However International Law does not forbid any State to claim rights, it only prohibits these rights to be enforced against other States if they do not conform with general principles of the Law of Nations. Several States have expressed the view, that they do not consider these provisions to be binding against their own nationals. We will review these protests in the next Section. In enumerating the Declarations and Decrees in this Section we have limited ourselves to those which were possibly related to the continental shelf and mentioned fishery rights expressis verbis. In Chapter III we will discuss other Decrees relating to an extension of territorial waters, which may of course limit fisheries on the high seas, if the countries concerned adopt exclusive fishery rights in their territorial waters.

Section 4. Protests

The Government of the United States informed the Governments of Chile and Peru [a], by means of notes dated 2nd July, 1948, that "it reserved the rights and interests of the United States so far as concerns any effects of the Declaration or the Decree in question, or of any measures designed to carry that Declaration (Decree) into execution". Taking the note to Chile as an example we quote:

---

[a] For the United States notes to Chile, Peru and Argentina see "Replies from Governments to Questionnaires of the International Law Commission" [82], p. 113–116.

"The U.S. Government notes that the principles underlying the Chilian Declaration differ in large measure from those of the United States Proclamations and appear to be at variance with the generally accepted principles of International Law ...." and "in particular that (1) the Chilian Declaration confirms and proclaims the national sovereignty of Chile over the continental shelf and over the seas adjacent to the coast of Chile outside the generally accepted limits of territorial waters, and (2) the Declaration fails, with respect to fishing, to accord appropriate and adequate recognition to the rights and interests of the United States in the high seas off the coast of Chile".

A similar note was sent on the same date to Argentina [a].

On 12 December, 1950, the United States Government sent a note to the Government of El Salvador [83], the last sentence of which reads, p. 24: "My Government desires to inform the Government of El Salvador, accordingly, that it will not consider its nationals or vessels or aircraft as being subject to the provisions of Article 7 or to any measures designed to carry it into execution".

On 7 June, 1951, the Government of the United States sent a note to the Government of Ecuador (Appendix No. 13[81]) which was very similar to the previous ones.

On 7 April, 1951, the Government of the French Republic sent a note to the Government of the United Kingdom (Appendix No. 14[81]) answering a request to inform the United Kingdom as to the attitude of the French Government towards the claims of certain Latin American countries concerning extension of their territorial waters.

The French Government replied as follows:

"Le Gouvernement français n'a jamais reçu, par la voie diplomatique, notification des résolutions ou propositions adoptées, de 1945 à 1950, par le Mexique, le Chili, le Perou, Costa-Rica et le Salvador, ayant pour effet de changer la limite de leurs eaux territoriales. Il n'a donc pas eu, dans ces cas précis, à formuler un avis.

Il estime cependant sur un plan général que de telles revendications ne sont pas recevables car elles lui paraissent en contra-

---

[a] For the United States notes to Chile, Peru and Argentina see "Replies from Governments to Questionnaires of the International Law Commission" [82], p. 113–116.

diction avec un principe de Droit International qui n'a jamais, jusqu'à présent, été contesté.

Les revendications contenues dans les décrets pris par les pays intéressés excèdent sans aucun doute l'étendue maxima des eaux territoriales admises en Droit International, même en tenant compte du fait que cette étendue est assortie parfois d'une 'zone contiguë' dans laquelle l'Etat adjacent peut exercer certains droits spéciaux (sûreté, police, douanes). Aucun Etat ne peut, par une déclaration unilatérale, étendre sa souveraineté sur la haute mer et rendre cette annexion opposable aux pays qui ont le droit d'invoquer le principe de la liberté des mers, tant que ces derniers ne l'auront pas formellement acceptée. Une renonciation à une règle de Droit International établie dans l'intérêt de la communauté des nations ne peut pas se présumer.

Telle pourrait être la position que le Gouvernement français soutiendrait si un quelconque pays lui notifiait officiellement sa résolution d'étendre la limite de ses eaux territoriales. Cette position n'a aucun caractère confidentiel puisqu'elle est fondée sur des principes universellement reconnus de Droit International''.

On 6 February, 1948, the United Kingdom Government delivered a Protest to the Government of Peru and on the same date to the Government of Chile (Anglo Norwegian Fisheries Case, British Reply, Vol. II[84], p. 27–28 and p. 29–30). Referring to our interpretation we have given about the purport of these claims, we find the same view in these notes for instance on p. 27: "The Peruvian Government's action .... in claiming that sovereignty may be extended over the large areas of the high seas above the continental shelf appears to be quite irreconcilable with any accepted principle of international law ....".

It is strange that the areas are described as lying "above the continental shelf", because most of the 200 miles is far beyond the continental shelf, especially in the case of Chile.

The notes then state that His Majesty's Government are prepared to enter into negotiations with the Peruvian (Chilian) Government "and with any other Government which may have an established interest in the waters concerned, in order to agree on such protection and conservation of the resources in the sea as can be proved to be necessary in the common interest". Reference is then made to the International Agreement for the Regulation of Whaling signed in Washington on 2nd December,

1946, by which (p. 30) "the contracting parties mutually imposed upon themselves certain restrictions directed towards the conservation of whales". "His Majesty's Government .... do not recognize and will not consider their nationals as being subject to any measures of restriction or control over the high seas outside territorial waters which the Peruvian (Chilian) Government may see fit to promulgate in pursuance of the declaration". A similar statement is made relating to British whaling vessels.

On 23 April, 1951, the Government of the United Kingdom sent a note to the Government of Honduras (Appendix No. 8[81]). A few passages will suffice from the paragraphs 4 and 5:

"4. The action of the Government of Honduras, moreover, in claiming that sovereignty may be extended to a distance of 12 kilometres from the coast of the Republic or alternatively over large and undefined areas of the high seas above the continental shelf, appears to be irreconcilable with the principles of international law governing the extent of territorial waters formerly recognized by the Government of Honduras and by the great majority of other maritime States. In this connexion His Majesty's Government in the United Kingdom wish to place it on record with the Government of Honduras that they do not recognize the claim of Honduras to exercise sovereignty over waters, outside a limit of 3 miles measured from the low-water mark along the coast.

5. His Majesty's Government in the United Kingdom recognize, however, that the protection of fisheries and the conservation of natural resources in the high seas outside territorial waters are a proper object of agreement between all interested States. They regard as a desirable model for this type of agreement the North-West Atlantic Fisheries Convention negotiated between no fewer than eleven States interested in developing and maintaining the fisheries in the North-West Atlantic and signed in Washington on 8th February, 1949 .... They note, however, with regret that Legislative Decrees Nos. 102 and 104 claim to establish sovereignty over the high seas without having obtained any agreement of this type and without providing any safeguards with respect to the established interests of other States. They therefore wish to place it on record with the Government of Honduras that, until such an agreement has been reached, they do not recognize and will not consider their nationals as being subject to any measure of restriction or control over the high seas outside territorial waters, which the Government of

Honduras may see fit to promulgate in pursuance of the above mentioned Legislative Decrees."

On 10 September, 1951, the Government of the United Kingdom sent another note to the Government of Honduras (Appendix No. 8[81]), referring to Decree No. 25, in which particular reference was made to the Preamble (quoted above).

This passage reads: "His Majesty's Government do not accept this statement as a correct statement of the international law bearing on this question".

The note ends by stating: .... "His Majesty's Government wish to bring it to the notice of the Government of Honduras that they cannot accept, as being in accordance with the principles of international law, Article III of Legislative Decree No. 25 of 17th January, 1951, in so far as it claims to extend the protection and control of the State in the Atlantic Ocean over the whole extent of sea composed within the perimeter formed by the coast and a parallel line 200 nautical miles distant from the north coast of the mainland of Honduras".

On 14 September, 1951, the Government of the United Kingdom sent a note to the Government of Ecuador (Appendix No. 9[81]). In paragraph 3 it is said:

".... His Majesty's Government wish further to emphasize that, in their view, Article 3 of the Decree of 21st February, 1951, is contrary to international law in that, not only does it claim a 12-mile limit, but it also fails to state that, subject to certain generally recognized exceptions, such as bays and islands, the outer limit of territorial waters must be measured from the low-water mark along the entire coast.

The formula indicated in Article 3 seems to envisage the drawing of baselines between the 'outermost promontories of the Ecuadorian Pacific Coast' regardless of the distance apart of such promontories and regardless of the fact whether the waters enclosed by the baselines drawn between successive promontories constitute a bay in law or not".

and in paragraph 4: ".... His Majesty's Government cannot, however, accept any Ecuadorian claim generally to control fishing areas outside the 3-mile limit of territorial waters. In this connexion His Majesty's Government wish to draw the attention of the Government of Ecuador to Article 3 of Part 1 of the Annex to the Report of the International Law Commission covering its third session, 16th May — 27th July, 1951 (U.N.

doc. A/CN.4/48 of 30th July, 1951, at p. 57) which, in their view accurately states the existing law on this subject. The article says:

'The exercise by a coastal State of control and jurisdiction over the continental shelf does not affect the legal status of the superjacent waters as high seas'.

In the conception of His Majesty's Government in the United Kingdom, there is no right under international law to control fishing outside the limit of territorial waters unless the right forms part of a historic claim to the regulation of sedentary fisheries, and even then such regulation does not affect the general status of the area as high seas".

It is also stated that Article 2 of the Decree of 22 February, 1951, is not in conformity with the rules of International Law. The note ends with paragraph 6:

"His Majesty's Government in the United Kingdom consider that Ecuador has no right to enforce and the United Kingdom would have no duty to acknowledge the enforcement of those portions of the Ecuadorian Decrees of 21st and 22nd February which His Majesty's Government have stated in this note that they are unable to accept, for the reason that such portions of the Decree are not in conformity with the rules of international law".

On 9 February, 1950, the Government of the United Kingdom sent a note to the Government of Costa-Rica (Appendix No. 10[81]) in which it was said that it was the understanding of His Majesty's Government that the rights and interests of United Kingdom nationals in the high seas will not be affected by the Decree No. 803 or by any measure taken by the Costa-Rican Government under it except as may be agreed with His Majesty's Government. "His Majesty's Government would be glad to receive a confirmation of this understanding".

On 12 February, 1950, the Government of the United Kingdom sent a note to the Government of El Salvador (Appendix No. 11[81]) Paragraph 3 reads:

"In the light of the foregoing considerations His Majesty's Government in the United Kingdom, while not opposed in principle to claims to the exercise of sovereignty over the sea-bed contiguous to the Salvadorean coast, are unable to recognize the claims set forth in the new Constitution''.

The note is further of the same tenor as the one to the Government of Honduras.

On 6 July, 1950, the Government of the United Kingdom sent a note to the Government of Iceland (Appendix No. 6[81]). This note refers to the Regulations published by the Icelandic Government on 22 April, 1950 (the same rules referred to by Mr. Fischer in the Copenhagen Conference of the International Law Association of 1950 mentioned above). We quote the following passages:

.... "His Majesty's Government cannot, however, agree that Iceland is entitled to apply a 4-mile limit within which United Kingdom vessels are excluded from fishing.
The baselines described in Section 1 of the regulations, are unacceptable to His Majesty's Government, being drawn in a manner which they consider contrary to international law. As the Icelandic Government will be aware, the question of the principles which should govern the determination of baselines for fishery purposes is at present under consideration by the International Court of Justice at The Hague in connexion with the rights which Norway is entitled to exercise in this matter. His Majesty's Government trust that the Icelandic Government will pay due regard to the ruling given by the Court and will, if necessary, amend their regulations to conform with that ruling".

In a subsequent note of 23 May, 1951 (Appendix No. 6[81]) the hope is expressed that as Iceland's denunciation of the Anglo Danish Fisheries Convention of 1901 [a] comes into force on 3 October 1951, the International Court of Justice not having decided in the Anglo Norwegian Fisheries Case, on that date, the status quo will be maintained and the new regulations will not be applied before there has been an opportunity for the British and Icelandic authorities to discuss the matter in the light of the judgement to be given by the Court.

On 1 September, 1951, the Belgian Government sent a note to the Government of Iceland. (Anglo Norwegian Fisheries Case,

---

[a] *See* [153], *p. 232.*

Verbatim Report, 18 October, 1951 [85], p. 16). We quote the
following passages:

".... cette nouvelle réglementation, si elle était admise,
constituerait une dérogation au principe de la limite de 3-milles
généralement respectée par les pays riverains de la mer du Nord
.... La Légation de Belgique se permet de demander aux Autorités islandaises de surseoir à la mise en vigueur du règlement
qu'il envisage jusqu'à ce que cette Cour se soit prononcée sur
ce cas" (referring to the International Court of Justice and the
Anglo-Norwegian Fisheries Case).

Finally the Netherlands Government sent a note to the
Government of Iceland (loc. cit. p. 17) in which it was stated:

"The Netherlands Government are of opinion that international law does not allow a State to take unilateral measures
prohibiting or regulating the fishing of foreign vessels in a certain
area of the high seas".

It continues stating that the Netherlands fishery interest

"would suffer from the coming into force of the regulations
in question, which imply by establishing a four-miles zone which
is moreover measured from unilaterally introduced long straight
baselines, an extension of the area in which a State may forbid
the fishing by vessels under a foreign flag".

SECTION 5. OTHER REASONS FOR DEALING WITH THE FISHERIES
INDEPENDENTLY OF THE CONCEPT OF THE CONTINENTAL SHELF

One reason may be derived from the opinion put forward
by François in the 63rd meeting of the International Law Commission [86], p. 10, namely that the Commission could not study
"a subject of such wide scope and which differed so much in its
various aspects from one part of the world to another that
regulations concerning it could not be embodied in a general
code; that question should be left for separate conventions
dealing, for example, with seals, the large cetaceans, and so
forth. A general codification could not include all the provisions
which would be necessary ....".
Where the continental shelf theory needs universal legal
provisions, the problem of marine resources does not seem to

be easily regulated by rules applicable the world over, and can therefore not be linked with the continental shelf theory.

This opinion of François has been expressed by Schücking in his notes on the amended draft Convention on territorial waters (see Report Committee of Experts, 1927 [52], p. 72–74). Article 2 of this amended draft reads: ".... Exclusive rights to fisheries continue to be governed by existing practice and conventions". Schücking explains in his notes (p. 74): "My colleague (Magelhaes) has shown the close connection which exists between the geographical condition of the littoral waters and the extension of fisheries. It is for this reason that a general uniform regulation of all fishery rights could not be applied to territorial waters. Clearly also it would be necessary to preserve all those rights which are essential for the conservation of fishery rights, for example the right to police fisheries".

The same vision appeared from the answers of Governments to the questionnaire, for instance, that of the United States (loc. cit. p. 161):

"1. International regulation of certain fisheries, such as those for whales is desirable and should be realizable;
2. ....
3. That in most cases particular fisheries may best be regulated by treaties between the nations most directly concerned".

Also Japan (p. 172) considered it preferable to regulate the exploitation of the products of the sea in bi- or plurilateral treaties.

Finally the answer of Great Britain (p. 146): ".... the exploitations of the products of the sea cannot be made the subject of any general convention, but should be attempted rather by particular conventions relating to particular products and particular areas between the Countries interested".

Sir Maurice Gwyer, one of the British delegates at the Conference for the Codification of International Law in the Hague, 1930 [61], p. 141, said: ".... with regard to fisheries I would only say this. This is a subject of primary importance to particular nations, but what is important for one nation in its relations to another nearby may be a matter of no importance at all to another nation on the other side of the globe. Fisheries on the coast of Norway, which have been a matter of interest and sometimes of difference of opinion between Dr. Raestad's country and my own, are, I should imagine, of no interest whatsoever to the delegate from Japan. I therefore suggest that the true line of approach for the solution of the problem of fisheries is rather that the individual nations concerned in the problems of particular fisheries should put their heads together and attempt to solve their domestic differences by means of separate Conventions, rather than that all nations of the earth should attempt to lay down one rule which would govern fisheries universally throughout the globe".

The International Law Commission although considering in its Report covering its second session [14], p. 22: ".... that protection of the resources of the sea should be independent of the concept of the continental shelf", obviously did not derive this conclusion from the view quoted above, because on the same page, "the Commission requested the special rapporteur to study the problem of protecting the resources of the sea for the benefit of all mankind by the *generalizing of measures* laid down in bilateral or multilateral treaties". (underlined by us)

At least for protection purposes general rules were not found impossible.

The view of the Commission concerning this independence of the protection of the resources of the sea from the concept of the continental shelf must therefore rest on a reasoning different from the one which we have given.

Several members expressed their views on this independence without giving a reason, for instance in the 69th meeting of the Commission, S.R. 69[67], p. 4, Hudson pointed out "that the right to regulate fishing was not related to the question of the continental shelf". Yepes (loc. cit. p. 5) and Sandström (p. 7) put forward the same opinion.

From the discussion which followed in that meeting, it became clear, that the Commission considered that the right of control and jurisdiction of the littoral State applied *only* to the sea-bed and the subsoil and that the littoral State would therefore have no special rights over the waters covering the continental shelf (loc. cit. p. 7).

This was the reason for the Commission's decision in its Report mentioned above.

Before we continue with the discussions in the International Law Commission, where, especially François referring to the declarations and decrees, several times put forward his opinion that fisheries had been linked with the conception of the continental shelf, we have to deal briefly with the important distinction between two kinds of fishery rights which has to be kept in mind.

The first is the right of a State to claim fisheries exclusively for its own nationals in a certain area. The second is the right of a State to enact and enforce regulations concerning the protection and conservation of fish in a certain area.

It is clear that this distinction plays no role in the area where a State has sovereignty rights, i.e. in its territorial sea. The difference does play a role as soon as the claim extends outside the area of sovereignty rights. In the latter area claims concerning protection, being of a more universal scope are far more acceptable than claims to exclusive fishery rights.

SECTION 6. EXCLUSIVE FISHERY RIGHTS

Most States claim exclusive fishery rights for their nationals in their territorial sea (sometimes also to domiciled foreigners like in Chile). This claim is based on their sovereign rights in these waters, and put forward because coastal fisheries form an important source of food and employment for the population. This right is for instance recognized in Art. 2 of the Convention for the Regulation of the Police of the Fisheries in the North Sea outside Territorial Waters [a] (in this Article the 3-mile limit is adopted) [b].

What is the exact nature of this right? In the Schücking-report (Report to the Council of the League of Nations, 1927 [52], p. 53) it is said that "In virtue of its right of dominion over the whole area of its territorial waters the riparian State possesses

---

[a] See [153], p. 179.

[b] Also recognized in Art. 6 of the Draft: "The legal status of the territorial sea," Appendix I (p. 212 et seq.) of the Acts of the Codification Conference at the Hague, 1930 [61].

for itself and for its nationals the sole rights of ownership over the riches of the sea, the fauna in the waters, every thing found above or below the subsoil of the territorial sea ....".

To call this right a property right seems somewhat unreal where the fauna is concerned. Even bottom-fishes migrate over considerable distances [a].

The only place where live fish can be the subject of ownership is in an aquarium, and for the rest the only way to become owner of fish is to catch it [b].

What one can say, is that, whereas the territorial sea is a part of the territory of the State, this State has the right to catch the fish in these waters and from this right results the power to give or lease this right to foreigners, but also to forbid foreigners to catch fish in these waters.

Early claims to this effect are mentioned by Fulton [31], p. 59, where he says, that "King William the Lion (A.D. 1165–1214) confirmed the grants given by King David I to the monks of the priory on the Isle of May in the Firth of Forth to whom the foreign fishermen had to pay tithes and addressed missives to all his good subjects and the fishermen who fish round the Isle of May commanding them to pay their tithes to the monks as they were paid in the time of his grandfather King David (A.D. 1124–1153); and he prohibited them from fishing in their waters or using the Island without license from the monks. This very early claim to the right of exclusive fishing in the sea is characteristic of the policy of all the Scottish kings", and p. 77: ".... the Scottish fishermen were .... little inclined to tolerate the intrusion of foreign fishermen within what they claimed as their 'reserved waters' that is the firths and bays and a distance along the coast described as 'a land-kenning' ...." that is (p. 84): ".... not nearer than where they could discern the land from the top of their masts. This distance was usually

---

[a] It is remarkable that the migration of fish was adduced by Plowden in the Sir John Constable case in 1575 (Moore [87], p. 229), as one of the reasons that the Queen (Elizabeth) "cannot prohibit any one from fishing there (the seas under her jurisdiction); and the water and the land under it are things of no value, and the fish are always removable from one place to another".
We will come back on this opinion later in another connexion.

[b] The Netherlands Civil Code for instance mentions under the means of obtaining property, Art. 641: "the right to appropriate the fish belongs exclusively to the owner of the water in which the fish is found ....".

placed at 14 miles, but sometimes a double land-kenning of 28 miles was claimed ....".

Again according to Fulton (p. 355), Welwood [a] answering Grotius' Mare Liberum, "was the first author who clearly enunciated, and insisted on, the principle that the inhabitants of a Country had a primary and exclusive right to the fisheries along their coasts—that the usufruct of the adjacent sea belonged to them; and that one of the main reasons why that portion of the sea should pertain to the neighbouring State was the risk of the exhaustion of its fisheries from promiscuous use".

Finally the first mention of 3-miles as a belt for exclusive fishing rights was contained in one of the papers in the volume provided for the use of the English Ambassadors to the Peace Conference in Cologne in 1673, according to Fulton, p. 499: "Ye Art of the Fishery as contained in ye Project 1673" concluding "with this sentence: 'In w[ch] fisheing y[e] said States shall oblidge themselves that their subiects shall not come wth in one league of y[e] shores of England and Scotland', which is the first mention of a 3-mile limit that has been discovered".

We mentioned a few milestones in the history of the exclusive fishing rights, but it would be out of place to follow up the interesting story of these claims, which were closely related to the varied aspects of the history of the territorial waters.

Not only that a tendency existed to extend the marginal zone for the purpose of exclusive fishing rights, but some countries actually enacted decrees to this effect.

In 1907 for instance the Minister of Agriculture of the Argentine Republic "issued a series of ordinances for the regulation of the fisheries (Reglementende la pesca y caza, Boletin oficial, 20th September, 1907) a zone of water up to a distance of 10 miles from high water mark on the land is under control of the State. The great gulfs and bays are moreover, included such as the Gulf of San Matias, the Gulf of St. George, and the Gulf of Nuevo,

---

[a] An Abridgement of all Sealawes, gathered forth of all writings and monuments, which are to be found among any people as nation upon the coasts of the greate Ocean and Mediterranean sea and specially ordered and disposed for the use and benefit of all benevolent sea farers, within the Maiesties Dominions of Great Brittane, Ireland and the adiacent Isles there of, London, 1613. Two years later he published: De dominio Maris Iuribusque ad Dominium praecipve spectantibus assertio brevis et methodica, Cosmopoli, 16th January, 1615.

the closing line in some cases considerably exceeding 100 miles from point to point and extending more than 70 miles beyond a 3-mile limit .... Within the declared limits the exercise of seafishing is free, provided that the regulations are adhered to ... Commercial fishing is forbidden within (that area) unless by vessels entered on the official list (matricula nacional) and foreigners are thus excluded". (Fulton, p. 661).

In France, according to Fulton, p. 657: "certain fisheries were allowed to be temporarily suspended beyond the 3-mile limit, if it was found necessary for the preservation of the bed of the sea, or of a fishery composed of migratory fishes (Art. 2 .... 'si cette mesure est commandée par l'intérêt de la conservation des fonds ou de la pêche de poissons de passage' ....)".

Norway (Fulton, p. 671–678) reserves the right in certain cases to exceed the limit derived from the general principle .... on some parts of the coast special laws regulate the extent of the sea in which the exclusive right of fishing is reserved to subjects.

"The first of these was a royal decree of 16th October, 1869, which prescribed that a straight line drawn at a distance of one geographical mile (4 miles) from and parallel to a straight line drawn between Storholmen en Svinö shall be taken as the boundary of the waters off the coast of the Söndmöre district, in which the fishing is entirely reserved for the inhabitants of the country .... it was necessary to consider local circumstances and what was natural, convenient and just. *The line* that had been drawn, they said, *coincided with a natural depression in the bottom of the sea* which separated the inshore from the offshore fishing banks, and it formed a *natural boundary* which could be readily ascertained by the use of a sounding lead" (underlined by us) [a].

We gave this example particularly because it shows that the idea to link exclusive fishery rights to the configuration of the sea-bottom has dawned on the human mind much earlier than the first mention of the continental shelf in connection with fisheries by Odón de Buen and his Portuguese adherents.

---

[a] This Decree has been discussed at length during the Anglo-Norwegian Fisheries case before the International Court of Justice. We only refer to the speech of Sir Eric Beckett, who said (Verbatim Report, Sitting 19 October, 1951 [85], p. 38): "I think that Norway did acquire as against the United Kingdom a prescriptive title to the waters enclosed by the Sunnmöre Decree".

We have to distinguish clearly between two kinds of extensions. One is a claim for a marginal belt as territorial water wider than the 3-mile limit, partly and sometimes mainly motivated by the desire to have a wider area where nationals of the country concerned have exclusive fishery rights. Examples are Spain (6 miles) *a* and Norway (4 miles). The other one is a claim for exclusive fishery rights *outside* the territorial sea adopted by the country concerned as part of their territory. The latter extension is a contiguous zone, either claimed in the form of national legislation or agreed upon between two States in bilateral Treaties.

François in his first Report to the International Law Commission [10], p. 30, says:

"As regards fishing, several national legislations lay down limits which differ from those of the territorial waters. This is in order to prevent fishing vessels of other countries from entering these zones. After noting the Government's replies to questions on this matter, the Preparatory Committee of the Codification Conference concluded that it would not be possible to arrive at an agreement establishing a contiguous zone for fishing and it merely proposed the establishment of such a zone exclusively for Customs, sanitation and security purposes" *b*.

In his second Report [26], p. 50, he gives the following list of exclusive fishing rights claimed (data provided by the Secretariat of the United Nations):

"Ecuador — fifteen miles.
Argentina, Canada, Colombia and Portugal — twelve miles.
Indo-China and Mexico — twenty kilometres.
Lebanon, Morocco and Syria — six miles".

The provisions concerned read as follows (quoted from the data provided by the Secretariat U.N.):

---

a At the Hague Codification Conference in 1930, the Spanisch delegate, Goicoechea, gave the following reason, Acts, 1930[61], p. 27: " .... the Spanish American countries, which have inherited the tradition of Spanish law, signed with Spain and Portugal, at the Madrid Congress in 1892, an agreement establishing a breadth of six miles. This was based on technical considerations, as a breadth of three miles would not include parts of the sea sufficiently deep to allow edible species to live in it".

b Art. 2 last paragraph of the Draft Convention amended by Schücking (see Report to the Council of the League of Nations 1927 [52], p. 72–73) reads: "exclusive rights to fisheries continue to be governed by existing practice and conventions".

Ecuador. Regulations concerning maritime fishing and hunting, Decree No. 607, 29 August, 1934, "Registro oficial", 31 August, 1934, No. 257, pp. 9, 15; Julio T. Torres, "Compilacion de Reformas al Código Civil. Leyes y Reglamentos Conexos" (Quito 1942), p. 280, 294. Translation by the Secretariat of the united Nations [153], p. 68.

Art. 78. "As territorial waters for fishing purposes shall be considered the waters contained within 15 miles from the low-water mark at the most salient points of the Islands (of the Colón Archipelago).

Art. 129. Fishing is in general free during the whole year, with respect to fish, molluscs and crustaceans, by any means whatsoever, provided that it is exercised outside territorial waters, more than 6 miles from the coasts, measured from the low-water mark; and subject to the prohibitions with respect to close seasons which are contained in these Regulations"...

The actual zone reserved for nationals seems to be 6 miles, whereas the 15 miles is a zone for protection purposes, which will be discussed in the next section of this Chapter.

Argentina. We mentioned the Decree of 1907 above [a].

Canada. Fisheries regulations for the Province of Prince Edward Island adopted by order in Council (P.C. 837) 11 May, 1927, "Statutes of Canada 1928", Prefix p. XXVII. [153] p. 57.

Section 19: "Trawlers prohibited (1) .... within the territorial waters of Canada.

(2) The master of every steam trawler at any port on the Atlantic seaboard of Canada, shall, before departure, come before the collector of customs .... and deliver to him a report outwards under his hand of the destination of such vessel ....
The report outwards shall also contain a declaration to the

---

[a] *The 12 miles mentioned in the Report of François is derived from the Fishing Regulations, enacted by Decree No. 148, 119, 19 April 1943. "Anales de Legislación Argentina ".vol. 3 (1943), p. 142. Translation by the Secretariat of the United Nations,* [153] *p. 51.*

Art. 4 .... ".... As coastal fishing shall be considered fishing conducted within the limits of a line running parallel to the coast at a distance of twelve marine miles, to be reckoned from the low-water mark. As high seas fishing (pesca mayor) shall be considered fishing conducted beyond that limit, as well as fishing at the mouth of the Plate Rivir if it concerns maritime and/or anadromous migratory species".

The same figure is mentioned in Article 3 of the regulations of 4 July 1909, issued by the Ministry of Agriculture with respect to fishing concessions along the maritime shore between the mouths of Rio de la Plata and of Rio Negro: "the concessionaires may only employ trawl-nets towed by steamboats in a zone at least twelve (12) miles distant from the low-water mark". (Ministerio de Agricultura, Leyes, decretos, etc., sobre pesca, caza maritima e industrializacion (1944), p. 29) [153] *p. 52.*

effect that the master of the steam trawler in consideration of the clearance granted by the officer of customs, undertakes and agrees, for one year after clearance (a) to restrict all fishing-operations by such steam trawler to waters which are at least 12 miles distant from the nearest shore on the Atlantic seaboard of Canada, between the first day of May and the thirty-first day of December; (b) and also to refrain from all fishing operations by such trawler in .... (certain waters) during the month of January. The penalties .... made by the Governor in Council shall apply in respect of such steam trawler and the master thereof, for non-compliance with the undertaking prescribed by this regulation".

"Identical provisions are contained in Section 24 of the Fishery regulation for the Province of New Brunswick adopted by order in Council (P.C. 837) 11 May, 1927, Statutes of Canada 1928, Prefix p. XXXIII".

The zone of 12 miles here referred to is not actually a zone for exclusive fishery rights, because the prohibition appears to be limited to trawlers, which call at Canadian ports. It seems that fishing outside the ordinary territorial sea, but inside the 12 mile belt, by trawlers or other fishing craft from outside which have not called at a Canadian port, cannot be interfered with on the ground of these regulations.

Colombia. Law No. 14 amending the law concerning deposits of hydrocarbons, 31 January, 1923. "Leyes expedidas por el Congreso Nacional en su legislatura de 1923" (segunda edición, 1941), p. 47. Translation by the Secr. of the U.N. [153], p. 62.

Art. 17. "For the purposes of Art. 38 of Law 120 of 1919, concerning deposits of hydrocarbons, and of Law 96 of 1922, relating to fishing in the seas of the Republic, the term 'territorial sea' shall be understood to refer to a zone of 12 marine miles around the coasts of the continental and insular dominions of the Republic".

Law 96 of 1922 (Leyes etc. .... de 1922 [seg. ed. 1940], p. 512) authorized the Government "to organize the renting of fishing in the seas of the Republic, in a manner which it considers as most appropriate from the point of view of national interest".

Indo-China. Presidential Decree determining the extent of the territorial waters of Indo-China for the purposes of fishing, 22 September, 1936 "Journal Officiel", vol. 68, No. 226 (26 September, 1936), p. 10192 [153], p. 75.

Art. 1. "For the purpose of fishing, the territorial waters of French Indo-China extend 20 kilometres from the shore at low-water mark" (bays .... 10 mile-rule).

Mexico. Fisheries Regulations, 5 March, 1927. "Diario Oficial", Vol. 41, No. 13 (15 March, 1927), section 2, p. 1 [153], p. 84.

Art. 2. "National fish resources shall include all products of aquatic life which have their origin or live in .... and all those which can be exploited in the maritime waters along the Mexican coasts to the extent provided for in treaties and laws on this subject; in absence of express rules or provisions this extent shall not be smaller than 20 kilometres as provided in Art. 5 of the Law of 18 December, 1902".

Lebanon and Syria. Order No. 1104 with respect to the policing of maritime fisheries, 14 November, 1921.
"Recueil des actes administratifs du Haut-Commissariat", Vol. 2 (1921), p. 412. Translation by the Secretariat of the United Nations [153], p. 83.

Art. 1. "In the coastal zone of Syria and Lebanon under French mandate the territorial sea extends, for the purpose of fisheries, to a distance of six marine miles from the coast or islands".

Morocco. Dahir relating to maritime fishing, 31 March, 1919. P.L. Rivière, "Traités Codes et Lois du Maroc" (Paris, 1925), Vol. III, p. 996 [153], p. 77.
Art. 2. "As regards the French zone of Our Empire the territorial waters extend, for the purposes of fishing six nautical miles from the shore at low-water mark" (for bays the 12 mile-rule).
Dahir to regulate fishing by fishing fleets in the territorial waters of Morocco, 25 March, 1922, Rivière, p. 901.

Art. 1. „For the purposes of fishing the territorial waters of the French zone of the Sherifian Empire ...." (as above).
"The exercise of the right to fish in these waters shall be subject only to payment of the charge in respect of a permit".

The wording of these decrees is somewhat confusing; "for fishery purposes the territorial waters shall extend to ....". The meaning, however, is to claim fishing rights and not sovereignty rights, which are usually embraced under the notion territorial waters.

Where Argentina, Colombia, Indo-China and Morocco are concerned, foreigners are not excluded altogether, but the permission to fish is dependent on a license, and can thus be rented.

For Treaties in this respect we refer to the examples given on p. 53 of the Report to the Council of the League of Nations, 1927 [52]. (Treaties reserving to nationals the right of fishing in territorial waters and Treaties in which States grant to the nationals of other contracting States fishing rights in their territorial waters with or without reciprocity) [a].

It is very well to extend the territorial waters, for fishery purposes, but one thing must not be forgotten. It has little use to proclaim exclusive rights, if these rights cannot be enforced.

For countries with long coast lines and in large archipelagoes like, for instance, Indonesia it is extremely difficult to prevent intrusion.

In the former Netherlands East Indies the exclusive fishery rights inside the 3 mile-limit were before World War II often violated by Japanese fishermen who were trying to catch small fry as bait for their tuna fisheries. This fry lives on the reefs, which were mainly inside the 3 mile belt. They often used ruthless methods like dynamite, with disastrous results for the natives' food supplies, the national population being for the greater part Mohammedans and therefore not allowed to eat pork, whilst beef was often too expensive, were dependent for their protein nutrition on fish. Moreover, it is a well known fact that the Japanese misused their fishing fleet, to practice hydrographical survey in areas which they intended to invade later.

General MacArthur imposed heavy controls on Japanese fisheries in 1946, which were, however, relaxed in 1949. The Nippon Times of 23 September, 1949, wrote: . . . . "The S.C.A.P. authorization for a 30 per cent expansion of the Nation's present fishing grounds is highly welcomed by the Japanese people to say the least . . . . W. C. Herrington, Acting Chief of the natural resources section . . . . said that the relaxation of control, imposed by General MacArthur in 1946, was authorized in recognition of the effective work being done by the Japanese Government and industry to control violations and overfishing . . . .".

" . . . . But Mr. Herrington has warned, future developments concerning the further relaxation of restrictions on Japanese

---

[a] *See also, for instance* [153], *p. 170 (No. 15), p. 174 (No. 20), p. 179, p. 232 (No. 3), p. 241 and p. 244 (No. 6).*

fishing activities will depend upon whether or not Japanese fishermen will continue to respect national and international fishing agreements and regulations.

Japans prewar reputation for violation of international agreements was notorious. The recent good conduct has not as yet erased all the ill-feeling created before the war. Even greater efforts are thus essential before the world will recognize willingly the allowing of freer access to more extensive fishing grounds to Japanese fishermen. It is imperative that the Japanese fishermen demonstrate they 'will make good neighbours upon the high seas'. The expansion of the fishing grounds must thus be considered another test to determine whether or not Japan is worthy of international trust. There is thus an even greater need now than ever before for the Japanese Government and the fishing industry to prohibit all illegal practises. The Nation's good name and its qualification to be included in the family of nations depend upon the continued success of that program".

In the Japanese Peace Treaty the regulation of fisheries is left to negotiations between Japan and the Allied Powers. Art. 9 reads:

"Japan will enter promptly into negotiations with the Allied Powers so desiring for the conclusion of bilateral and multilateral agreements providing for the regulation or limitation of fishing and the conservation and development of fisheries on the high seas" [a].

Finally we like to mention a passage from the Memorandum on the Regime of the High Seas [22], which remarks not without wit on p. 40: "With regard to their legal policy concerning areas of the sea in connection with fishing, States with a well-developed fishing industry may be divided into two groups. Those whose fishing grounds are in the immediate vicinity of their own coasts favour the maximum extension of their territorial waters; those whose fishing grounds are in the immediate vicinity of the coasts of other countries to limit, as far as possible, the extent

---

[a] Allen [90], p. 177–178, expressed his anxiety for the case that no provisions for the protection of American fisheries would be incorporated in the proposed peace treaty with Japan. He said: "If the Department (of State) should grasp the opportunity to solve the acute Pacific Ocean fishery problem in connection with the negotiation of a peace treaty with Japan, the necessity for unilateral action (i.e. in accordance with the Proclamation of 1945) might be happily avoided . . . .".

of the territorial waters of the State in the immediate vicinity of whose coasts they carry on fishing, and oppose any extension of its powers beyond its territorial waters".

We have only given a brief survey of exclusive rights in order to explain their scope.

From the proclamations and decrees mentioned earlier in this Chapter it is clear, that the Proclamation of President Truman does not claim exclusive fishery rights although it remains to be seen, what would happen if a stray fisherman operated in the conservation zones not in accordance with the prevailing regulations.

The Chilian and Peruvian instruments, however, do claim in our opinion exclusive rights.

We end this section by saying that the International Law Commission did not deal with exclusive fishery rights. In the 119th meeting (S.R. 119[89], p. 16) François pointed out that .... "for the time being the Commission did not wish to make any change in the limits within which fishing could be carried on under existing international law, when not only the nationals of a single State were so enjoyed".

SECTION 7. PROTECTION AND CONSERVATION

Early in history it was recognized that the marine resources should be protected against depletion.

Some authors were of the opinion that the sea could not be exhausted by fishing, like Ferdinand Vasquez or Vasquius [a], quoted by Fulton [31], p. 339, and by Grotius [91], p. 253: "For it is generally agreed that, if a great many persons hunt or fish upon some wooded tract of land or in some stream, that wood or stream will probably be emptied of wild animals or fish, an objection which is not applicable to the sea". Others on the contrary did believe that exhaustion was possible like Welwood, quoted above, who mentioned the risk of exhaustion as one of the reasons why a portion of the sea should pertain to the neighbouring State.

We find during the reign of Charles I of England a revival

---

[a] Controversiae illustres, Venice, 1564.

of preservation laws, which were apparently of even older date, Fulton, p. 213: "The old laws for the preservation of the spawn and brood of fish, which had fallen into disuse, were put into force; proclamation appeared prohibiting wasteful fishing; a vigorous effort was made to suppress the use of injurious appliances . . . .".

Fulton also gives in Appendix C (p. 750) the text of an earlier proof of the existence of preservation regulations. It is a "License For Fishing at the 'Zowe' Bank in the Channel" given by James I: " . . . . To all to whom theis presentes shall come, Greeting, Knowe ye that I, according to the auntient ordinances and rules hertofore established and lately revived for the preservation of the fishing betwixt the subiectes of the Easterne coast of the kingdome of great Britayne and the frenche Fishermen accoastinge those partes, ... etc.". " . . . . for all sortes of Fishe without restrainte of season, soe the same be done and performed with nettes and engines lawful and accustomed by the English subiectes of that coast . . . . This license is to endure but untill the first daie of August wch shall be in the yeare of our Lord God 1616 . . . .".

We do not intend to give a complete history of the protection of fish as our aim is only to draw attention to the two types of fishery claims. Therefore one more example from history, this time from 1681 will suffice.

Valin [23], p. 622 et seq. (Liv. IV Tit. X) discusses a regulation concerning the cutting of a sea-weed called "varech" (or varec or kelp being the name for several kinds of Laminaria) used as fertilizer (and later also for extraction of iodine).

Article 1 reads: "Les habitans des Paroisses situées sur les côtes de la mer, s'assembleront le premier Dimanche du mois de Janvier de chacune année, à l'issue de la Messe paroissiale, pour régler les jours auxquels devra commencer & finir la coupe de l'herbe appellée varech & vraicq, sart ou gouesmon, croissant en mer à l'endroit de leur territoire".

Apart for reasons of equal distribution among the coastal population, to prevent a "rush and grab", this regulation also intended to protect the fry which frequented these weeds as a shelter: " . . . . pour conserver le frai du poisson, aussi-bien que le petit poisson, qui, trop foible pour lutter contre les vagues de la mer, lors même qu'elle n'est pas agitée, ou pour en soutenir la trop grande fraicheur, trouve sous le sart un abri qui tout à la fois le met hors d'insulte de la part des vagues, de même que des gros poissons, & lui fait ressentir une chaleur douce, causée par le soleil qui échauffe, dans la belle saison, la partie du rivage que la mer découvre pendant le reflux ou le temps qu'elle employe à se retirer".

In "Livre cinquième", p. 635 et seq., he deals with the liberty of fisheries and the regulations relating to protection, mainly prescribing the measurements of the meshes of the nets and periods in which fishing was forbidden, p. 642: "Rien, après tout, de plus naturel & de plus conforme au bon ordre pour l'amélioration & la conservation même de la pêche dont sans cela la source tariroit en peu de temps; car enfin que deviendroit la pêche, s'il étoit permis de la faire avec de filets, d'où le petit poisson, le frai même ne pourroit s'échapper"?

States are of course free to give regulations for the protection of different species of fish in their own territorial waters, or outside these waters to their own nationals.

One of the reasons, that States want to extend their territorial waters, is to be able to lay down rules for protection in a wider area.

In so far as only the coastal State lays down provisions for protection, it has been felt that a maritime belt of 3 miles was insufficient. This was one of the reasons that the Institute of International Law adopted in 1894 [a] the following Rules on the definition and regime of the territorial sea, Scott [92], p. 113–114:

"The Institute, considering that there is no reason to confound in a single zone the distance necessary for the exercise of sovereignty and for the protection of coastwise fishing and that which is necessary to guarantee the neutrality of non-belligerents in time of war;

That the distance most generally adopted of 3 miles from low-water mark has been recognized as insufficient for the protection of coastwise fishing;

That this distance moreover does not correspond to the actual range of guns placed on the coast;

---

[a] See also Annuaire [93], Vol. 13, p. 328 .

Has adopted the following provisions: ....

Article 2. The territorial sea extends six marine miles (60 to a degree of latitude) from the low-water mark along the full extent of the coasts".

For this purpose also contiguous zones have been claimed. It is, however, difficult to see how a State could enforce such provisions against nationals of other States outside its territorial waters.

It is true that States have enacted national laws concerning contiguous zones, that so to speak the principle of the freedom of the high seas has been nibbled at in these contiguous zones.

But the rights claimed in these contiguous zones are of a very limited nature, and are meant as a protection against infringements of customs or sanitary regulations or interference with the security of the State by foreign ships.

Laws to this effect rest- upon the principle (as Sir Charles Russell said in the course of his argument before the Behring-Sea Arbitral Tribunal in 1893) "that no civilized State will encourage offences against the laws of another State the justice of which laws it recognizes. It willingly allows a foreign State to take reasonable measures of prevention within a moderate distance even outside territorial waters" (see Report to the Council of the League of Nations, 1927 [52], p. 70).

But contiguous zones established for protection of fish are quite a different matter. Fish in the high seas are nobody's property. Grotius [91], p. 232, says: ".... those (things) which are also res nullius but which have not been assigned for common use: e.g. wild beasts, fish and birds.

Items belonging to the latter class can be made subject to private ownership, *provided* that some one does take *possession* of them ...." (underlined by us) and (same page) .... ".... the sea is common property, whereas the fish become the property of the persons who catch them ....".

Protection is highly necessary, but that is a matter which does not concern the coastal State alone, but all States which have fishery interests in a certain area.

International Tribunals denied States the right to exercise protective jurisdiction over fisheries on the high seas contiguous to their territories. We referred already twice to the Behring-Sea Arbitration. The American claim to exclusive rights in the seal fisheries beyond the ordinary limit of the territorial waters was not recognized by the Tribunal. (See amongst other documents Leonard [94], p. 80). Leonard, p. 86, draws attention to the remarkable fact that in the diplomatic correspondence preceeding submittance to the arbitrator Asser in September, 1900, of the case of the American vessels "Cape Horn Pigeon", "James Hamilton Lewis", "C. H. White", and "Kate and Anna", seized by the Russians for illegal sealing, but outside Russian territorial waters, the Russians argued on the same principles as "had been espoused by the American Government" in the Behring-Sea Arbitration. These principles involved the right of a nation to undertake protective measures in the waters beyond the 3-mile limit. Leonard, p. 87: "The United States took the position which Great Britain had taken several years before; Russia became the defender of the sea products for all humanity, just as the United States had been in 1893".

The decision was a confirmation of the award of 1893: ".... the seizures by the Russian Government did not conform to international law 'since the jurisdiction of a State does not extend beyond the limits of the territorial sea, unless this rule has been derogated by a special convention' "(Leonard [94], p. 89).

It is also significant that neither the Preparatory Committee of the Codification Conference nor the Conference itself in 1930 arrived at an agreement establishing a contiguous zone for fishing, even not for the protection of fish. The Preparatory Committee merely proposed the establishment of such a zone exclusively for customs, sanitation and security purposes (see quotation of François First Report [10], p. 30, above) and as François says (same page): "The Conference (the Hague, 1930) was not prepared to accept the contiguous zone even in this limited form". François said in the 69th meeting of the International Law Commission, S.R. 69[67], p. 23, that "the Hague Conference of 1930 had not even discussed the establishment of a contiguous zone within which fishing rights would be regulated by the littoral State....", but this statement cannot be meant to cover rights of protection because several speeches during that Conference did discuss protection [a].

However, no decision was taken. In Annex V, in the Report of the Committee on April, 10th 1930, Acts 1930 [61], p. 212, the following recommendation was submitted to the Conference: "Taking into consideration the importance of the fishing industry to certain countries;

Recognizing further that the protection of the various products of the sea must be considered, not only in relation to the territorial sea, but also the waters beyond it;

And that it is not competent to deal with these problems nor to do anything to prejudge their solution;

Noting also the steps already initiated on these subjects by certain organs of the League of Nations,

Desires to affirm the importance of the work already undertaken or to be undertaken regarding these matters, either through scientific research or by practical methods; that is, measures of protection and collaboration which may be recognized as necessary for the safeguarding of riches constituting the common patrimony".

What sort of provisions are usually laid down for protection? To mention some examples, the size of meshes of nets, to prevent immature fish to be caught and to be imported or to be sold; certain months or places in which fishing is prohibited to allow fish to spawn;

forbidding certain destructive methods as dynamite;

prohibiting to sell undersized fish on the market;

prohibiting trawling in areas, frequented by small fry, etc.

Especially trawling, if practised indiscriminately, can do a lot of harm. Fulton [31], p. 704, mentions, that "at a conference of practical fishermen held in 1883 in connection with the International Fisheries Exhibition at London, statements were made by 'trawlers' as to the enormous destruction of under-sized fish, and the depletion of the grounds".

---

[a] The Portuguese delegate Magelhaes (Acts [61], p. 18) and the Danish delegate Lorck (p. 25) submitted proposals to add to the text of basis of discussion No. 5, dealing with contiguous zones, a provision concerning fishery protection, and François says in his Second Report [26], p. 47: "certain delegations pointed out how important it was that the coastal State should have in the contiguous zone effective administration of its fishery laws and the right of protecting fry".

It is imperative that these measures should be taken soon, but we believe that the road taken by certain governments to achieve this aim is not the right one. We do not believe in unilateral proclamations or declarations because they do not seem to us to be the most suitable means to solve international problems. The only psychological effect they may produce could perhaps be compared with a shock therapy. Governments may become conscious of the fact, that half-hearted measures do not suffice and that intensifying the fisheries and industrializing the fishing methods make drastic measures for protection more essential than ever before, and that these measures can only work if taken and enforced through joint action of all the States interested in the particular kind of fishery concerned.

That more than ever protection is essential is well explained in the Memorandum on the Regime of the High Seas [22], p. 38: ".... fishing is continually becoming more destructive. There are many reasons for this; fishing vessels are driven by mechanical power instead of sails; fishing tackle which formerly had to be man-handled is now mechanically operated; refrigeration plants increase the time that fishing vessels can remain at sea; in certain types of fishing, a factory ship sends out catching vessels and the fish they bring back to it is processed quite independently of the land; detailed information is available on the habits of the various species, their migrations, their seasonal assembly points, the nature of the fishing grounds and the temperature of the water; finally, instruments have been adapted for fishing purposes of which one type (asdic) makes it possible to detect shoals of fish and another (radar) enables vessels to continue full-scale operations in spite of such conditions as fog, which formerly prevented them".

We will now continue with the work of the International Law Commission.

Section 8. Discussion on fishery rights in the International Law Commission

The subject was discussed in the second session of the International Law Commission. In the 69th meeting (Summary Record [67], p. 25), el Khoury said: ".... As for the control of fishing, the Commission should not make a distinction between the contiguous zone and the high seas".

François on the contrary thought (p. 26): .... "that fish protection measures were particularly necessary near the coasts. The waters near the coasts contained the young fish and that was why protection was required. At the present time there were no rules of international law on the subject".

We believe that the opinion of el Khoury and of Cordova who said, p. 26, "that fish protection and conservation both in the coastal areas and on the high seas were of the greatest importance to humanity" is more in accordance with the biological facts, (whaling, pelagic sealing, interception of salmon or cod on their way to their spawning places, etc.).

Spiropoulos touched another important facet of the problem, p. 27, in saying "that the Commission, ... could not accept such rules (the Proclamation of President Truman and the Chilian Declaration) as standards of international law for the protection of particular interests".

The motive for protection is primarily the national interest, although really is it the interest of humanity in general. But because of the more obvious national motives, the psychological effect is a reluctance to submit to rules laid down by a particular State.

The whole subject is in our opinion a typical example of the difference between the scientific development of the human race and the level of culture we have been able to reach, the latter being far behind the scientific level. Science has told us years ago, that protection is highly necessary to prevent exhaustion of marine resources, but the human character has on an average not reached the necessary development to be considerate to its fellow men or to sacrifice immediate gain in order to keep the supplies going for future generations.

Kerno made an interesting remark (p. 28), saying, that whereas there was an "extensive State practice" as to contiguous zones for "fiscal, statute customs and sanitary matters" where the Commission could carry out a task of codification proper, "in the case of fish protection the situation was different, and the Commission was faced instead with a question of progressive development. The Rapporteur could therefore examine the question and make suggestions the following year. One point to be established was whether littoral States should be given a kind of 'trustee' role which they would perform in the general interest. It should be added that, in principle, the need for the protection of marine resources was felt everywhere, even outside a 'contiguous' zone".

François suggests the following regulations, Second Report [26], p. 37:

"Every coastal State shall be entitled to declare, in a zone 200 sea-miles wide contiguous to its territorial waters, the restrictions necessary to protect the resources of the sea against extermination and to prevent the pollution of those waters by fuel oil.

The coastal State shall endeavour to enact such rules in agreement with the other countries interested in the fisheries in those waters. The rules shall not discriminate in any way between the nationals and vessels of the various States, including the coastal State; they shall, in all respects other than protection of the resources of the sea and repression of pollution of the sea, observe the regime of the high seas.

If a State considers that its interests have been unfairly injured by a restriction of the kind provided for in the first paragraph, and if the two States are unable to reach agreement on the subject, the dispute shall be submitted to the International Court of Justice".

He emphasizes that this zone is only meant for protection of the marine resources and not for exclusive fishing rights. We will deal with the pollution of waters in a later Section.

It seems to us that the word "endeavour" is too weak. As follows from what we said above we think that such a zone should only and alone be established by multilateral Conventions, which would of course only be applicable to the Parties concerned. A stray fisherman from a non-party State could not be forced to adhere to the regulations. If this would lead to frequent misuse the State whose nationals are involved can always be invited to adhere to the Convention. But if these intrusions remain limited to a few, there is no reason to interfere more than necessary with the freedom of the seas. We should not try to be perfect, because such an attempt fails. If a multilateral Treaty for a certain area is established between the most interested parties, we will have achieved an enormous improvement, and a *few* intruders do less harm than the conflict which would be the result of forcing them to obey a law which is not applicable to them.

We therefore think that if more States are interested there should be an obligation on the State most concerned to call a Conference together or bring the subject into the Assembly of the United Nations or the Food and Agriculture Organization, or the Economic and Social Council with the request that a Convention be drafted.

Although we are not in favour of giving a coastal State such a power as in the proposal of François, a point which we deal with later, the width of the area, 200 miles, is certainly an improvement compared with the 6 miles claimed by Magelhaes in 1930.

It is, however, not wide enough. François explains in his Report that he had two reasons for his choice of 200 miles, one being the fact that this figure appeared in a number of proclamations (p. 37), the other that the draft Convention of Washington of 1926, concerning the pollution of sea-water mentioned zones of considerable extent, (p. 38). In the 117th meeting of the International Law Commission the proposal of François was discussed.

Referring to recent Conventions (which will be the subject of our next Section) Hudson made the following counter-proposal (S.R. 117[34], p. 25): "The States whose nationals are engaged in fishing in any area of the high seas may regulate and control fishing activities in such area for the purpose of conserving its resources against extermination.

If only the nationals of a single State are thus engaged in an area, that State may take such measures in that area. If the nationals of several States are thus engaged in an area, such measures shall be taken by those States in concert. In any case, however, no area may be closed to the entry of nationals of other States to engage in fishing activities".

In the 118th meeting, S.R. 118[95], p. 4, "the Chairman (Brierly) pointed out that neither Mr. Truman's proclamation nor Mr.

Hudson's proposal stated specifically what would happen if the nationals of a third State engaged in fishing without conforming to the established rules. It was, doubtless, to be assumed that they would be forced to conform", and Hudson agreed, but he said: "it would then be necessary to include the States in question in the agreement".

We have already said that this "forcing" of nationals of a third State would be the wrong thing to do. It would seem to be "ultra vires" and quite against the most elementary principle of International Law.

Several members of the Commission had misgivings about a single State, the coastal State, having the right to make regulations concerning the high seas and Scelle (loc. cit., p. 9) even expressed his fear that adopting that principle "would amount to a return to the days when the republics of Genoa and Venice claimed possession of the seas just as far as they could impose their sovereignty".

Referring to Hudson's proposal, he said (p. 6) that it should be supplemented by the idea expressed in the third paragraph of the proposal of François. "Again it was not sufficient to say that the States should come to an agreement; provision should be made against their not so doing. It would be usual to provide, in such a case, for recourse to arbitration or the International Court of Justice".

Other members stood up for the coastal State and a vote on the two proposals resulted in 6 votes for the principle of François and 6 votes for that of Hudson (S.R. 118[95], p. 20).

This vote may well be considered as characteristic for the dilemma, which has so far been responsible for the small achievement in the field of protection and conservation. The most obvious solution being agreement needs the will to agree which is only too often forestalled by egoistic and short sighted motives. Fulton [31], p. 737, writes: "The second probable reason that nothing has yet been done to arrive at an international understanding (the first being that the international investigations are still going on) appears to be that the representatives of the great trawling industry have changed their minds .... They fear that if the question of fishery regulation beyond the ordinary 3-mile limit is opened up with foreign Powers in the interest of the North-Sea fisheries; proposals may be made, as a *quid pro quo*, by some of the other Powers for similar regulations on their coasts; and it is evident from the statements made in Parliament that this view has hitherto prevailed". Although he published his book in 1911, the situation has not changed much.

The other alternative: extension of territorial waters or establishment of contiguous zones for this purpose, infringes on the principle of the freedom of the seas and is bound to give rise to conflicts between States who so extend their conservation area, and States who do not consider conservation measures taken in such area enforceable against their nationals.

Moreover the contiguous zone for conservation of fisheries may well gradually become an extension of the territorial sea (see François Second Report [26], p. 46). François gives as one of the obstacles to the acceptance by other States of the legislation enacted for that purpose by the coastal State: "the fact that such unilateral legislation by the coastal State with respect to a portion of the high seas does not offer sufficient guarantees; it does not in practice always take into account the general interest of all who use those waters, and it sometimes favours intolerably the special interests of the coastal State" (Second Report [26], p. 36).

Against such extension also egoistic motives may be at work. The Memorandum of the Regime of the High Seas [22], p. 39: "The general criticisms based on the principle of freedom of the seas as traditionally understood, which are levelled against any extension of State powers by a littoral State beyond the limits of the territorial waters, gain a special hearing from the corporate organizations of the fishing industries, which can count on the support of their local representatives in political assemblies. Governments themselves lend a willing ear to the complaints of sympathetic sections of the population, in which the best naval recruits are found". This is true, but more so for Europe than for America, we believe.

Of course the whole problem is yet another example of the struggle between private interests and the general interest, and as the latter is on the winning side inside the States, we believe that it will eventually win as between the States.

We quote two passages of Riesenfeld demonstrating how difficult the choice is. Riesenfeld [96], p. 282: "Therefore if it can be assumed justly that coastal fishing grounds, owing to their primordial importance for coastal States and owing to the very imminent danger of their complete destruction resulting from the employment of piratical techniques by distant nations, can be adequately preserved only by control and exclusive exploitation by the coastal State, international law must and does recognize the right to such control and exploitation by the coastal State, unless the vested, long standing rights of other nations are thereby infringed".

But the same author writes on p. 277: "Where a satisfactory settlement by multilateral agreement can be reached this method is naturally the preferred answer to the problem. Thus far, however, the nations have succeeded in concluding only a few general agreements".

Hudson then proposed as an addition to his text, S.R. 118[95], p. 20: "If any part of an area is situated within X miles of the territorial waters of a coastal State, that State must be permitted to take part in any concert adopted even though its nationals are not engaged in fishing in that area".

He wanted to meet François' idea "that conservatory measures, or the absence thereof in an area situated within X miles of the territorial waters of a State could affect the fishing within those territorial waters".

This was adopted with a change in the wording: the coastal State is "permitted to take part", being changed into: "is entitled to take part"; together with the first paragraph of Hudson's proposal (S.R. 119[89], p. 11).

In the meanwhile a "new" idea took shape during the discussions. In the 117th meeting (S.R. 117[34], p. 25) Kerno had already hinted in that direction. He said that "rules for the protection of the resources of the sea should be established in the general interest of mankind". He suggested that international rules established by States should be brought to the notice of the Economic and Social Council, albeit for information only.

In the 118th meeting Scelle (S.R. 118[95], p. 6) said that it was "unfortunately, not possible at the moment to appeal to an international administrative authority, the need for which had already made itself felt, but which could not come into existence under existing world conditions".

Spiropoulos (p. 7) thought that "the best solution would be the establishment of an international board for the protection of the resources of the sea", and François referred to a similar proposal by Schücking to the League of Nations Committee of Experts.

Amado suggested as regards Spiropoulos' idea, "to consult the Fisheries Division of the United Nations Food and Agriculture Organization, which was doing important work on the subject". In a document of the Regime of the High Seas, dealing with questions under study by other Organs of the United Nations or by Specialized Agencies [97], p. 3, the following passage about fisheries is written:

"3. *Fisheries*. The Food and Agriculture Organization of the United Nations is empowered by Article 1 of its Constitution to promote and, where appropriate, to recommend national and international action with respect to 'the conservation of natural resources and the adoption of improved methods of agricultural production'. Article XVI of the Constitution interprets the term 'agriculture' as including fisheries and marine products. FAO established a special division on fisheries which made a survey of world fisheries".

Hudson eventually submitted a third paragraph of his text, S.R. 119 [89], p. 3: "The FAO should confer competence on a permanent body to conduct continuous investigations of the world's fisheries and the methods employed in exploiting them, and such body should make recommendations of conservatory measures to be adopted by the States whose nationals are engaged in fishing in various particular areas".

As an alternative text he proposed that "such body should have power to notify to the States whose nationals are engaged in fishing in various particular areas the conservatory measures to be adopted in each area. Any notification should come into force in the area to which it relates within X months, provided no such State makes objection within that period".

The alternative text was based on Art. 1, paragraph 9, of the Convention of Mexico City of January 25, 1949 (see below).

Spiropoulos, p. 6, thought that the best solution would be "regulation by an international body", ... "rules would first be established and there would then be set up an international institution which, in the last resort, could take decisions in case of dispute". What we see here is a proposal for an international legislative body and an International Court. This proposal goes much further than that of Hudson, because there is nothing in his draft which compels interested parties to agree, nor to accept a judicial or arbitral decision, as Scelle remarked (p. 7).

This difficulty was solved by adoption of el Koury's amendment (p. 12), to substitute in the first text of the third paragraph of Hudson the word "recommendations" by "regulations" and "adopted" by "applied".

However, two points remained to be answered. One, which we have discussed earlier, is the question whether a new State coming into an area where other States have been engaged in fishing and have adopted certain rules, has to conform to the rules or not. François (p. 13) asked for an answer.

Hudson thought, that the new situation would necessitate a new agreement with the new-comer. Scelle's idea was that the latter "would be joining an international organization having a rule of law which those acceding to it must observe. On the other hand, as soon as it joined the organization, the new State could propose changes and the Food and Agriculture Organization would decide whether, in that particular instance, it should recognize the objection".

Cordova wanted to give the international organization powers to intervene, when a State joined those already engaged in fishing.

The other question was, *when* should an international organization intervene? Of course, there would be no reason for such intervention when the States agreed upon the regulations to be laid down, and therefore the international organization should only make regulations when the States were unable to agree amongst themselves.

"The principle that failing agreement, an international organization should intervene, was unanimously adopted", (S.R. 119, p. 11). The Chairman (Brierly) eventually suggested the text (S.R. 132[98], p. 8) which was adopted (see here under).

Here follows the text and comments adopted, and reproduced in the Report of the International Law Commission, covering its third session [36], Part II, Related Subjects, Resources of the Sea, p. 59–61:

Article 1.

"States whose nationals are engaged in fishing in any area of the high seas may regulate and control fishing activities in such area for the purpose of preserving its resources from extermination. If the nationals of several States are thus engaged in an area, such measures shall be taken by those States in concert; if the nationals of only one State are thus engaged in a given area, that State may take such measures in the area. If any part of an area is situated within 100 miles of the territorial waters of a coastal State, that State is entitled to take part on an equal footing in any system of regulation, even though its nationals do not carry on fishing in the area. In no circumstances, however, may an area be closed to nationals of other States wishing to engage in fishing activities.

Article 2

Competence should be conferred on a permanent international body to conduct continuous investigations of the world's fisheries and the methods employed in exploiting them. Such body should also be empowered to make regulations for conservatory measures to be applied by the States whose nationals are engaged in fishing in any particular area where the States concerned are unable to agree amongst themselves.

1. The question of conservation of the resources of the sea has been coupled with the claims to the continental shelf advanced by some States in recent years, but the two subjects seem to be quite distinct, and for this reason they have been separately dealt with.
2. Protection of marine fauna against extermination is called for in the interests of safeguarding the world's food supply. The States whose nationals carry on fishing in a particular area have therefore a special responsibility, and they should agree among them as to the regulations to be applied in that area. Where nationals of only one State are thus engaged in

an area, the responsibility rests with that State. However, the exercise of the right to prescribe conservatory measures should not exclude newcomers from participation in fishing in any area. Where a fishing area is so close to a coast that regulations or the failure to adopt regulations might affect the fishing in the territorial waters of a coastal State, that State should be entitled to participate in drawing up regulations to be applied even though its nationals do not fish in the area.

3. This system might prove ineffective if the interested States were unable to reach agreement. The best way of overcoming the difficulty would be to set up a permanent body, which in the event of disagreement would be competent to submit rules which the States would be required to observe in respect of fishing activities by their nationals in the waters in question. This matter would seem to lie within the general competence of the Food and Agriculture Organization of the United Nations.

4. The pollution of waters of the high seas presents special problems, not only with regard to the conservation of the resources of the sea, but also with regard to the protection of other interests. The Commission noted that the Economic and Social Council has taken an initiative in this matter. (Resolution 298 (XI)C, of 12 July, 1950).

5. The Commission discussed a proposal that a coastal State should be empowered to lay down conservatory regulations to be applied in a zone contiguous to its territorial waters, pending the establishment of the body referred to in the previous paragraph. Such regulations would as far as possible have to be drawn up in agreement with the other States interested in the fishing grounds in question. They would make no distinction between the nationals of the various States, including the coastal State. Any disputes arising out of the application of the rules would have to be submitted to arbitration. The figure of 200 sea-miles was suggested as the breadth of the zone. In view of the fact that there was an equality of votes concerning the desirability of this proposal, the Commission decided to mention it in its report without sponsoring it".

After having followed the genesis of these texts fairly closely, a few remarks will suffice.

The system of a contiguous zone for conservation purposes has been dropped. Only in the note No. 5 on Art. 2 such a zone s envisaged as long as the establishment of the international body mentioned in Art. 2 is pending. (In note 5 it says "the body referred to in the previous paragraph", but this must be an error, the previous paragraph being note 4, deals with pollution of waters of the high sea). This means that the Commission accepts the system of the Proclamation of President Truman, only as a temporary one. Besides it introduces in that case compulsory arbitration for disputes arising out of the application of the rules, a provision *not* contained in President Truman's Proclamation.

Finally the Commission introduces an international body which will act as international legislator, in cases where the States are unable to agree. Of course the power of this body to legislate would be a delegated power, derived from a Convention whereby such a body would be established, and the duties of that body would be specified. Nevertheless, we believe that we may qualify such delegation by States to an international body as an improvement in the field of international organization. The idea belongs to the field of the functional organization.

In this connection we may draw attention to the fact that compulsory submission to the International Court of Justice in case States are unable to reach agreement, as suggested in the last paragraph of François' proposals, has not been inserted in the draft article. This is only natural, because the other alternative has been chosen. We should not forget that jurisdiction and legislation are very much akin and were in primitive stages of development even performed by the same body.

We believe that in the very first stages of international legislation of this kind, a legislation which makes its first timid appearance in our generation, the difference with international jurisdiction or arbitration is not great. As the decision of the International Court would be binding upon the parties, so would be the provisions laid down by the international body. Both are limited to the parties concerned, in the area in which the rules will be applicable. This kind of legislation is of a technical rather than of a political nature. This legislation has for these reasons only a limited scope, but being operative possibly for several States, it is international in character.

The legislative activities of the international body will, however, only be invoked, when the States concerned are unable to agree. Agreement being thus the primordial way of creating the regulation for protection. The idea to give an international body the power to lay down technical provisions is not new. To mention an example. A very limited power of this kind was given to the International Commission for Air Navigation, in Article 34 of the Convention for the Regulation of Aerial Navigation of 1919 [99], regulating the duties of this body. This article refers for instance to Article 14 according to which the Commission has the power to determine the methods of employing wireless apparatus and can extend the obligation of carrying these apparatus to all other classes of aircraft, etc. According to Article 34 (c) the Commission has the power to amend the provisions of the Annexes (A) to (G) dealing with markings of aircraft, certificates of airworthiness, rules as to lights and signals, rules of the air, etc. [a]. Further examples will be met in recent Treaties concerning protection and conservation.

We shall now discuss a few recent Treaties, some of which were concluded after the Proclamation of President Truman of 1945, in which the United States were a Party, thus showing that so far it has been possible to find co-operation between States willing to submit to conservation measures agreed upon in a certain area.

Section 9. Treaties concerning protection

The object of this Section is not an enumeration of Treaties, but mainly a comparison of the principles on which they are built with the principles adopted and proposed by the International Law Commission.

As we have said, the International Law Commission did not differentiate as to species; they envisaged special agreements in particular areas.

Typical examples of Treaties concerning protection of particular species, which were threatened with depletion are (1) the Convention for the protection of fur seals in the waters of the

---

[a] By Protocol of June 1, 1935 [100], Art. 34 was modified and the powers of the Commission relating to Art. 14 and Annex H (Customs), were abrogated.

The International Civil Aviation Organization, the successor of the Air Navigation Commission, in the Convention on International Civil Aviation of Chicago [101], 1944, although its functions are of a wider scope, has less legislative power, because States are not bound by its decisions (Art. 90 and also Art. 38).

North Pacific, Washington, 7 July, 1911 [102] [a]; (2) the Convention revising the Convention of May 9, 1930, for the preservation of halibut fishery of northern Pacific Ocean and Behring-Sea, Ottawa, January 29, 1937 [103].

(3) The protection of whales was lastly regulated by the International Convention for the Regulation of Whaling, signed at Washington, 2nd December, 1946 [104].

There is good reason to believe that these Conventions did help to prevent a threatening extinction of the species for the protection of which they were concluded.

As another example of an international body with legislative or regulating power we draw the attention to the International Fisheries Commission of the Convention for the preservation of halibut of 1937.

Article 1, 2nd paragraph of this Convention reads: "The International Fisheries Commission provided for by Article III is hereby empowered, subject to the approval of the President of the United States of America and the Governor-General of Canada, to suspend or change the closed season provided for by this article, as to part or all of the convention waters .... and to permit, limit, regulate and prohibit in any area or at any time when fishing for halibut is prohibited, the taking, retention and landing of halibut caught incidentally to fishing for other species of fish, and the possession during such fishing of halibut of any origin".

Article 3 .... "The high contracting parties agree that for the purposes of protecting and conserving the halibut fishery.... the International Fisheries Commission with the approval of the President of the United States of America and of the Governor General of Canada, may, in respect of nationals and inhabitants and fishing vessels and boats of the United States and of Canada from time to time,

(a) divide the convention waters into areas;
(b) limit the catch of halibut to be taken from each area within the season during which fishing for halibut is allowed;
(c) prohibit departure of vessels from any port or place, or from any receiving vessel or station, to any area for halibut fishing, after any date when in the judgment of the International Fisheries Commission the vessels which have departed for that area prior to that date or which are known

---

[a] Japan abrogated the Treaty by note of October 23, 1940 (Riesenfeld [96], p. 201, note 84).

to be fishing in that area shall suffice to catch the limit which shall have been set for that area under section (b) of this paragraph;

(d) fix the size and character of halibut fishing appliances to be used in any area;

(e) ....

(f) close to all halibut fishing such portion or portions of an area or areas as the International Fisheries Commission find to be populated by small, immature halibut".

Also the International Whaling Commission established according to Art. III of the Convention of Washington, 1946, has certain regulating powers.

Art. V: "The Commission may amend from time to time the provisions of the Schedule by adopting regulations with respect to the conservation and utilization of whale resources, fixing (a) protected and unprotected species; (b) open and closed waters" .... etc.

We now turn to more recent drafts and Conventions. We will compare their provisions with some characteristic points of the International Law Commission's draft, in particular the power of an international body, the limitation of agreements to certain areas, the participation of coastal States etc.

The draft Convention proposed by the International Fisheries Conference, London, 1943 [105], contains a provision which gives certain powers to a permanent Commission to be set up. Paragraphs 1, 2 and 6 of Art. 32 read:

Article 32 (Chapter III).

1. "The Contracting Governments undertake to set up a permanent Commission, to which each of them will appoint one delegate.

2. It shall be the duty of this Commission to consider whether the provisions of Annexes I and II to this Chapter should be extended or altered. For this purpose the Commission shall consult the International Council for the Exploration of the Sea". (Annex I deals with minimum sizes of mesh for nets in different areas and Annex II gives the sizes of different species of fish below which such fish may not be landed, carried on board or sold.) ....

6. "The Contracting Governments *undertake to give effect*, on the date stated in the Report, *to any recommendation* of the Commission for the extension or alteration of such Annexes which has been carried unanimously at a meeting of the Commission and accepted by any delegate who was not present at the meeting" *a*.

The Convention signed at the International North-west Atlantic Fisheries Conference, Washington, 8th February, 1949 [106], contains no provision concerning delegation of legislative or regulating powers. The Commission to be established under Art. II, which shall be known as the "International Commission for the North-west Atlantic Fisheries" can merely, according to Art. VIII:

"transmit to the Depository Government (the United States) proposals for joint action by the Contracting Governments designed to keep the stocks of those species of fish which support international fisheries in the Convention area at a level permitting the maximum sustained catch by the application, with respect to such species of fish, of one or more of the following measures:

a. establishing open and closed seasons;
b. closing to fishing such portions of a sub-area as the Panel concerned finds to be a spawning area or to be populated by small or immature fish;
c. establishing size limits for any species;
d. prescribing the fishing gear and appliances the use of which is prohibited;
e. prescribing an over-all catch limit for any species of fish".

Such proposals shall only become effective for all Contracting Governments, according to paragraph 8 of the same Article, four months after the date on which notifications of acceptance shall have been received by the Depository Government from all the Contracting Governments participating in the Panel or Panels for the sub-area or sub-areas to which the proposal applies, (compare with the alternative text of paragraph 3 of Hudson's proposal, S.R. 119 [89], p. 3).

An other characteristic of the International Law Commission proposal is the regulation per *area*.

---

a A proposal by the Canadian delegation gave powers of a much wider scope to the organizations to be established according to their draft (see Annex II, Art. 52 A) (loc. cit. p. 20–21).

It has been said before that conditions differ so widely in different parts of the world, that it is impossible to make universal regulations on this subject.

We referred already to the Canadian proposal at the Conference in London in 1943, relating to Special Areas .One of the principles which applies within such Special Areas, according to Art. 52 A (1) is that "(a) the conservation of the marine resources of such area, although of concern to all nations, may be of particular interest to certain countries having a special relationship thereto....".

According to par. 6 of this Article the North-west Atlantic should constitute such a Special Area and as we have seen this idea materialized in the 1949 Convention.

In a Memorandum of 7th May, 1947 to member Governments the Food and Agriculture Organization gave notice of the intention of the Director-General to invite member Governments to establish by Convention under the auspices of the F.A.O. Regional Councils for the study of the sea.

The Memorandum then mentions recommendations made at the first session of the F.A.O. (Report of the first session of the Conference, p. 35) and continues:

"4. The Standing Advisory Committee on Fisheries, at its meeting at Bergen in August, 1946, recommended the establishment at as early a date as possible of regional bodies for starting and developing intensive programs of co-operative research and conservation of fisheries by the member nations concerned.

5. The regional method of approach has certain obvious advantages over the all-inclusive method.

Although the seas are confluent, ecological zones within them differ and the problems of conservation and development differ correspondingly ......

7. The functions of the proposed regional councils for the study of the sea would be:

(a) to formulate the oceanographic and biological aspects of the problems of conservation and development of the resources of the high seas ...., etc. ....

10. The establishment of a series of regional councils for parts of the high seas not now actively served by existing bodies is an aim that must be attained by stages".

As a provisional delimitation the Memorandum then suggests the following regions:

a. North-western Atlantic;
b. South-western Pacific;
c. Mediterranean Sea and contiguous waters;
d. North-eastern Pacific;
e. South-eastern Pacific;
f. Western-south Atlantic;
g. Eastern-south Atlantic and Indian Ocean (African Area).

It was suggested to give immediate consideration to the establishment of Councils for the first 3 regions (a, b and c).

In the Convention for the North-west Atlantic Ocean of 1949, covering a limited, though fairly big area described in Art. I this area is even divided into five sub-areas, each with their own Panel (Art. I, par. 3 and Art. IV).

Treaties of this kind should of course be founded on scientific data. Several countries replied on the Questionnaire No. 7, concerning the "Exploitation of the products of the sea", to the Commission of Experts in 1926–1927 (Report to the Council of the League of Nations, 1927 [52], p. 120 et seq. and p. 129 et seq.) that scientific data should be made available before Treaties could be concluded.

The first big step in this direction had been taken already with the establishment of the International Council for the Exploration of the Sea at Copenhagen, which held its first meeting in 1902. Its main field of scientific investigation is the North-Sea, the Baltic and the Arctic. Similar, but mostly smaller organizations exist for the Mediterranean and the Atlantic. Some of these organizations are dealing with a particular species only, like "The International Halibut Commission" and "The International Whaling Bureau". Recently new steps have been taken.

At the Fisheries Conference held at Singapore from 6–8 January 1947, the desirability was stressed to establish a permanent Eastern Fisheries Council. It was also recommended that Governments should give full consideration to the acceptance of the principle of *international collaboration* (underlined by us), in respect of the fisheries of the areas defined. Such a Council should be established under the F.A.O.

On the F.A.O. Fisheries Meeting at Baguio, Philippines, from

25–28 February, 1948 [107], the agreement was signed for the establishment of the Indo-Pacific Fisheries Council. This is a purely scientific and technical body. Its functions, described in Art. III, being mainly research and dissemination of information and reporting to the F.A.O.

On January, 25, 1949, the Convention for the establishment of an International Commission for the Scientific investigation of Tuna was signed at Mexico City [108]. Art. 1, par. 2, reads:

.... "The Commission shall submit annually to the respective Governments a Report on its findings, with appropriate recommendations, and shall also inform them, whenever it is deemed advisable, on any matter relating to the objectives of this Convention".

When Hudson proposed his alternative paragraph in the 119th meeting of the International Law Commission (quoted above) he referred to (S.R. 119 [89], p. 4) paragraph 9 of Article I of this Convention which reads:

"9. The Commission shall be entitled to adopt and to amend subsequently, as occasion may require, by-laws or rules for the conduct of its meetings and for the performance of its functions and duties. Such by-laws, rules or amendments shall be referred by the Commission to the Governments and shall become effective thirty days from the date of receipt of notification unless disapproved by either of the two Governments within that period".

The functions of this Commission (Art. II) are very similar to those of the Indo-Pacific Fisheries Council, although here limited geographically to the "Pacific Ocean off the coasts of both countries and elsewhere as may be required" and as to the species to tuna, tuna-like fishes and tuna-bait.

The International Commission for the North-west Atlantic Fisheries, mentioned in Art. II of the Convention for the North-west Atlantic, 1949, is again a scientific and technical Commission.

Art. VI enumerating its functions starts as follows:

"1. The Commission shall be responsible in the field of scientific investigation for obtaining and collating the information necessary for maintaining those stocks of fish which support international fisheries in the Convention area ...."".

The functions are similar to the other Commissions we have discussed above.

In Art. XII the parties give a sort of promise to co-operate:

"The Contracting Governments agree to take such action as may be necessary to make effective the provisions of this Convention and to implement any proposals which become effective under paragraph 8 of Article VIII: Each Contracting Government shall transmit to the Commission a statement of the action taken by it for these purposes".

whilst Article XIII tries to solve the problem of stray fishermen belonging to a non-party country:

"The Contracting Governments agree to invite the attention of any Government not a party to this Convention to any matter relating to the fishing activities in the Convention area of the nationals or vessels of that Government which appear to affect adversely the operations of the Commission or the carrying out of the objectives of this Convention".

We see that none of these *scientific* Commissions have any more powers than to recommend certain measures to be taken. This is only natural, they consist of oceanographers and biologists who judge the problems from the scientific angle and not from the legal one. It is, however, quite a different proposition, if like in the proposal of the International Law Commission an *administrative* body like the F.A.O. should have the power to make regulations if the Parties concerned cannot arrive at an agreement.

Agreement is, also in our opinion, the best way to achieve the desired results.

The original proposal of François to give the coastal State a primary say in the matter was replaced by the clause, that the State whose territorial waters are situated within 100 miles from the area concerned, is entitled to take part in the regulation in that area. This clause finds its parallel in Art. IV, par. 2, of the North-west Atlantic Convention of 1949, though without the 100 mile margin, where it says that "each Contracting Government with coastline adjacent to a sub-area shall have the right of representation on the Panel of the sub-area"; that is to say even if the reason for representation on that Panel, as mentioned in the Article, namely "current substantial exploitation in the sub-area concerned of .... (certain species)", does not exist.

The International Law Commission arrived at the conclusion, that fishery protection and the continental shelf theory were quite distinct, as we have mentioned, because of the status of the overlying waters. We came to the same conclusion on biological facts. But it cannot be denied that certain items of the continental shelf theory and certain branches of fisheries are uncomfortably intermixed and that certain activities in the exploitation of the mineral resources of the continental shelf will certainly affect fishery activities in the neighbourhood.

We will discuss these points now.

SECTION 10. BOTTOM-FISHERIES

As far as bottom-fisheries are carried out by trawling, they are linked up with the shelf, as described in Chapter I (the sea-bottom which is not deeper than 100 fathoms), because trawling is mostly carried out up to this depth. We mentioned this fact already above, quoting a statement to this effect on page 19 of the Report on a Survey of the Fishery Resources of the United States [56], and may add a statement in the Report of the Commission of which Govare was chairman [29], page 3:

"Les instruments les plus modernes utilisés (chaluts et palangres) pour la pêche sur le fond ne dépassent guère une profondeur de six cents pieds et se trouvent donc limités au plateau continental".

As far as declarations or decrees do claim the sea-bottom or sea-bed of the continental shelf to belong to the coastal State, or being under the control and jurisdiction of the coastal State, it is obvious that trawling by foreign fishermen on that sea-bed will be objected to.

It would be closing our eyes to reality if we tried to argue that bottom-fish are not fixed to the sea-bed but move freely and only live on or near the sea-bed. The trawlnet scrapes the sea-bottom very closely. Many bottom-fishes like plaice, sole, weever etc. burrow beneath the sand to be less conspicuous to their enemies. Of course they will be caught as well when swimming around near the sea-bottom, but it would be quite impossible to make out which one was caught free swimming and which one when dug in. All this may seem hairsplitting, but even when this claim of the sea-bed is made for the purpose of having the exclusive right to build installations for oil exploitation, to drive piles into it, we are not so certain that this right once established will not be misused for other claims and would be adduced to prohibit non-nationals to carry out trawling on the sea-bed of the shelf.

Hudson uttered a similar warning, when during the discussion on sedentary fisheries, he said in the 114th meeting of the International Law Commission [109] (S.R. 114, p. 7):

"Hence the paragraph would have to be drafted very carefully so as to ensure that sedentary fish, pearls etc. as well as bottom-fish, were not placed under the control and jurisdiction of the coastal State".

He referred to the draft Article 4 of François in his Second Report [26], p. 69: "The waters covering the continental shelf outside the territorial waters remain within the regime of the high seas", for which Article Hudson had proposed a slightly differently worded text.

If the waters covering the continental shelf remain under the regime of the high seas, a prohibition to fish in those waters, as we imagined above, would be contrary to this principle of freedom. The International Law Commission expressly provided that "There could be no question of such right of control and jurisdiction over the waters covering those parts of the sea-bed" (Second Report François, p. 65). Article 3 of the draft Articles on the continental shelf, in the Annex of the Report of the International Law Commission, covering its 3rd session [36] (p. 57) reads:

"The exercise by a coastal State of control and jurisdiction over the continental shelf does not affect the legal status of the superjacent waters as high seas".

Leaving apart the chances of adoption of this draft Article (and we think of the declarations and decrees in which sovereignty over these waters is claimed) our worries, expressed above, are not taken away even if the Article was adopted.

It is noteworthy that some instruments we discussed in this Chapter (and some which will be discussed in Chapter IV) give an assurance that "the character as high seas of the waters above the continental shelf and the right to their free and unimpeded navigation are in no way thus affected" (to take the wording of the Proclamation of President Truman with respect to the resources of the continental shelf). Although the clause concerning the unimpeded navigation is actually superfluous, being already comprised in the statement that these waters keep their character of "high seas", it was apparently thought necessary to stress this fact in particular.

The other most important right derived from the notion "high seas", the right of unimpeded fishery, however, is *not* mentioned in any of these instruments.

In spite of the fact that the right to fish is comprised in the notion "high seas" and should therefore in view of this statement not be interfered with in these waters, the suspicious reader may well ask why one of the main rights inherent in the notion "high seas" is especially mentioned again and the other not. As to other instruments we refer to the last paragraph of the Proclamation of President Truman concerning Fisheries (wording as quoted above). Article 6 of the Treaty between the United Kingdom and Venezuela of 1942 [110] reads:

"Nothing in this Treaty shall be held to affect in any way the status of the waters of the Gulf of Paria or any rights of passage or navigation on the surface of the seas outside the territorial waters of the Contracting Parties".

A source of conflicts seems to be opened if we proceed in this train of thought.

Anticipating our suggestions in Chapter IV dealing with mineral resources, in which we propose to divorce the sea-bed from the subsoil, we believe that the possibility of interference with fisheries is one of the reasons to consider seriously our suggestion of giving the sea-bed another status than the subsoil and to sever the link which the continental shelf theory emphasizes so strongly, as we will see.

Section 11. Sedentary Fisheries

Sedentary fisheries are generally understood to concern those products of the sea which are permanently or semi-permanently attached to or at least constantly living on the sea-bed, like edible oysters, pearl-oysters, chanks (a shell of the gastropod mollusc "Turbinella rapa" used for decorative purposes and as a "musical" instrument in temples, according to Thurston [111] (p. 55), trepang (the Malay word for sea-cucumber or sea-gherkin, "Holothuria marmorata", sometimes called bêche-de-mer, a quasi French corruption of the Portuguese word: bicho do mar, sea-worm), coral, sponges etc.

Gidel [112] (p. 488), however, comprises under sedentary fisheries also, the fishing of pelagic (swimming) fish carried out with permanent installations fixed on the sea-bottom, like weirs or pound nets, in Malay: "sero's". Although this method could well be called sedentary fisheries, we believe that this word is usually understood to denote the species concerned and not so much the equipment employed. François uses the word in both meanings (First Report [10], p. 31).

In both meanings the sedentary fisheries become important for our subject, when exercised outside the territorial sea.

The legal difficulty arises from the fact, that the coastal fishermen, and often the coastal State, have claimed for many centuries the exclusive fishing rights on the banks or reefs adjacent to their coasts.

According to Gidel [112], p. 489, these legal difficulties have not been many, and when they arose they were easily solved. He mentions for instance a case (p. 492) of a Greek captain and a French merchant who tried to protest in 1875 against the lease of sponge fisheries on the Tunisian coast, instituted by legislation as public revenue for the security of the debts to European Powers. They invoked the principle of the freedom of the seas against this act of sovereignty, but by judgment of their respective consuls their claims were dismissed.

The difficulties between French and British fishermen concerning oyster fisheries in the Bay of Granville were solved by the Convention of Paris, August 2, 1839.

Fulton [31], p. 612, writes: ".... a very considerable stretch of water containing oyster-beds, in the Bay of Granville on the French coast, between Cape Carteret and Point Meinga, south-east of Jersey, and extending far beyond the 3 mile-limit, was reserved exclusively for French fishermen, the boundaries being minutely defined and laid down on a chart annexed to the Convention; and the British fishermen were prohibited from carrying on any kind of fishing, even for floating fish, within this area. The bay thus appropriated is over 17 miles in breath, and the closing line passes in some places about 14 miles from the shore. (The greater part of the closing line, however, is, curiously, within the 3 mile-zone). This concession to France was a recognition of the principle that fisheries of this nature — that is for objects which are attached to or stationary on the bottom — require special treatment". (See also Gidel, p. 489).

This Convention contained in Article 11 (Gidel, p. 423) an agreement to establish a commission whose duty it was to draft regulations for the fisheries. This code of regulations, confirmed by the respective Governments in June 23, 1843, contained several provisions concerning oysters and laid down a minimum size for oysters (Gidel, p. 447).

This code was in Great Britain embodied in an Act to carry into effect the Convention concerning the Fisheries in the Seas between the British Islands and France of 22nd August, 1843 [113] [a]

To give an idea about its contents, we quote:

"Art. XLV, Oysterfishings shall open on the First of September and shall close on the Thirtiest of April.

Art. XLVI, From the First of May to the Thirty-first of August no boat shall have on board any dredge or any Implement whatsoever for catching oysters.

Art. XLVII, It is forbidden to dredge for oysters between Sunset and Sunrise.

Art. XLVIII, The fishermen shall cull the oysters on the fishing-grounds and shall immediately throw back into the sea all oysters less than two and a half inches (six centimètres, French) in the greatest diameter of the Shell ....".

So far a few disputes.

---

[a] For later Treaties between France and Great Britain on this subject (June 23, 1843, November 11, 1867, not ratified, and September 29, 1923) see Gidel, p. 489–491.

How far beyond the three mile-limit do these fisheries extend?

This question was asked in the British Parliament in 1923, regarding Ceylon. The Under Secretary of State for the Colonies answered (column 993) [114]: "Under Ordinance 18 of 1890 fishing for chanks is prohibited within the limits bounded by a straight line drawn from a point six miles westward of Talaimannar to a point six miles westward of the shore two miles south of Tailavilla" [a].

In an answer to a further question (colomn 1417) [114] the Under Secretary of State for the Colonies said that "it is estimated by rough measurement that the line in the Gulf of Manaar (see previous reply) is at its furthest point 25 miles from the coast of Ceylon. The prohibition applies to all vessels of all nationalities and is imposed for the protection of the pearl-banks ....".

We have seen already, that the boundary line delimiting the exclusive fishery rights for French fishermen, by the Treaty of 1839, on the oyster-banks in the Bay of Granville, was at some places 14 miles from the shore. In Australia the delimitation is not so clear. The Pearl-shell and Bêche-de-mer Fishery Act of 1881 [115] (p. 777–790), was amended several times (in 1886, 1891, 1893, 1896 and 1898 [b].) The Amendment Act of 1891 (55 Victoria No. 20) section 6 (loc. cit. p. 784) reads: "An Inspector may within the limits of the territorial jurisdiction of Queensland, exercise any of the following powers .... etc.".

According to Jessup [116], p. 16: ".... judging from the official map of New Guinea and Papua (at. p. 969 Official Year Book of the Commonwealth of Australia, No. 15, 1922) the British claim to jurisdiction in these regions is very extensive indeed, being measured by a line drawn frequently more than 100 miles from shore to embrace the numerous scattered islands. The Queensland boundary is shown in part as extending more than 100 miles out to sea" [c].

---

[a] See Schedule B of the Chanks Ordinance, 1890 (No. 18–1890) in Thurston [111], p. 58–62.

[b] In the Second Report of François, p. 55, two more amendments are mentioned: 1913 and 1931.

[c] This map [117] shows a dotted line, marked "Boundary Queensland" running along the outside of the Reefs in the Torres Strait, crossing over the Torres Strait and turning to the West including the Warrior Reefs and following closely the south-coast of Papua (New Guinea), in short very much the line described in François' Second Report [26], p. 56, quoting the Federal Legislation, i.e. the Queensland Pearl-

Article 31 of the Oyster Act of 1886 (an Act for the protection
of oysters and the encouragement of oyster fisheries, 50 Victoria,
No. 22, loc. cit. p. 772–777), reads: "This Act shall extend to
and be in force, only in such ports and parts of the Colony ....
as the Governor in Council shall from time to time by Proclamation declare to be and to come within the operation thereof".

Here the delimitation cannot immediately be found in the
Act. For other Australian Acts see Second Report of François [26],
p. 55–57. He concludes by saying: "The present Australian
Constitution continues in the Commonwealth Government
the power to regulate fisheries beyond territorial limits
(Section 51, x)".

In Tunisia, according to François, loc. cit. p. 58: "The bed
extends as far as 17 miles from the mainland".

Finally Section 67 of the Act to carry into effect the (non
ratified) Convention of 1867 [118], mentioned above, reads: "The
Irish Fishery Commissioners may from Time to Time lay before
Her Majesty in Council By-laws for the purpose of restricting
or regulating the dredging of Oysters on any Oyster Beds or
Banks situated within the Distance of Twenty miles measured
from a straight Line drawn from the Eastern Point of Lambay
Island to Carnsore Point on the Coast of Ireland, outside the
exclusive fishery limits of the British Islands ...." and Fulton [31],
p. 621, points out that this is "an area of nearly 1300 square
(geographical) miles outside the 3 mile-limit, including the
Arklow and Wexford Banks and stretching from 12½–19 miles
beyond the ordinary limit".

This will suffice to point out that the areas, where sedentary
fisheries take place, are in many cases situated outside the
territorial waters.

A curious fact which we mentioned already in Chapter I
is, that for the delimitation of these areas or delimitation between
parts of these areas for different purposes, the criterion of depth
is used.

---

shell and Bêche-de-mer Fisheries (Extra Territorial) Act of 1888 (51 Victoria, No. 1).
There is also under the Federal Legislation a similar Act for Western Australia. But
as both these Acts apply only to British ships (loc. cit. p. 57) they are not useful to
show the existence of exclusive rights or conservation rights applicable to foreigners
outside the ordinary 3 mile-limit.

Where the first delimitation is concerned, so to speak the staking of a claim as against outsiders, François, loc. cit. p. 54, says that the Pearl Fisheries Ordinance of 12 February, 1925, of Ceylon "delineates as a pearl-bank an area between the three and five fathom-lines on one hand and the 100 fathom-line on the other hand" [a]. He further mentions, p. 57–59, the reserved zone off the coasts of the Regency of Tunis in which the Tunisian Government "regulates fishing". The outer limit of this zone in the Gulf of Gabes is the 50-metre isobath (this boundary-line has been laid down in Article 29 of the Instruction of 31 December 1904, relating to the Navigation and Fishery Service, see François, p. 58).

These boundary-lines have a connection with a delimitation, which we got acquainted with in Chapter I, namely the depth where exploitation was possible. François, p. 58, says: "As fishing by diver and trawl was carried out in the past at depths not exceeding 50 metres, the supervising authorities adopted that depth as the practical boundary of the Tunisian beds".

As to the second kind of delimitation, inside the outer boundary, we give the following examples. As a limit for the possibility for exploitation can be described the depth of 20 metres in Tunisia, mentioned by François, p. 58: "As sponges cannot be fished by trident at a greater depth than 18 or 20 metres, the depth of 20 metres has been chosen as the inner limit of fishing by diver and trawl ....".

Sometimes delimitation is made between an area reserved for the population of the land, who are often too poor to afford expensive equipment and are left to work the more shallow parts of the banks, whereas the deeper parts are farmed out. An example is to be found in the Chanks Ordinance 1890 (No. 18, 1890) in Thurston [111], p. 61: ".... Provided that nothing in this section contained shall prevent any person from collecting coral or shells from any portion of the said seas in which the water is of the depth of one fathom or less ...." [b].

Another example is the Ordinance for Netherlands South New Guinea, of 15 January, 1905 [119], concerning pearl, mother

---

[a] *See also* [158], *p. 61, (Note under No. 5 (c))*.
[b] *The same wording in Chanks Ordinance, 30 June 1891, as amended by Ordinance No. 2 of 1929* [158], *p. 59, under (a)*.

of pearl and trepang fisheries. Article 2 of this Ordinance states that the old rights of the local population remain secured, with the exclusion of everybody else, for all places not deeper than 5 fathoms (9 metres) at low water. These rights are inalienable. For the rest no fishing is allowed without a permit, costing for the season per ship 60 guilders and per diving-apparatus 40 guilders.

The same provision is to be found in Article 2 of the Ordinance for the former Netherlands East Indies, of 29 January, 1916 [120], applicable in the rest of the Archipelago, and regulating sedentary fisheries as mentioned above, and the sponge fisheries. This Ordinance is based on a lease-system. However, both Ordinances are only applicable within the 3-mile limit.

Sedentary fisheries are carried out in many places. We mentioned Ceylon, where we find pearl- and chank fisheries north and south of the Adams bridge, in Palk's Bay and in the Gulf of Manaar, on the Ceylon side (Dutch Bay) and the opposite side near Tuticorin.

In Dutch Bay, according to Thurston [111], p. 33, the yield in the season (March) in 1889, with a fleet of maximum 193 boats, was from well over 1 million to 2 million oysters daily, with a topscore of nearly 2,5 million on the 19th March. We further mentioned Ireland, France, Tunisia, Indonesia and Australia.

During the Codification Conference of the Hague, 1930, (Acts [61], p. 150) the Columbian delegate, Urrutia, pleaded for a wide belt of territorial sea, for the pearl fisheries in the Atlantic and for the exploitation of "our other natural resources such as petroleum"! This is a remarkable example of linking up sedentary fisheries with mineral resources in a claim for at least a part of the continental shelf, at an early date.

In the Report on a Survey of the Fishery Resources of the United States [56], p. 2, we read: "The United States sponge fisheries is centered on Florida's west coast ...." .... "the canning of shrimps and oysters are the most important shore industries (in the South-Atlantic and Gulf States)".

Fulton [31], p. 697, further mentions the coral beds in the Mediterranean off the coasts of Algeria, Sardinia and Sicily, regulated by French and Italian laws beyond the 3-mile limit.

Finally François describes (loc. cit. p. 59) the pearl fisheries
in French Oceania, New Caledonia and French Somaliland,
however, usually confined to territorial waters; in the Persian
Gulf, p. 59, where these fisheries are governed by customs and
usages and not by national legislation. "Intrusion by outsiders . . .
has been discouraged by the British, who have long exercised
powers of maritime police in the Gulf. This British protection
of pearling has been based on British political and naval pre-
dominance in the Gulf rather than on any legal authority";
and the pearl fisheries of Venezuela (p. 60) and Panama (p. 61).

After this orientation we arrive at the main questions, what
is the difference with ordinary fisheries, what right or title has
the coastal State on sedentary fisheries?

Westlake [121]. p. 190–191, writes:

"The case of the pearl fishery is peculiar, the pearls being
obtained from the sea-bottom by divers, so that it has a physical
connection with the stable element of the locality which is
wanting to the pursuit of fish swimming in the water. When
carried on under state protection, as that off the British Island of
Ceylon, or that in the Persian Gulf which is protected by British
ships in pursuance of treaties with certain chiefs of the Arabian
mainland, it may be regarded as an occupation of the bed of
the sea. In that character the pearl fishery will be territorial
even though the shallowness of the water may allow it to be
practised beyond the limit which the state in question generally
fixes for the littoral sea . . . . and the territorial nature of the
industry will carry with it, as being necessary for its protection,
the territorial character of the sea at the spot".

He calls it occupation. His last remark relating to the waters
above the pearl-banks, outside the territorial sea, seems to us
difficult to accept, although, as we will point out, freedom of
fisheries above these banks will not always be allowed, for
instance trawl-fishing.

In his notes to Chapters VIII and IX (referring to pages
190–191), p. 203, Westlake criticizes Oppenheim, who does not
believe that a part of the sea-bed under the open sea can be
occupied. Oppenheim [122], p. 10: "Obgleich die Freiheit des

offenen Meeres gewohnheitsrechtlich allgemeine Anerkennung
gefunden hat — ein Entwicklungsprozess, der erst im Viertel
des neunzehnten Jahrhunderts zur vollen Ausreife gelangte —
so blieb doch die staatliche Herrschaft über die Perlen fischerei
in der nähe der Bahraininseln und der Insel Ceylon vollkommen
und allseitig anerkannt. Wir hatten es also um eine historisch
und gewöhnheitsrechtlich anerkannte Anomalie zu tun".

Oppenheim [42] gives the same explanation in his book, p. 568
(note), namely: "the fact that the freedom of the open sea was
not a rule of International Law when these fisheries were taken
possession of".

We should have put it differently. The freedom of the seas
has, we believe, always existed. In the olden days people took
this right for granted. Assuming that we may believe Jessup [116],
who says, p. 15: "The history of the Ceylon chank fisheries was
traced from the 6th century B.C. to modern times", we believe
that in those days the density of navigation and fishing was so
little, that human activities, like fishing pearl-oysters, could
go on for some considerable time unnoticed by other than the
coastal fishermen and fishermen of the neighbourhood. It is
noteworthy in this connection that in the Ordinance of Netherlands New Guinea of 1905 [119], these "orang laoet" (seamen, here
fishermen), "who dwell on these coasts from time to time", are
expressly put on the same footing as the local population, where
traditional rights are concerned.

We believe, that prescriptive rights could develop quietly,
and had existed long enough to be respected when people became
conscious of the freedom of the seas. That the freedom of the
seas has always existed is our assertion, which we base on the
fact that in spite of extravagant claims and except small areas,
where the claiming State had the (naval) power to enforce its
claims, in practice the sea has been and has remained free
throughout human history. After all, the great advocate for this
freedom, Grotius, who had to fight one of these extravagant
claims, did not invent that freedom, but based it on many
witnesses, who stated the same truth centuries earlier.

Hurst [123], p. 40, refers to Hall [71], who holds the view that
appropriation of larger sea areas was given up through abandonment and that maritime occupation must be effective in order

to be valid (p. 189). Hurst then says: "If it is disuse and disuse alone which has led to a restriction of the rights of the Sovereign in the bed of the sea, it follows that in cases where there has been effective occupation of a portion of the bed of the sea within the meaning of the principle enunciated by Hall and such occupation still continues, there has been no abandoning of the rights of ownership, and consequently the ownership still continues. Assuming that this proposition is sound, it removes a difficulty which has found expression in writings on international law as regards sedentary fisheries".

Hall, however, speaks about the sea and Hurst about the sea-bed.

It is true, as Oppenheim [42], p. 575, says, that "there has been a tendency in the past to assume that the surface of the bed upon which the open sea rests must be likened in legal condition to the waters of the open sea themselves". But Oppenheim criticizes this view when he continues to say that the reasons for "the abandonment of the former claims to occupy the waters of the open sea .... do not apply to the sea-bed ....".

On p. 576 he says that "although it is traditional to base some of these cases on the ground of prescription .... that a State may by strictly local occupation acquire, for sedentary fisheries and for other purposes, sovereignty and property in the surface of the sea-bed, provided that in so doing it in no way interferes with freedom of navigation, perhaps we should add, with the breeding of free swimming fish".

The tendency that the sea-bed and the waters above it have the same legal status can be found for instance in Moore [87], p. 180 et seq., in Chapter X: "The origin in the prima facie title of the Crown", where he refers to a treatise of Thomas Digges, "Proofs of the Queens Interest in Lands left by the Sea and the Salt shores thereof", written in 1568-9. We must, however, not exaggerate the property rights of the Crown as meant by Digges. He deals with tide-lands, what he calls the "salt shore". He even says (p. 185 Moore) that the sea is the property of the King, but he does not say anywhere how far those property rights extend.

He does, however, delimit the "salt shore" (p. 191): "So that not only from the lowe watermarck downward but also upwarde to the full sea, so highe as naturally by ordinary course euery mooneht wthout rage of the wynde or wether or other rare accident, so farre I saye as the sea hath by sutche his naturall course flowde; ....", and p. 199: "....; but perhappes a question may bee mooued howe farre this salt shore doothe extend and howe yt is to bee limited. To this I answere, although by the rigour of the lawe the salt shore dooe reche as farre as the gretest winter ffloodd dooth flowe, yet dooe I suppose the lawe and lawe-full custoome of the realme will thus interprete the same, that yt bee taken no farder then the sea by Naturall course not enforced by rage of wether doothe rise".

The margin is thus meant to be between high and low water spring tide. A very small claim of the sea-bed indeed. The only wider rights on the sea-bed he mentions are those in harbours and roadsteads and "shelves", on p. 189 he says that it appears manifestly that the Prince has greater property in the ground of his ports and harbours than any borderer or Lord of the soil adjoining to any fresh river has in the banks of the said river .... (because) .... no man can let fall an anchor upon the Kings ground in any port without permission and paying therefore to the Kings officers for that purpose .... that it is against all reason to .... deny his (the Kings) interest in islands, shelves and the salt shore; and on p. 190 that precedents are to be seen in records of sundry princes in divers places since the Conquest... (for instance) the custom and usage of the land that strangers pay to the prince "anckorage" for breaking (brekinge) of his ground. Finally on p. 203 that also to make it manifest that the soil of the seas is also "intirelie the King's, no man can let fall any anchore in any roade aboute this Realme but he paiethe for breakinge the Kings grownde to the officers of the Kinge".

The same vague claim we find in the lectures of Serjeant Callis on the Statute of Sewers (23 Henri VIII cap. 5), (Moore [87], loc. cit. p. 252) according to Fulton [31], p. 363, delivered at Gray's Inn in August, 1622.

Callis discusses the King's ownership of the sea and says (Moore, p. 252) "that the King has therein these powers and properties, viz.:

1. Imperium Regale.
2. Potestatem legalem.
3. Proprietatem tam soli quam aquae.
4. Possessionem et proficuum tam reale quam personale".

The extent of these rights is vague. Fulton, (loc. cit.) says that the King is full Lord and owner of the seas and that the seas are within "the realm of England".

One gets the impression, however, that these property rights on the sea-bed were thought of in areas not too far from the shore, for instance again in the suit relating to Sutton Pool, near Plymouth, during the reign of Charles II in 1662, Moore, p. 314: "It is an information by the Attorney General against Oliver Ceeley and others for making encroachments on the soil by wharves and buildings erected below high-water mark." The information recites that "whereas our Sovereign Lord King Charles .... is seised in his demesne as of fee in right of His Crown of England of and in the ground and soil of the coasts and shores of the seas of and belonging to his Kingdom of England, and of and in the ground and soil of all and every the ports, havens, arms of the sea, creeks, pools, and navigable rivers thereof into which the sea doth ebb and flow or hath used to ebb and flow, and in particular of and in the ground and soil of all that pool, haven, creek, and arm of the sea commonly called Sutton Pool, near unto the port and town of Plymouth, and of the fishery, wreck of the sea, wharfage, tolls, &c".

We are coming nearer our subject in the case of the Duchess of Sutherland v. Watson in 1868, an action for taking mussels from mussel scalps (banks) in Cromartie Bay. Of course at that time the extravagant claims on the sea had been abandoned but we see a more real relation between the "foreshore" and the waters near the coast in this case. Moore [87], p. 578: "Lord Neave held that 'the foreshore is in the Crown as a patrimonial right of property subject to public rights. Mussel scalps are partes soli because, although the mussel has powers of locomotion, and uses them in early life, when it once settles down and fixes its domicile it seems to do so animo remanendi, and, becoming part of the soil, it is a trespass to take it. Mussel scalps are the subject of exclusive grant by the Crown to an individual. The Crown may grant the right of using these scalps and taking the mussels from them in two ways. It may give the solum and fundus of the sea as a feudal estate, and, if it does so, it gives the scalps along with it' ".

In Hall's Essay on the rights of the Crown [87] we find the same statement, p. 668–669:

".... the writers on the common and municipal law of England, as well as the decisions of our judicial courts, all speak the same language and appropriate the dominion of the British seas *tam aquae quam soli*, to the King".

One who was more specific in describing distances was Plowden in the Sir John Constable Case, 1575 (Moore [87], p. 227–229). But he lived in the time of Queen Elizabeth who sponsored the freedom of the seas.

Nevertheless he states that the Queen had jurisdiction over the sea between England and France, and the Irish sea and in the other seas only "the moiety".

But he says "although the Queen had jurisdiction in the sea adjoining her realm, still she has not property in it, nor in the land under the sea, for it is common to all men, and she cannot prohibit anyone from fishing there ....".

We believe, that, even if theoretically a property claim may have existed on the sea-bed, in reality the writers and the judicial decisions were actually dealing only with practical cases, like building wharves below high-water mark, encroaching on mussel- or oyster-banks, questions of aluvion, wreck and flotsam, a.s.o.

We suggest that the tendency, Oppenheim refers to, only concerned parts of the sea-bed which were in the picture of human activities and that further claims, if at all meant seriously, and not just to please a monarch, were of no significance whatsoever. In other words the similarity in legal status of the sea and the sea-bed was only a limited and local one. Therefore we beg to disagree with Hurst, that there even may be question of disuse which has led to restriction of the rights of the Sovereign in the bed of the sea, how convenient such a thesis may be to explain away the difficulties as regards sedentary fisheries.

These ideas are somehow connected with the theory, that the territorial sea is a sort of residue of wider claims in former times. We will deal with this view in Chapter III.

Hurst [123], dealing with the property rights of the Crown

in the bed of the sea, just before he starts to discuss sedentary fisheries, ends up by saying, p. 39:

"The question, however, remains to what distance do these property rights of the Sovereign in the bed of the sea extend? [a] So far as the law of this country is concerned, the rights of the Crown were fixed long before the doctrine of the 3-mile limit was thought of, and yet it seems to be agreed that nowadays these property rights do not in general extend beyond the 3-mile limit". We may perhaps even say that property rights in the sea-bed have rather increased than decreased.

Hurst makes another remark which we find difficult to accept. On p. 40 he says: "the exclusive right to the produce to be obtained from these fisheries may be based on their being a produce of the soil" [b]. The biological facts are quite different. We cannot compare the sea-bed where mussels or oysters or sea-cucumbers or sponges happen to be with a garden and the animals with real cucumbers, pumpkins or other fruit growing in that garden. The animals do not grow in the soil with roots, they do not derive food from the soil, but from the water of the sea covering the sea-bed, they are not even in all cases permanently attached to the soil. We quoted on purpose the case of the Duchess of Sutherland v. Watson concerning the taking of mussels from mussel scalps, because the reasoning was, biologically speaking, correct. Mussels do move about as long as they are not attached to a cluster of other mussels. Oyster-larvae, when trying to settle down, do detach themselves, when the spot is not to their liking and can repeat this several times, until they "anchor" forever, showing their "animus manendi", and thus obtaining "domicile". The sponges of course are similarly attached to the sea-bed. The sea-cucumbers or trepang on the contrary move freely about without ever settling down. Therefore there is no question of "products of the soil". In the

---

[a] We noticed an unusual lapse of his argument, when Hurst first said on p. 39: "Lord Hale's writings give no indication as to the distance to which he considered the property of the Crown in the bed of the sea to extend", and then (same page): "If the rights of the Crown to the ownership of the bed of the sea are now more restricted than they were at the time at which Lord Hale was writing, it can only be that these also have been narrowed by disuse".

[b] Smith [41], p. 11, makes the same incorrect statement, which he may have quoted, however, from Hurst, as the wording is practically the same.

case of corals we may even say that the soil is a product of the corals.

Hurst makes a clear distinction between the sea-bed and the waters above, where he says: "the exclusive right to the pearls to be obtained from the banks flowed from the ownership of the bed of the sea where the banks were situated, and not from any claim to maritime jurisdiction over the waters".

Is this statement tenable?

On p. 43 he says: "It cannot be too strongly emphasized that the recognition of special property rights in particular areas of the bed of the sea outside the marginal belt for the purpose of sedentary fisheries does not conflict in any way with the common enjoyment by all mankind of the right of navigation of the waters lying over those beds or banks. Nor does it entail the recognition of any special or exclusive right to the capture of swimming fish over or around these beds or banks".

Gidel very rightly remarks, p. 500, that the principle of the freedom of the seas involved for all the right to fish freely on the high seas. But sedentary fisheries limit this rights as far as the "sedentary" species are concerned.

We would like to go even further. Trawling on those banks, even with the exclusive object to catch bottom-fish, will certainly be objected to by the coastal State as being destructive for the oysters or other "sedentary" species.

Gidel says that the freedom of the high seas and sedentary fisheries are incompatible. To recognize sedentary fisheries as International Law does, means that on these places outside the territorial waters, an exception is made to the general principle of the freedom of the seas. Gidel wants to limit this exception, and the conditions on which the sedentary fisheries can be held as legitimately carried out are the effective and continued use of a part of the high seas for that purpose, other States having special interests because of their geographical situation not objecting.

What sort of rights does the littoral State usually claim?

As example we mention Article 8 of the Chanks Ordinance of 30 June, 1891, as amended by Ordinance No. 2 of 1929 (quoted from the Second Report of François [26], p. 53 [a]: "It shall not be lawful for any person to fish for, dive for or collect chanks, bêche-de-mer, coral, or shells in the seas within the limits defined in Schedule B except in accordance with the rules for the regulation, supervision, protection or control of such operations which may be made by the Governor and published in the Gazette, ..." [b].

Article 4 of the Pearl Fisheries Ordinance of 12 February, 1925 (quoted from François, loc. cit. p. 54) [c]: "1. No person shall fish, or dive for, or collect, pearl oysters on or from any pearl-bank, or use a vessel for any such purpose, unless he holds a licence (in this Ordinance referred to as a pearl fishery licence) authorizing him so to do" [d].

The same system we meet in the Australian Act we mentioned above and in the Pearling Act (No. 45 of 1912) of Western Australia (François, loc. cit. p. 55) and in Tunisia. These provisions come under the category of exclusive control. The littoral State prohibits fishing without licence and farms out fisheries which means that it acts as proprietor. Although the coastal State has thus the power to prevent anybody especially foreigners to fish, by refusing a license or in the handling of the lease system, the regulation has been liberal enough to admit fishermen without distinction of nationality, according to Gidel, p. 501 [e]. Next thereto we see the right to lay down conservation regulations, as for instance in Article 8 of the Chanks Ordinance of 1890: "8. It shall not be lawful for any person to use any dredge .... for the purpose of fishing or collecting chanks", which was made to protect the oyster-beds. Provisions fixing the dates for the pearling season and protection of the oyster-beds are found in the Venezuelan Pearl Fisheries Act No. 19143 of 22 July 1935, according to François (loc. cit. p. 60); and rules establishing

---

[a] *See also* [158], *p. 58.*

[b] The demarcation in Schedule B has been given above in the answer of the Under Secretary of State for the Colonies in the British Parliament in 1923 [114], (column 993).

[c] *See also* [158], *p. 60.*

[d] An example of farming out of these fisheries by the Government can be found in correspondence relating to an angreement for lease of pearl fisheries on the coast of Ceylon [114].

[e] In the same sense Hudson in the 66th meeting of the International Law Commission (S.R. 66[7], p. 12).

zones and periods within which the taking of mother-of-pearl shell with mechanical devices is permitted, are found in Chapter III of Title V of the Codigo Fiscal of Panama (Official Edition 1931), (François, p. 61).

As the coastal State acts as proprietor, the question may be asked where this right came from. The Under Secretary of State for the Colonies, answering a question to that effect regarding the chank fisheries of Ceylon, in the British Parliament in 1923 [114], (column 993) said: "The claim of the Ceylon Government to this fishery is based on immemorial user by successive sovereigns of the island", and again (column 1418): "The justification for this prohibition is based on rights over the fisheries enjoyed in uninterrupted and undisputed proprietorship by successive rulers, native, Portuguese, Dutch and British since period prior to the development of the doctrine of the 3-mile limit".

About the "age" of the pearl fisheries in the Gulf of Manaar, Thurston [111] writes on p. 9:

"Tuticorin has been celebrated for its pearl fishery from a remote date, and as regards comparatively modern times, Friar Jordanus, a missionary bishop, who visited India about the year 1330, tells us that as many as 8000 boats were then engaged in the pearl fisheries of Tinnevelly and Ceylon. In more recent times the fishery has been conducted successively by the Portuguese, the Dutch and the English". We mentioned already the age of chank fisheries according to Jessup, but even if this cannot be proved for other fisheries the age given above for the pearl fisheries seems quite sufficient for obtaining a prescriptive title.

As most writers speak of occupation, does that mean that sedentary fisheries are quite different in character from other fisheries?

Sedentary fisheries, says Fulton [31], p. 697: "have always been considered on a different footing from fisheries for floating fish .... they are looked upon rather as belonging to the soil or bed of the sea than to the sea itself. This is recognized in municipal law, and International Law also recognizes in certain cases a claim to such fisheries when they extend along the soil under the sea beyond the ordinary territorial limit".

The Under Secretary of State for Foreign Affairs said about
this question in the British Parliament in 1923 [125], (column 1261):

".... pearl fisheries stand on a different footing to the ordinary
kind of fishing in the waters of the sea, because the banks where
the pearl oysters lie must be treated as part of the bed of the
sea. For many centuries the pearl-banks off the coast of Ceylon
have been claimed as subject to sovereignty of the rulers of the
neighbouring territory and subject therefore also to their control ...." and (column 1262): ".... where they are situated
under the high seas the claim to sovereignty and control is
limited in extent to the area of the banks, and does not affect
the rights of navigation or of ordinary fishing in the waters over
the banks". To conclude the British view, we quote the British
answer, comprised in the "General observations submitted by
certain Governments" on Point II:

"Application of the rights of the coastal State to the air above
and the sea-bottom and subsoil covered by its territorial waters",
in Bases of Discussion [126], p. 28: "(c) No claim is made by His
Majesty's Government in Great Britain to exercise rights over
the high seas outside the belt of territorial waters.

There are certain banks outside the 3-mile limit off the coasts
of various British dependencies on which sedentary fisheries
of oysters, pearl oysters, chanks or bêches-de-mer on the sea-
bottom are practised, and which have by long usage come to
be regarded as the subject of occupation and property. The
foregoing answer is not intended to exclude claims to the sedentary
fisheries on these banks. The question is understood to relate
only to claims to exercise rights over the waters of the high
seas".

The object of the fisheries is the animal, which is usually found
on certain banks. But this is not characteristic for oysters or
sea-cucumbers. The New Foundland Banks, or the Dogger Bank
may be safely visited with a good chance of a rich yield of bottom-
fish. The fact that oysters are attached to the bottom once
they have settled down, does not guarantee a good yield every
year. Thurston [111], p. 54, remarks what a spasmodic industry
the pearl fisheries are:

"Writting in 1697, for the instruction of the political council
of Jaffnapatnam, the then commandant of that town justly

remarked that the pearl fishery is an extraordinary source of revenue, on which no reliance can be placed, as it depends on various contingencies, which may ruin the banks, or spoil the oysters. And this remark holds good after the lapse of two centuries".

We said that the larvae, which are free swimming, are rather particular in their choice of domicile. But of course if certain banks are traditional places for oyster-fisheries, we may assume that the conditions to settle down are favourable. However, as long as the larvae are free swimming, even if they should have wanted to settle down on the grounds of their parents they do not always get the chance. Strong currents may sweep them away before they have "anchored" and if in the season where the youngsters are going to change their life habits strong currents occur on the banks, the banks will be empty and this has happened many years according to Thurston. Either they are swept to other places favourable for a "sedentary" life or they die.

Nevertheless, once settled down we can say, that, because of the place where it settled down the oyster becomes the property of the coastal State, which had a prescriptive right.

Here is the difference with bottom-fish: we can say that an oyster or sponge attached to a bank *is* somebody's property, whereas the fish can only *become* somebody's property when caught.

Does this property right derive from the property of the oysterbank? This is not an easy question. Ceylon is not a very good example.

Hurst, p. 40, says: "The instances where ancient usage justifies a claim to sedentary fisheries outside the 3-mile limit do not seem to be numerous, and of those which are known some appear to be situated in bays or gulfs which are claimed as part of the national territory by the State contiguous to whose shore they lie", and p. 41: "Both the Gulf of Manaar and Palk's Bay would probably be claimed as part of the national territory, and not part of the high seas. Palk's Bay at any rate has now been held by the Madras Courts (Annakumaru Pillai v. Muthupayal, 1903, 27 Indian Reports, Madras Series 551) to be an integral portion of the British Dominions, and if the question arose a similar decision might possibly be given as to the Gulf of Manaar".

In other words the ownership of the oyster-banks in Ceylon is not necessarily derived from occupation because of the oysters, but may be founded on a historic bay claim, the banks are within territorial waters [a].

"Another instance" says Hurst, p. 41, "where there is no doubt that the site forms part of the high seas is that of the oyster-beds off the east coast of Ireland".

It may well be that Hurst was led to his theory by this example. But even if ever an *effective* occupation of the bottom of the Irish sea had taken place outside the oyster-banks and outside the 3-mile limit (which would have been abandoned later but for the oyster-banks and the bottom under the 3-mile belt) a statement which we cannot accept as depicting a historical situation with any measure of resemblance, even then, this example would not be sufficient a basis for a theory. There is of course no question of remnants of sea-bottom properties in the sponge-banks of Florida or Mexico.

We may perhaps distinguish between a claim on certain fisheries and a claim on the sea-bed in certain areas as Oppenheim [122] does, p. 10: "Staatliche Herrschaft über bestimmte Arten der Fischerei einerseits, und andererseits staatliche Okkupation des Bettes des offenen Meeres sind zwei sehr verschiedene Dinge".

A fishery right, however, cannot be thought of apart from a place. It is difficult to imagine a right to sedentary or other bottom-fisheries without claiming some sort of right on the banks. Even if no exclusive rights are claimed, but only conservation rights, the coastal State must be able to enforce its regulations, for instance to prevent fishermen to dredge or use a trawl on the banks, before the season for oysters is open. We therefore believe that "a sort of property right" on the banks must be admitted.

The Sedentary Fisheries had much attention in the International Law Commission. We will deal with the discussions briefly. In the Second Session, especially in the 66th meeting,

---

[a] It is, however, not altogether impossible that the bays were considered to be historic because of the pearls.

different opinions were expressed as to the nature of the rights of the coastal State, some members preferred to call it occupation, some prescription and some "usucapio" (S.R. 66[7], p. 6–15). We may add to the confusion the notion "usufruct" from the common property of marine life in the high seas. We could, however, accept prescription, as the nearest term to cover the nature of existing rights.

As Kerno remarked (p. 15), sedentary fisheries are carried out on the continental shelf. Several members, however, were of the opinion that sedentary fisheries were quite separate from the question of the continental shelf.

We are inclined to agree with them, and to think that sedentary fisheries come under the realm, although not under the regime of the high seas, because as we have said, the animals concerned live *in* the water, take their food from the water and only happen to be found on the sea-bed, just as bottom-fish happen to be found there; besides the banks occupy usually a small part of the shelf.

The only reasons to recognize certain rights of the coastal State are (1) the fact, that unscrupulous fishing for bottom-fish with trawlnets would be disastrous for the sedentary fisheries, and therefore a restriction on the freedom of the seas should be allowed here; and (2) the fact that some coastal States can claim acquisitive prescription.

To continue the activities of the International Law Commission, François made the following proposal in his Second Report [26], p. 62:

"Sedentary fisheries characterized by the effective and continued use of a part of the high seas without any formal and repeated protests against such use having been made by other States, and particularly by such States as, by reason of their geographical situation, could have put forward objections of particular weight, shall be recognized to be lawful, provided that the rules governing them allow their use by fishing craft irrespective of nationality and are limited to maintaining order and conserving the beds in the best interests of the fisheries by means of duties fairly assessed and collected".

This proposal, as François said in the 119th meeting (S.R. 119[89], p. 17) maintained the status quo. The Commission decided

that the question of sedentary fisheries should *not* be connected
with that of the continental shelf, with 11 votes to 1 (S.R. 119,
p. 20).

A dispute developed in that meeting (the 119th), whether
only the coastal State or all States engaged in sedentary fisheries
in a certain area, should lay down conservation regulations.
Hudson explained in the 120th meeting (S.R. 120 [127], p. 4),
that in the Persian Gulf, the inhabitants of Kuweit and Bahrein
exploited sedentary fisheries in waters contiguous to the territorial waters of Iran. François (p. 6): "considered that, in the
case of the Persian Gulf, the term 'occupation' could not be
applied".

Hudson (p. 11): "The Commission should state the principle
justifying regulation by a State whose territorial waters were
contiguous to the areas of the high seas in which sedentary
fisheries were exploited; such regulation, however, could not
exclude the nationals of other States from using the sedentary
fisheries, unless such nationals failed to comply with the rules
established by the regulating State".

The following text is included in the Report of the International
Law Commission Covering its Third Session [36], p. 61–62.

"Sedentary Fisheries

Article 3

The regulation of sedentary fisheries may be undertaken by
a State in areas of the high seas contiguous to its territorial
waters, where such fisheries have long been maintained and
conducted by nationals of that State, provided that non-nationals
are permitted to participate in the fishing activities on an equal
footing with nationals. Such regulation will, however, not affect
the general status of the areas as high seas.

1. The Commission considers that sedentary fisheries should
   be regulated independently of the problem of the continental
   shelf. The proposals relating to the continental shelf are
   concerned with the exploitation of the mineral resources of
   the subsoil, whereas in the case of sedentary fisheries, the
   proposals refer to fisheries regarded as sedentary because
   of the species caught or the equipment used, e.g. stakes

embedded in the sea-floor. This distinction justifies a division of the two problems.

2. Sedentary fisheries can give rise to legal difficulties only where such fisheries are situated beyond the outer limit of territorial waters.

3. Banks where there are sedentary fisheries, situated in areas contiguous to but seaward of territorial waters, have been regarded by some coastal States as under their occupation and as forming part of their territory. Yet this has rarely given rise to complications. The Commission has avoided referring to such areas as 'occupied' or 'constituting property'. It considers, however, that the special position of such areas justifies special rights being recognized as pertaining to coastal States whose nationals have been carrying on fishing there over a long period.

4. The special rights which the coastal State may exercise in such areas must be strictly limited to such rights as are essential to achieve the ends in respect of which they are recognized. Except for the regulation of sedentary fisheries, the waters covering the sea-bed where the fishing grounds are located remain subject to the regime of the high seas. The existing rule of customary law by which nationals of other States are at liberty to engage in such fishing on the same footing as the nationals of the coastal State, should continue to apply".

From this text it is clear, that the British view, which was defended by the Chairman (Brierly), not as British view, but as principle, namely, the right of appropriation of an area, which the (coastal) State should regulate without being obliged to treat the nationals of other States on the same footing as its own nationals (S.R. 120 [127], p. 17) was not accepted. Nor is any provision included, concerning the rather exceptional customs in the Persian Gulf.

Concluding we like to say:

1. That new discovered banks would not come under the proposed article. The Chairman (Brierly) said (S.R. 119[89], p. 23) that "sedentary fisheries in which no rights had been acquired, by prescription would be subject to the general regime of fisheries".

Lindley [128] referring to Hurst writes, p. 69: "It thus appears that .... certain .... sedentary fisheries .... can be sustained only on the prescriptive ground that they date

back to a time when dominion over the open sea was recognized. They cannot, therefore, be regarded as precedents that
would justify a new occupation of a portion of the surface
of the bed of the sea outside territorial waters". As said
before we do not agree with the reasoning, although we do
agree with the final result. The same result could be based
on the fact that there is a tendency to respect old rights
and. therefore this exception to the regime of the high sea
is accepted, but no more "old" rights can be established,
now that the status quo is going to be "codified".

2. That exclusive fishery rights are ruled out and only conser-
    vation rights can be exercised by the coastal State. This,
    however, is necessary for instance to be able to prohibit
    trawling.

3. That sedentary fisheries is one of the few restrictions on the
    freedom of the seas which was recognized in existing Inter-
    national Law.

As a curious note we may add a point of resemblance to the
continental shelf theory, which is the necessity to make a
division by agreement between countries adjacent to the same
shelf. The solution, which was adopted for the Ceylon- and
Tuticorin pearl-banks, was not a division in space, but one in
time, Thurston [111], p. 35: "In view to the possibility of clashing
of the fisheries in future years, a mutual agreement relating to
the division of the pearl fishery season between the Ceylon- and
Tuticorin pearl-banks, has been come to between the Madras
and Ceylon Governments; and the proposal of the Madras
Government that the Ceylon fisheries should begin in February
and close at the end of March, leaving April and May for the
Tuticorin fisheries, met the wishes of the Government of Ceylon".

The other form of sedentary fisheries, where constructions
are used made of stakes or piles driven into the sea-bed, has
a great similarity to constructions used to extract oil from the
subsoil, in so far as they obstruct navigation. In the Strait of
Malacca such "sero's" exist in considerable numbers and far
outside the territorial waters of the littoral States concerned.
There are even huts built on piles for watchmen. They are a
nuisance for smaller ships, bigger ships being usually not affected

because the constructions are mostly built in relatively shallow waters.

According to an Appendix to the Statute Book of the former Netherlands East Indies [129], several Ordinances (i.e. Statute Book 1877 No. 107, 1878 No. 118 and 1899 No. 114) prohibited to place sero's in the shipping canals indicated in those Ordinances. These regulations could be remembered with profit when regulating the building of oil-installations.

This kind of sedentary fisheries is completely different from the foregoing kind. With these permanent constructions free swimming fish is caught and it was an error concerning the facts that the Chairman (Brierly) (S.R. 65 [130], p. 11) said that for sedentary fisheries "both conditions had to be fulfilled, i.e. an exploitation of sedentary species by means of static equipment". Sedentary fisheries in the first sense are never carried out with permanent constructions.

This kind of sedentary fisheries has a closer connection with the continental shelf because of the permanent attachment to the sea-bed. On the other hand the resources which are exploited have nothing to do with the sea-bed or subsoil and are pure products of the sea as in the other kind of sedentary fisheries. As the constructions are only occupying shallow waters and not very wide spread in the world, we do not believe that they should give international lawyers sleepless nights.

SECTION 12. POLLUTION OF SEA-WATER

One of the possible consequences of extracting oil from the subsoil of the continental shelf by means of installations in the sea, will be the possibility of pollution of the sea-water. It does not need much imagination to realize, that oil leakages may occur. This may not have serious consequences, but the same cannot be said about a "blow-out". It should be admitted, that a blow-out, not an unusual occurrence in the early days of petroleum exploitation, has become an exceptional mishap, thanks to special devices made for its prevention (blow-out preventers). Nevertheless no human work being perfect, the possibility of such a blow-out to happen cannot be excluded, and the results will be disastrous. It must be kept in mind, that patches of oil on the sea are not stationary but will move by current and wind.

Pollution of sea-water caused by ships discharging oil or oily mixtures has had the attention of Governments for some time. On 1 July, 1922, a Joint Public Resolution was passed by the United States Congress, requesting the President to call a Conference of maritime nations with a view to the adoption of effective means for the prevention of pollution of navigable waters by oil-burning and oil-carrying ships. (Bishop [131], p. 336). One of the considerations in the Preamble reads: "Whereas most serious is the destruction of ocean fisheries resulting from the constant discharge into territorial waters of the waste products of the oil used for fuel on many steamers in place of coal, which threatens to exterminate the food fish, oysters, clams, crabs and lobsters, which are a vital part of our various national food supplies; . . . .".

Bishop, p. 337, pointed out that the Resolution "clearly sets out that the remedy against this evil is of a double nature. Oil is found as well on the high seas which are not subject to any national control as within territorial waters where the national legislator can lay down binding measures for the prevention of wanton acts of a nuisable character".

In 1926 a Conference was convened in Washington which made a draft Convention on Oil Pollution of navigable waters [132]. The following passage may be quoted: "The Conference therefore agreed to recommend that a system of areas should be established on the coasts of maritime countries, and on recognized fishing grounds, within which no oil or oily mixture, which constitute a nuisance, should be discharged.

Each country can determine what the width of the area off its own coasts should be, in the light of its own special circumstances and conditions, such as prevailing winds, currents and the extent of its fishing grounds, and after consultation with its neighbours where this appears necessary. The general rule in the case of coasts bordering the open sea should be that the width of the area should not exceed 50 nautical miles, but that in exceptional cases, where the peculiar configuration of the coast or other special circumstances render such a course necessary, the width might be extended to 150 nautical miles".

Whether the proposed system would work satisfactorily is a difficult question. As said before, oil patches move and to trace the culprit who discharged the oil, perhaps months ago and miles away, will be in most cases quite impossible. Oil remains well-nigh perpetually on the water. It is, we believe, more a question of education of ships-engineers and captains. They should feel their responsibility as good world citizens.

Again, as we remarked before, civilization has not kept pace with the technical development. We don't give a driving licence to a boy of 12 years old because he has not yet developed the sense of responsibility towards his fellow citizens. Unfortunately many adults still lack this sense of responsibility.

Of course there are also technical devices i.e. oil-separators, which if supplied to all ships and in the harbours, would help to prevent pollution of the sea.

It is curious to note, that these zones of 50 or 150 miles have been mentioned by the American delegate Miller at the Hague Codification Conference in 1930 (Acts [61], p. 127) in connection with contiguous zones. He remarked (p. 128) that the British delegate (Sir Maurice Gwyer) had suggested that the difficulties (which gave rise to the desire of having contiguous zones) should be met entirely by special conventions between particular States. But, he said, in the case of oil pollution, it makes no difference what flag the ships fly, "it is not a question of an agreement between neighbouring States".

No further steps were taken as to the draft Convention.

In 1934 "The United Kingdom Government having considered that the pollution of the sea owing to the discharge of oil or oily water was increasing, decided to submit this matter to the League of Nations" [a]. In an oral explanation the British delegate Shakespeare in the Second Commission [134] (p. 8) drew attention to the damage to fisheries: ".... En second lieu, il y a des plaintes, emanant des administrations intéressées aux pêcheries, plaintes appuyées par les avis d'hommes de science éminents au sujet des dommages causés aux prairies sous-marines par les déchets d'huiles et de pétrole".

---

[a] See Note of the Secretary General of the United Nations on pollution of sea-water by oil [133], p. 2. This Note gives a comprehensive survey of the steps taken on this subject.

In January 1935 the Council of the League adopted the following resolution (see Note Secretary General [133], p. 3–5):

"The Council

Authorizes the Communications and Transit Organization to make all the necessary preparatory studies with a view to facilitating the future conclusion of an international convention in regard to the pollution of the sea by oil".

In order to give effect to the resolution adopted by the Council, the Secretary General, on 23 January 1935, addressed a circular letter together with a questionnaire to all States members of the League and to non-member States as well, a total of 69 Governments.

A brief survey of the general trend of opinion expressed by Governments in their replies on some of the main points of the above mentioned questionnaire is given below:

"1. *Damages caused by oil pollution of the seas are*:

(a) *birds*: slight to more definite bad effect according to regions and circumstances;
(b) *fish and fishing*: same as above; on some occasions acute bad effects have been noticed;
(c) *seaside resorts*: in some cases the beaches were so polluted that bathing was prohibited, and on one occasion the effect lasted for a very long period (several years);
(d) *ports*: (especially in respect to danger of fire) tendency to increase danger of fire.

2. *Damages caused to fish and fishing industry*:

(a) *fish* (high seas): doubtful effect, except in the opinion of one Government which believed that fish eggs are very harmfully affected by oils;
(b) *fishing* (coastal): affects fish and shell fish (and sea weeds); serious damages have been noticed; in some instances fish were no longer edible, or died;
(c) *fishing* (trade): on some occasions fishing districts had to be abandoned; oil can affect condition of fishing nets.

8. *Scientific or other findings concerning the extent of sea pollution by oil*: ....

Some countries, which had made a special study of the question, reached the conclusion that oil can remain on the surface for a very long time, almost indefinitely. Even if the volatile elements of oils disappear by evaporation, the quantity which evaporates represents only a very small proportion of the total volume, and only a small amount of the oil founders.

Regarding the drifting of oil, many experiences have proved that a drift of more than fifty miles is not uncommon. In one instance, oil had been detected five hundred miles away from its point of discharge" ....

The Assembly adopted a resolution on 24 September, 1935, by which the Council was requested to instruct the Communications and Transit Organization to take the necessary steps to complete the preparation of a draft Convention. A committee of experts prepared a new draft Convention and a draft Final Act which were sent to the Governments. Nearly all the replies received from States with sea-coasts were in favour of concluding such a Convention.

The Council adopted the following resolution on 10 October, 1936 (Journal Official, 1936 [135], p. 1196): "Le Conseil décide de convoquer pour une date qu'il fixera ultérieurement .... une Conférence dont l'ordre du jour comprendra la conclusion d'une convention et d'un acte final relatifs à la pollution de la mer ...." (p. 1197: "La résolution est adoptée").

On p. 1391, Annex 1626 (No. official C 449, M 235, 1935, VIII) the draft Convention is given [a].

Article I of this draft: "(1) The High Contracting Parties agree .... to take the necessary measures to render illegal and punishable by adequate penalties the discharge by such vessels (specified in Article IV) of oil or oily mixtures, as defined in Article III (1), within any zone established as provided in Article II ...."

Article II gives the same extent for zones as in the draft Convention of Washington of 1926, mentioned above. This Conference never took place.

Most countries with sea-coasts have made laws to prevent oil pollution of the waters of their coasts and harbours.

To mention a few examples:

---

[a] See also for the resolution of the Council the Note of the Secretary General [136], Annex III on p. 20 and for the text of the draft Convention and the Final Act, Annex I and II on p. 10 and 18.

a. the United States: Oil Pollution Act, 7 June, 1924 (43 Stat. 604–605) (International Law Documents [72], p. 182).
Section 3: "…. it shall be unlawful for any person to discharge …. oil …. into or upon the coastal navigable waters of the United States from any vessel using oil as fuel for the generation of propulsion power, or any vessel carrying or having oil thereon in excess of that necessary for its lubrication requirements …." etc.

b. the Cuban General Law on Fisheries, enacted by Decree Law No. 704, 28 March, 1936, [a]
Article 46: "The masters of ships shall in no case allow ashes, rubbish, fluid flushed from oil, molasses or petroleum tanks or waste matter of any kind to be discharged within the confines of bays and harbours.
Such ashes etc. shall be discharged into the sea offshore at a distance of not less than 5 miles from the coast".

c. Japan, Port Regulations, enacted by Law No. 174 of 1948, as amended by Law No. 98, 24 May 1949, "Official Gazette" (English edition) Extra No. 47, (24 May 1949), p. 2, [b]
Article 24: "(1) Any person shall not throw or discharge ballast, waste oil, cinder, ashes, dirt and other refuge matters, without permission, into the waters in a port or within 10,000 metres from the boundaries of a port …."

In the meanwhile, after the war, the thread was taken up again by the United Nations. The problem had become more serious because the total world merchant fleet according to the Note of the Secretary General [133], p. 6 (quoting from Lloyd's List, 23 March, 1949) shows an increase from June 1939 to June 1948 of 11,634,529 tons, being on the last date 81,074,188 tons. Moreover, whereas in 1914, 89 per cent of the world merchant fleet were coal burning vessels, at present only 22 per cent depend on coal for fuel, and the number of oil tankers has greatly increased (loc. cit. p. 6–7).

In the Note of the Secretary General, p. 8–9, it is suggested that the Transport and Communications Commission might deem it useful and appropriate: (1) to draw the attention of the Inter-governmental Maritime Consultative Organization, once the latter has started functioning, to this particular problem and (2) pending the coming into force of the Convention establish-

---

[a] See [153], *p. 65.*
[b] See [153], *p. 83.*

ing that organization, to recommend to the Economic and Social Council that Governments be requested to submit their views. The Resolution No. 3 adopted by the Transport and Communications Commission at its fourth session reads:

"*Taking into account* that the Inter-governmental Maritime Consultative Organization (IMCO), when it has started functioning, would be the competent agency to handle this subject,

Recommends to the Economic and Social Council

1. That the Secretary General be instructed to request the views of Member Governments on the following points:
   (*a*) Whether, pending the establishment of IMCO, preliminary action should be taken;
   (*b*) If so, what should be the best procedure to follow;
   (*c*) In particular whether the convening of a meeting of competent governmental experts would be the appropriate procedure;
   (*d*) Whether the draft Convention on the Pollution of Sea-water by oil prepared in 1935 under the auspices of the League of Nations could serve as a working basis for the consideration of the problem;
   (*e*) Whether the scope of the examination of the problem should be extended to cover the possible future pollution of sea-water by atomic waste from fuel which may be used by ships;
   (*f*) Whether they would wish to give priority to the consideration of any one of the several consequences of the pollution of sea-water;
2. That the Member Governments which possess the technical facilities to do so be requested to undertake research studies on this problem forthwith, and to establish between them such collaboration as might be useful and possible".

The resolution was adopted by the ECOSOC on 12 July, 1950 (see E/CN. 2/100 [136], p. 1–2).

The answers of the Governments can be found in the document last mentioned. In the United Kingdom reply (p. 17) under the main consequences of the pollution of sea-water is mentioned (II) the damage to fish-life, shell fish, birds and the pollution of fishing gear [a].

---

[a] Further replies are contained in E/CN. 2/100 Add. 1, 1 March, 1951, and Add. 2, 14 March, 1951, and Add. 3, 14 March, 1951.

As to the alleged damage to fisheries a thorough scientific research will be necessary. Some research work has already been done. We mention the following two reports on this subject. Leenhardt [137] says, p. 56: ".... nous avons été chargé de faire quelques études expérimentales sur la toxicité du mazout pour les principaux coquillages comestibles marins". He found, p. 57: ".... le mazout contient un corps nocif pour les animaux marins, au moins pour les mollusques. Ce corps doit agir par sa dissolution dans l'eau puisque la mortalité se faisait sentir dans les bassins où le mazout ne venait pas en contact avec les coquillages. Ce poison tue indistinctement et de la même façon les moules, les gryphées et les huitres. Son action se fait sentir à la longue pour de faibles concentrations et plus rapidement à mesure que les proportions augmentent" .The concentrations he used would, however, hardly be met in nature. He ends up by saying, p. 58: ".... Si l'on ajoute à cela le fait que la marée brasse ces eaux et les renouvelle à chaque instant, on peut conclure que le mazout, bien que contenant un poison pour les animaux marins, ne peut, vu les proportions, être incriminé dans les mortalités ou disparitions d'espèces signalées par les pêcheurs ou ostréiculteurs".

Where the effect of oil-pollution upon certain forms of aquatic life is concerned, Roberts [138] reports the following results of research in this matter, on p. 267:

"Weigelt found that fish completely disappeared from pools and lakes into which mineral oil had been discharged. The same worker states that ulcerations and attacks of disease follow exposure to petroleum.

According to Thomas the emulsions formed from a petroleum distillate and from a light fuel-oil coated the gill membranes of fish and caused death .... Mussels and oysters were apparently unharmed by contact with fuel-oil .... In general, Orton is of the opinion that the oils discharged at sea are comparatively harmless ....", and on p. 268: ".... Planktonic organisms were killed when they came into contact with the oil but otherwise they lived for several days; ....", and finally the summary on p. 273–274:

"1. Oil-films slow down absorption of oxygen from air, but in very thin films, likely to be met with at sea, no slowing down was appreciable.

2. Agitation, such as would be met with at sea, markedly increases the rate of absorption through an oil-film . . . .

4. All the oils tested were found to be toxic to fish. It is believed that this toxicity is due to both soluble toxic substances and to emulsions . . . .

The adverse effects of fuel-oils on fish, shell fish, plankton and plants, are not very marked, but these oils are the cause of death of countless sea-birds.

It is not contended, however, that oil-pollution in its effects upon marine life need cause no anxiety. On the contrary, it is felt that it will, if unchecked, become a menace in the future".

That our fears are not personal ones, may be concluded from the fact that where offshore drilling was going to take place, special regulations for oil-pollution have been laid down.

Firstly we mention Article 7 of the Treaty between the United Kingdom and Venezuela relating to the submarine areas of the Gulf of Paria [110]:

"Each of the High Contracting Parties shall take all practical measures to prevent the exploitation of any submarine areas claimed or occupied by him in the Gulf from causing the pollution of the territorial waters of the other by oil, mud or any other fluid or substance liable to contaminate the navigable waters or the foreshore and shall concert with the other to make the said measures as effective as possible".

Secondly: Article 5 (2) of the Submarine Areas of the Gulf of Paria (Annexation) Order in Council, 6 August, 1942 [153], p. 47:

"5. The Governor of the said Colony shall, as soon as may be after the date of this Order, make regulations to ensure: . . . . (2) that all practicable measures shall be taken to prevent the exploitation of any of the said submarine areas from causing the pollution of coastal waters by oil, mud or any other fluid or substance calculated to contaminate the seawater or shore-line". On this Order in Council is based:

Article 21 of the Submarine (Oil Mining) Regulations, 22 May, 1945 of Trinidad and Tobago (Government Notice No. 87; Proclamations, Orders, Regulations, etc. 1945, p. 101), [153], p. 36.

Similar Regulations are contained in Article 15 (*a*) of the Oil Mining Regulations, 2 September, 1949 of British Honduras

(Statutory Instruments, No. 56 of 1949), [153] p. 32 and in Article 26 of the Petroleum Act 1945 of the Bahamas [139] [a]:

"26. The holder of a prospecting licence or a mining lease shall adopt all practicable precautions to prevent pollution of the coastal waters by oil, mud or any other fluid or substance which might contaminate the sea-water or shore-line or which might cause harm or destruction to marine life".

Thirdly, we quote (*a*) the Preamble and Article V of the Rules and Regulations governing geological, geophysical and other explorations in areas within tidewater limits belonging to the State of Texas .... of 16 June, 1949 [140]:

"Whereas, House Bill 665, passed by the 51st Legislature of Texas, authorized the Commissioner of the General Land Office to issue permits for geological, geophysical, and other surveys and investigations of areas within tidewater limits belonging to Texas for exploration for oil and gas, subject to such rules and regulations as the Commissioner may prescribe.
Now therefore, I, Bascom Giles, Commissioner of the General Land Office, do hereby prescribe, ..., the following rules and regulations to govern explorations in the areas within tidewater limits as said term is defined herein" [b], and Art. V:
"If any person, firm or corporation should drill a well or wells in any of the bays or in the Gulf of Mexico, he shall drill said well or wells in such manner as will so far as practicable prevent the pollution of said waters, and in such manner as to interfere as little as possible with the fishing and/or shrimping industries. Upon the abandonment of such well all of the rigging and material used shall be removed, and the bottom of the Gulf or Bay where said well was drilled shall be restored to its former condition as nearly as possible";

(*b*) the Preamble of the Rules and Regulations governing drilling and producing operations in coastal waters of Texas of 6 April, 1948 [141]:

---

[a] *See also* [153], *p. 30.*
[b] The term "tidewaters" has nothing to do with the tide of the sea, i.e. flood and ebb. Article 1 of the Rules and Regulations explains: "the term 'areas within tidewater limits' means .... that portion of the Gulf of Mexico within the jurisdiction of Texas". To find the limit of this jurisdiction one has to consult the Act declaring the sovereignty of Texas along its sea-coast, 16 May, 1941, as amended by Act of 23 May, 1947. We will postpone this complicated (and controversial) matter, however, to Chapter IV.

"Pursuant to the provisions of Article 5366 of the Revised Civil Statutes of Texas, 1925, I, Bascom Giles, Commissioner of the General Land Office of Texas, have made extensive investigations and have consulted numerous engineers and experts who have investigated these matters, and find that there are certain hazards incident to drilling for and producing oil and gas in the areas of this State covered by the coastal waters; that there are great pressures in reservoirs containing oil and gas; that on occasions hurricanes occur; that there are consequent dangers of blow-out and cratering; that if a well should get out of control there is danger of the connections being cut out; that it will be necessary to avoid as far as possible the corrosive effects of salt water; and that the rules and regulations hereinafter set out are necessary and are reasonably calculated to prevent pollution of such coastal waters as a result of operations therein for the discovery and production of oil and gas";

Then follows a number of technical provisions all made for the purpose of preventing pollution of the sea-water through leakages, breakdowns and blow-outs. We mention only the 4 following rules:

Rule O. "The movement of oil from wells to storage shall be made in such manner as to prevent the escape of oil ....".
Rule P. "No slush or mud materials shall be disposed of in water before all oil has been removed therefrom. Likewise no salt water shall be disposed of in water before all oil has been removed therefrom".
Rule R. "Without limitation upon the foregoing rules, all reasonable precautions shall be exercised to prevent such pollution of the waters as will destroy fish, oysters and other marine life·"
Rule T. "It is recognized that the game, Fish & Oyster Commission of Texas is charged with the duty under Article 5366 of the Revised Civil Statutes of Texas, 1925, of enforcing rules and regulations and that said Commission and its representatives shall also be privileged to exercise all of their lawful powers in order to prevent pollution";

and lastly (c) the Louisiana Oil and Gas Conservation Law (revised Statutes Title 30 Sections 1–20, inclusive) [142]:

Par. 1. ....
"There is established the State Department of Conservation, ..., which is directed and controlled by a commissioner of conservation ....

Par. 4. . . . .

c. The commissioner has authority to make . . . . rules, regulations, or orders for the following purposes . . . .

(1) To require reasonable bond with security for the performance of the duty to plug each dry or abandoned well.

(6) To prevent blow-outs, caving and seepage in the sense that conditions indicated by these terms are generally understood in the oil and gas business . . . .".

To mention an example of the practical application of these regulations: Kastrop [143], p. 145, describes a safety device which prevents a stock tank from overflowing. In this stock tank the oil from 13 marine wells is gathered in Galveston Bay, Texas coast, before it is pumped through a pipeline to the shore tank. This precaution is of course necessary to prevent pollution of bay-waters.

The Committee on rights to the sea-bed and subsoil, suggested that the Copenhagen Conference of the International Law Association in 1950 [24] should adopt the following solutions, p. 13–14: "I. The Conference affirms that the following principles are existing international law: . . . . (4) The exploration and exploitation of the resources of the sea-bed and subsoil of the continental shelf outside territorial waters is permissible only in so far as it does not substantially interfere with shipping and fisheries, e.g. in so far as it does not constitute an obstruction of traffic routes, a pollution of fishing waters, or their disturbance by seismic operations . . . .".

We will now look into the discussions and decisions of the International Law Commission in this matter. In the 65th meeting [130] (p. 22) it was decided that the Commission should not broach this subject because it was being considered by other bodies and there was no international legislation in connection with it. In his Second Report [26] François, p. 37, proposed the regulation for the protection of marine resources, quoted above in Section 7 in which the coastal State was entitled to prevent pollution in the area of 200 miles contiguous to its coastal waters. As the suggestion of a regulation by the coastal State was not accepted, the prevention of water-pollution moved to Article 2 of the preliminary draft made by the Commission.

In the 132nd meeting [98] (p. 7) Hudson proposed the following text: "..... Competence should be conferred on a permanent international body to conduct continuous investigations of the world's fisheries and the methods employed in exploiting them, and to make regulations for conservatory measures to be applied by the States whose nationals are engaged in fishing in any area of the high seas, and for the prevention of water-pollution, unless such States reach agreement otherwise" .... but after some discussion the Commission eventually decided in view of the action taken by the Economic and Social Council, to mention pollution only in the comment (see par. 4 of the comment, quoted in Section 8) [a].

It may be noted that the reasons of the activities of the Economic and Social Council are not oil-pollution caused by offshore oil-exploitation, but pollution caused by ships. In the 1926 and 1935 draft Conventions the States were given "contiguous zones" in which their regulations concerning prevention of pollution would be applicable, based on the principle of self-protection.

Here we feel, that another element enters the picture. The coastal State, under whose supervision oil-exploitation on the continental shelf adjacent to its coasts takes place, takes on an international responsibility to prevent oil-pollution in the interest of other States, whose fisheries or shores may be damaged by the oil. At the same time it is of course in its own interest to prevent this pollution. If the activities of the Economic and Social Council should remain limited to pollution caused by ships, we may expect that the International Law Commission will take up the prevention of pollution caused by offshore drilling again in its next meeting. As the two causes of pollution are so different we believe that the original idea of the International Law Commission should be considered again.

It should, however, not be forgotten that due to the drift of oil patches, it may well happen, that fisheries are effected in an area where the State whose oil-installations have caused the pollution has no fishery interests. This consideration may

---

[a] For the discussion on the drafting of this comment, see Summary Record of the 133rd meeting [144], p. 4–5.

make a redraft of Article 2 of the proposals of the International Law Commission necessary.

Rules could also be embodied in a Convention amongst all maritime States, concerning the exploitation of mineral resources of the continental shelf, to the effect that if the coastal State does not take efficient measures to prevent pollution either obligatory arbitration or jurisdiction of the International Court of Justice or an International Maritime Court will be accepted or powers will be given to an international body to intervene and lay down the proper regulations.

SECTION 13. SEISMIC EXPLORATIONS

Another cause of damage to fisheries, connected with offshore oil-exploration, is the method used for geophysical survey work, carried out for the exploration of the oil bearing geological formations, known under the name of seismic explorations or seismic operations.

These operations entail the setting off of an explosive charge in the water, which is destructive to marine life (see Fitch and Young [145], p. 53).

The destructive effect of explosions in the water on fish has long been known, and as a matter of fact misused as a ruthless method of fishing.

It is ruthless, because all fish, big and small are killed within a certain radius of the explosion, in contrast with electrical fishing methods, where the size of fish to be caught can be regulated. In an article called "Explosieven in de visserij" of June, 1950 [146] (Explosives in fisheries), an English extract is to be found on p. 61–62, the results are given of a research carried out by the Royal Siamese Navy into the effect of explosions on fish. It says that this method of fishing has nearly become a habit along the coasts of Thailand. Looking for methods to suppress this practice, for instance by confiscating the ship and prosecute the persons who are guilty of using explosives, it was necessary to know how to detect this practice. In other words did the fish so caught show any signs which would prove the use of explosives. A "Plastic C-2" charge of 600 grams was used 4 times and one of 900 grams the fifth time. The charges exploded at the bottom at different depths between 4 and 7 metres. The total weight of dead fish, which partly floated to the surface, varied between 2–56,9 kilogram. But a great deal of bodies remained at the bottom and could not be collected because of sharp corals. Fish at 5 metres from the explosion were killed at once, the ones at 10 and 15 metres from the spot died soon afterwards. The external signs were very clear for fish killed near the explosion and for the rest detectable after section (mostly internal hemorrhage and burst air-bladders).

The charges used in seismic operations are much heavier. For a good understanding a few words may be said about these seismic operations. Fitch and Young [145] give the following description, p. 53: "To obtain accurate records, it is necessary for the geophysicists to set off charges of explosives approximately every 250 to 1,000 feet on predetermined lines running north and south and east and west. The sound waves from the explosion travel down through the various strata of rock, and as they bounce back from these strata they are picked up by a number of geophones floated on the surface. An electrical impulse generated by the sound waves then passes from the geophones along a cable to the survey ship, where a permanent record is made on light-sensitized photographic paper. These graphs are later worked out in detail in the offices of the exploring company and anticlinal structures located accurately in various areas of the ocean-floor. The weight of explosive in a single shot has varied from 10 pounds to 160 pounds. In most instances the weight of each individual charge was 40 or 80 pounds for 'open' shots and 20 pounds for 'jet' shots". ("open" shots are charges exploded a few feet below the surface, "jet" shots are charges exploded when buried under the floor of the ocean). In the seismic refraction measurements carried out by Ewing, Worzel, Steenland and Press [147], from the coast-line to the edge of the continental shelf on the Atlantic coast of the United States, shots were fired *on* the bottom in most cases. Geophones and hydrophones were placed directly on the ocean-bottom (p. 879). In 3 lines of traverse .... an unconsolidated layer .... a semi-consolidated layer and a layer considered to be the basement (cristalline rockfloor) were traced across each traverse. On the Cape May traverse the thickness of the sedimentary column runs from about 5,000 feet near the beach to about 16,000 feet near the edge of the shelf, (p. 877).

Now back to the research reports of Fitch and Young [145], p. 53: "In an attempt to throw more light on this problem the Bureau of Marine Fisheries, California Division of Fish and Game, has closely observed the operations of the oil-companies engaged in geophysical survey work in our coastal waters;" p. 54: "When an explosion takes place in the water the number of fish killed depends upon the number present in the area. Most of the data collected by the authors concerns only those fish which floated to the surface when killed"; p. 55: "The largest fish observed killed by a shot was a 365-pound black sea-bass (Stereoleopis gigas) and the smallest were larval anchovies (Engraulis mordax). The greatest estimated quantity killed as a result of one explosion was approximately two tons of rockfish (Sebastodes), but there have been numerous times when not a single fish was killed in a whole day's operation .... The damage that occurs to fish-life is, at the present time, thought to be largely limited to those species possessing an air-bladder .... The ability of an underwater explosion to inflict damage decreases inversely as the cube of the distance. A fish 10 feet away from an explosion will receive approximately eight times the force from the shock wave as a fish 20 feet from the same explosion .... The great speed of the shock wave, approximately 4,940 feet per second (Gowanloch and McDougall, 1944), exerts terrific pressure on a fish that presents a broad surface to the direction of wave travel .... Fish presenting a broadside are literally plastered up against an incompressible wall of water. The body wall and air spaces within the fish are compressed instantaneously, allowing no opportunity for physiological adjustment to the sudden pressure"; p. 58: "A record of the estimated quantity of fish killed as a result of geophysical survey work conducted on days when an observer from the Bureau of Marine Fisheries was present is given in Tables 1–6 .... During the period of this study jet shots killed an estimated 0,23 pound of fish per pound of explosive used and 4,93 pounds of fish per shot. On the other hand, open shots killed 0,47 pound of fish per pound of explosive, double that of jet shots, and 31,56 pounds of fish per shot, more than six times that recorded for jet shots. This represents 41,000 pounds of fish killed by open shots and 1,000 pounds killed by jet shots". Table, p. 65, giving a summary of all fish killed by all methods, gives on 1,499 shots with a total weight of 90,940 pounds of explosives 41,963 pounds of fish killed. They end up by saying, that "Much more work and experimenting must be done in this line before a true picture can be had of those fish killed which do not float".

The danger has been detected in the Report laid before the Conference of the International Law Association at Copenhagen in 1950 [24]. In the proposals of the Committee on the rights to the sea-bed and its subsoil, under I (4) it is said that "The exploration .... of the continental shelf is permissible only .... in so far as it does not constitute .... disturbance (of fisheries) by seismic operations". (The article is fully quoted near the end of the previous Section on Pollution).

It is to be regretted that the International Law Commission has treated the problem rather stepmotherly in their 115th meeting (S.R. 115 [148], p. 7–8).

It is true of course that a provision as adopted by the Commission in Article 6 (1) (Report Third Session [36], p. 58): "The exploration of the continental shelf and the exploitation of its natural resources must not result in substantial interference with navigation or fishing. Due notice must be given of any installations constructed and due means of warning of the presence of such installations must be maintained", would comprise the hindrance to fisheries caused by seismic operations.

It would, however, in our opinion have been clearer if in the comment, seismic operations would have been mentioned.

Coastal States have already laid down rules to lessen the danger.

In the Rules and Regulations governing geological and geophysical and other explorations in areas within tidewater limits belonging to the State of Texas, of June 16, 1949 [140]:

Article I reads: "Whenever the word 'explorations' is referred to in these rules, it shall mean geological, geophysical, and other surveys and investigations including seismic methods for the discovery and location of oil and gas prospects, and which may or may not involve the use of explosives. The word 'seismic explorations' shall mean any geophysical exploration method which involves the use of explosives. The word 'shot' as used in these rules shall mean the use and detonation of powder, dynamite, nitro-glycerin, or other explosives ....".

Article II: "No shots shall be discharged except during daylight hours. No shots shall be used in excess of forty pounds. The Commissioner of the General Land Office may, upon application and after investigation, permit the use of larger shots in areas specified by the Commissioner and upon such terms and conditions as will be in keeping with the intent and purpose of these rules ....".

Article III: "Shots discharged in the waters of the Gulf of Mexico shall be suspended at a depth not greater than one-half the distance from the surface to the bottom and in no event nearer to the bottom than 5 feet or buried at least 10 feet below the bottom ....".

Article IV: "Before any shot is discharged in either the Gulf of Mexico or the bays, the exploration party shall employ methods approved by the Commissioner of the General Land Office to frighten or drive away the fish and/or marine life which may be in the area where the shot is to be discharged. If there is a school or schools of fish in the area to be shot, operations must be suspended in that particular area until said school or schools of fish have been driven away".

Article VI: "At least seven days before conducting explorations each exploration party shall notify the Commissioner of the General Land Office in writing of the time and place where explorations will be made ....".

Article VII: "Each exploration party shall furnish representatives of the Commissioner of the General Land Office and the Game, Fish and Oyster Commissioner with transportation facilities upon reasonable notice at any time they desire to enable them to visit the working area ....".

Articel X: "No shot shall be detonated within one mile of a shrimping fleet previously and in good faith operating in the area. By shrimping fleet is meant ten or more boats dragging for shrimp within an area of not more than one mile in diameter".

Article XI: "No shots shall be detonated during the months of May, June, July, August and September within 3 miles of a major resort beach".

Article VIII contains penalties for violation, ending up with this clause: ".... Further, the person, firm or corporation violating any of the provisions of said Act or these rules after due notice has been given shall be prohibited from further exploring the public lands covered by these rules except upon such terms and conditions as the Commissioner of the General Land Office may expressly stipulate".

To finish this Section we quote a few rules adopted on the Pacific coast. Fitch and Young [145], p. 68–69:

"The following rules and regulations were adopted by the California Fish and Game Commission at the meeting on August 22, 1947, and became a part of the Fish and Game Code effective September 19, 1947. As provided by Section 480, the following are the regulations under which permits are granted to use explosives in the waters of this State inhabited by fish, in so far as such explosives may be used for seismic exploration:

(a) Permits shall be issued for such areas and seasons as will result in a minimum of destruction to marine life and fisheries.

(b) No blasts shall be set off in waters of less than one hundred feet (100) (seventeen (17) fathoms), except as they are placed below the surface of the ocean-floor, unless prevented from so doing by rock formations.

(c) An employee of the Division of Fish and Game shall be permitted to accompany the boat or crew which is conducting the exploratory work, as an observer to determine if any damage is done to the fisheries. This observer shall have the authority to stop operations in any given area if damage to marine life seems too great ....

(h) In order to subject the fish-life to a minimum of disturbance, no permit shall be issued to any applicant after October 1, 1948, for any area until twelve (12) months after the last previous issuance of a permit for that area to the same or any other applicant, providing such permit was exercised ....

(i) Any permit granted by the Fish and Game Commission to conduct seismic operations grants permission only in so far as the Fish and Game Commission is concerned. No permit is valid nor shall be exercised unless at the time thereof the permittee has in force and effect a permit covering such operations issued by the State Lands Commission of the State of California" ....

SECTION 14. PRELIMINARY CONCLUSIONS

Reviewing what has been said in the preceding Sections we arrive at the following preliminary conclusions.

1. The best way to arrive at a satisfactory protection and conservation of marine resources is agreement between States interested in fisheries in a specific area and to give powers to an international body to make provisions for that area in as far as the States concerned are unable to agree, as

proposed by the International Law Commission. The areas concerned should be chosen independent of the limits of the continental shelf *a*. Mapping out areas and laying down provisions concerning protection and conservation should be based on scientific data.

It seems that these lines have been followed lately and that anxiety as to the possible implications of the Truman Proclamation concerning third States has been proved unfounded, so far. In the Treaties concluded up to now no difficulties with non-party States has been encountered and as Hudson said concerning the Northwest-Atlantic Treaty of 1949 (S.R. 65 [130], p. 11): "No attempt had been made to forbid nationals of other States access to these fisheries". Also on the European side of the Atlantic, where proposals for protective regulations have always been met with a certain reluctance, a better spirit may prevail in the future *b*. It seems that the United States have so far succeeded in finding States willing to co-operate.

We mentioned the Northwest-Atlantic Convention of 1949 and the Convention with Mexico of 1949. The latter is even more important as it may have checked a development on the lines of the original Declaration by the President of Mexico which could have led to departures from the principles of International Law. The same can be said about the Convention between the United States and Costa Rica for the establishment of an Inter-American Tropical Tuna Commission, signed in Washington, May 31, 1949 [150]. This treaty may well be a test, as it includes problems of interest to a number of countries besides the signatories (loc. cit. p. 766).

2. The principle of freedom of fishing outside territorial waters should be maintained and threatening depletion should not be solved by diversions from this principle as for instance in the bill introduced in the House in the United States on

---

a Leonard [94] gives in his book a map, opposite p. 180, showing the principle fishing areas of the world and also the outlines of the continental shelf (presumably the 200 metre-isobath). This map confirms our thesis, that fishing areas stretch far beyond the continental shelf.

b The last regulations were laid down in the Convention drafted at the International Overfishing Conference in London, 1946 [149].

15 November, 1937, by Mr. Dimond, delegate of Alaska. Section 2 of this bill provided (Jessup [151], p. 130): "the salmon which are spawned and hatched in the waters of Alaska are hereby declared to be the property of the United States, and it shall be unlawful for any person .... to fish for, take, or catch any of the said salmon in the waters adjacent to the coast of Alaska ....".

3. No attempt should be made concerning international regulation of exclusive fishing rights at least before the decision of the International Court of Justice in the Anglo-Norwegian Fisheries Case. It may be that that decision gives us a firm foundation for a new attempt to arrive at some sort of (not necessarily uniform) delimitation of territorial waters. Outside territorial waters no exclusive fishing rights should be recognized except those which can be proved to be in a historic bay or in historic waters or at least are based on prescription. The International Court of Justice or a special international body could be asked for an advisory opinion in cases of doubt.

4. Where sedentary fisheries are concerned, only the existing ones should be allowed to form an exception to the principle of the freedom of the seas. This may involve a historical research in cases of doubt [a].

5. Pollution and Seismic Exploration regulations should be made to prevent or limit in a satisfactory manner possible damages. Only in cases where States would be lacking in fulfilling this international duty, the decision of an international body should be called for.

6. Although for the regulation of conservation the Food and Agriculture Organization would be the most competent one, we doubt whether this Organization would have any compe-

---

[a] In this connection we may mention the idea expressed by Hatschek [152], p. 210, that the reason for the exception to the principle of the freedom of the seas has to be sought in the Islam rule of law according to which a tax called "haraǧ" was imposed on the sea-bed and its products. We are not sure, however, whether this interesting remark solves the problem of the history of sedentary fishery rights, or only shifts it to a further place. We would not like to agree with the idea without further investigation, where Ceylon is concerned. He mentions his idea in connection with Ceylon, but Ceylon has been for centuries a stronghold of Buddhism and we do not know whether the original claims were made by an Islamic majority or in a part where Mohammedans prevailed (see literature mentioned by Hatschek).

tence for dealing with matters like intervention in the field of pollution or seismic operations if necessary. Therefore we may combine the care of these subjects with other duties, yet to be discussed in an international body to be established for these special purposes.

CHAPTER III

THE REGIME OF THE HIGH SEAS AND NAVIGATION

INTRODUCTION

It has been said, that through the exploitation of the mineral resources of the continental shelf by means of installations built in the sea, navigation will be hampered. For instance in the Brussels Report of the International Law Association [44], p. 172, 177 and especially on p. 201, where the question is put by the rapporteur Feith: "When the interests of international shipping come to be weighed against America's national exploitation of submarine petroleum sources, will shipping come out on the winning side?". Our first reaction on this question is that we are dealing with a problem of International Law and not with power politics.

We have to distinguish between two kinds of obstructions. One is the pure and simple physical obstruction which would be laid in the path of shipping by erecting an installation in the sea. The second one is a possible extension of rights, which may be claimed or may be recognized or agreed upon in a general Convention on the subject, in waters outside the territorial sea of a coastal State exploiting the resources of the shelf adjacent to these waters.

In both cases shipping may be involved. We will deal with these two kinds of obstructions separately. Before doing so, however, we believe it to be useful to analyse the right of free navigation and this will necessitate a short discussion of the regime of the high seas and on the territorial seas and the contiguous zones.

SECTION 1. THE HIGH SEAS

Gidel [112], p. 125, gives the following definition of the high seas: "La haute mer (high sea, open sea, offenes Meer) se définit par opposition à la mer territoriale. Elle comprend donc tous les espaces maritimes qui se trouvent au delà de la limite vers le large des eaux territoriales".

Higgins and Colombos [154], p. 38, give the same sort of definition: "so much of the ocean as is exterior to .... the territorial waters ....", and finally Art. 4 of the draft Convention prepared by the Research in International Law of the Harvard Law School [155], p. 265, reads: "The high sea is that part of the sea outside marginal seas".

It seems necessary to define the territorial waters first. This, however, as we will see in a following Section is not easy because neither the extent nor the legal nature of these waters are generally accepted notions. We can take for the time being an extent of 3 miles and a description of the territorial waters as coming under the "territory" of the coastal State, the latter having sovereignty over these waters. The definitions given above make it clear that such sovereignty does not exist in the high seas.

The comment on the Harvard Art. 4 reads: "The term high sea is here defined for the purpose of its use in this Convention. It seems essential to employ a terminology which accurately differentiates the seas over which no state has sovereignty, and the term high sea is the most convenient term to be employed. It is recognized, however, that this term is often used to include the marginal sea ....". This does not mean that the States have no rights at all in the high seas. The contrast is not so sharp as the definitions could make us believe. In the first place there are some rights of an administrative character which the coastal State can exercise outside its territorial waters. It is fairly generally recognized (but *not* without exceptions) that the coastal State may exercise a certain jurisdiction, usually limited to the prevention of infringement of its customs or sanitary regulations. We could call this an "individual" right of a coastal State. It will be established by Convention or by the national legislation. It is limited in scope and in extent. For the time being we will take as a maximum the distance of 12 miles from the coast, following the Basis of discussion No. 5, laid before the Codification Conference in the Hague (Acts 1930 [61],

Annex I, p. 179). Can we say that outside a belt of 12 miles there remains the vast domain of the high seas where no State has any right?

Obviously not, but the rights of States outside this belt of 12 miles are not individual rights in the sense of particular rights of one State only and different from those of other States. The rights in that vast domain of the high seas are common rights. If in an area no particular rights can be exercised, no particular sovereignty prevails, we can speak about freedom. Freedom is the characteristic notion used in connection with the high seas. Gidel [112], p. 125, says: "Le régime juridique de la haute mer est caractérisé par l'idée de liberté; d'où l'emploi fréquent de l'expression: mer libre, comme synonyme de l'expression: haute mer".

What does freedom mean in this connection? Is everybody allowed to do everything on the high seas? Of course, this domain is not one without law and order. The word freedom means that everybody, i.e. the States or their nationals, are free to use the high seas for different purposes without, however, causing unreasonable hindrance to other people who are enjoying the same freedom. Now Gidel enumerates these ways of use as follows: ".... tous les Etats et les ressortissants de tous les Etats, s'ils satisfont aux conditions posées par leurs autorités nationales respectives, peuvent se servir de la haute mer pour les diverses utilités qu'elle comporte: navigation de surface et sous-marine; pêche ou chasse, ou plus généralement capture ou récolte de la faune et de la flore marines; immersion de câbles télégraphiques; survol des surfaces maritimes".

Is this enumeration of Gidel exhaustive? Our impression is, that Gidel summed up the different kinds of use which, at the time he wrote his book (Preface dated August, 1931, published 1932), had been made of the high seas.

We do not believe that no other means of using the high seas would be permitted if such a use could be made under the same conditions as the existing ways of using the high seas. What are these conditions? The different kinds of use do not exclude each other, but they may well limit each other. In order to be able to use the right to fish in the high seas, one usually has to use the right to navigate the high seas at the same time.

The same can be said about laying telegraph cables. But the use which one makes of the high seas limits the use of somebody else. The right to navigate is not completely free, it is limited by the use of that right by other persons and by their use of other rights like fishing or laying telegraph cables. A ship finding a fleet of drifters in its course may have to make a considerable detour to avoid the driftnets.

Therefore if yet another way of using the high seas has been discovered, we cannot reject such a use by saying that it interferes with existing kinds of use of the high seas. Using the high seas naturally and logically limits the use some other person makes of it if the place where we want to make use of the high seas happens to be the same one chosen by somebody else. It is also natural, that the one who can move easier, will have to give way for the one whose kind of use of the high seas involves a limitation in his possibilities to move or perhaps makes any move impossible. Hence the rule that a steamship which can easier be manoeuvred than a sailing ship, has to give way for the latter. All ships have to give way for a fisherman laying behind his nets or for a ship at anchor. These are established rules, which are perfectly logical. Therefore there is nothing new in the fact that the freedom of navigation is a "limited" freedom, limited because other people may be exercising their rights on the high seas, at the same time and at the same place (either the same right, i.e. that of navigation, or another right, e.g. fishing).

If this view could be accepted it may also give a solution of the problem of conservation and protection of certain species of fish. Jessup [151] discussing the "factoryships", writes, p. 133: "Just as weapons of offense and of defense develop to counteract each other in the field of military activity, science seems to perfect more or less concurrently the means for destroying and conserving the food supplies of the seas. The difficulty lies in the legal bases for making the measures of conservation operative on the high seas where no State has sovereignty".

Perhaps we could say that certain ways of exercising the right to fish in the high seas do not fulfil the condition we have put to the use of the high seas, i.e. that the use made should not cause unreasonable hindrance to other people who are enjoying the same freedom. If therefore a nation is sending factory-ships to the fishing grounds, where the States whose nationals have fished there for many years have agreed upon protective regulations, and that State is not willing to submit to these regulations, we could say that this refusal is an unreasonable frustration of a wise and careful way of exploitation of the marine resources by other users.

We may say the same about intercepting salmon on a scale which would lead to certain extinction or using methods like trawling just outside the territorial waters of a coastal State, which may be very harmful for the fisheries inside the territorial waters.

Of course an impartial body should in these cases decide whether such an unreasonable use exists and we meet again the international body proposed by the International Law Commission.

We made these remarks only because it is an approach from another angle which may be helpful for the solution of this very difficult problem.

Coming back to the definition of the high seas, we have seen that Gidel on one side defines the high seas by saying that it is the ocean outside the territorial waters and that he on the other side characterizes the high seas as the domain where the freedom of the seas is applicable. He says that "haute mer" and the free seas are synonyms. One of the oldest, as a matter of fact the oldest use of the high seas, i.e. navigation is, however, not limited to the high seas in the sense indicated by Gidel. The right of innocent passage, and we limit our whole argument for the moment to peace time, is a generally accepted right, which a coastal State should grant any merchant vessel wanting to traverse its territorial sea. To find out more about this inconsistency we have to go back and see what this right of navigation actually means. We think that the great advocate of the freedom of the seas may help us. Grotius [91], p. 216, says: "Access to all nations is open to all, not merely by the permission but by the command of the law of nations".

The main reason of the freedom of the seas is that it is the common highway, which connects the countries and continents. Of all the arguments used by Grotius to prove the freedom of the seas and to prove that the seas could not be appropriated we believe that this one has remained unshaken. Grotius has been attacked and some of his arguments were not unimpeachable. However, as we said, the one we mentioned has not lost any of its strength and therefore we believe that we can say that we are not far from the truth if we adopt this argument as an axiom to build upon. If navigation is the oldest and most important way of using the ocean, and we think that we can say that without making any serious mistake, it strikes us immediately that this use is not limited by a marginal belt, by the territorial waters of any State. We do admit that the right to fish is a very important one, and we do know that the coastal State can prohibit the exercise of this right to non-nationals.

On the other hand the remaining kinds of use of the high seas being laying telegraph cables and using the air above the seas are, we believe, less important in this connection.

The definition given, at least in connection with the notion of the free seas, a "synonym" seems therefore not quite satisfactory, because concerning one of the principal rights, i.e. navigation it cannot stand.

The territorial waters of course are of relatively recent times. Is it necessary to tie a definition of the high seas to a relatively young notion? Can we say that the high seas existed before the territorial waters and what definition has been given to the high seas before the territorial waters existed? It is a well-known fact that certain particular rights have been claimed on the seas in olden times. These rights have always been limited in extent, although often the extent of the *claim* was considerable. "Venice attributed to herself the sovereignty of the Adriatic, whilst Genoa and Pisa claimed the Ligurian Sea" (to quote Higgins and Colombos [154], p. 40). We all know about the division of the Oceans by Pope Alexander VI, confirmed by the Treaty of Tordesillas of 7 June, 1494. Later the Danes and the British put forward claims of sovereignty of certain areas of the seas.

But were these claims important, can we speak about a practice of States here as a source of our knowledge of International Law? Can we deduce from that practice a definition, for instance by saying, that the high sea is that part of the ocean which has not been claimed by any one State? We are afraid that this definition would not have been accepted in the times when these claims were made, for the simple reason that these claims have never been recognized by other States. Of course it was partly a question of naval power in a certain area and the claims could by lack of power of resistance be realized for some time. But when one subject of law forces the other one to do something and this practice goes on for a limited stretch of time, can we deduct from all this that the subject of law forcing the other has obtained a right? Of course not. In most places and times these so called rights have been hotly contested and in practice been often difficult and even impossible to realize. Therefore we are inclined to believe that the notion of the high seas and the freedom of the seas has always existed and has not been seriously shaken by the theoretical and certainly not by the practical denial of its existance.

Of course the definition of Gidel still holds good where the other rights are concerned and we may adopt the definition as meaning that the high seas are that part of the ocean where no State has sovereign rights. This will be further explained in the section on territorial waters.

It is only lately that doubts about the usefulness or about the exact scope of the principle of the freedom of the seas has arisen.

Lauterpacht [156] has dealt with the subject on p. 98 et seq., but we believe that the ideas about the freedom of the seas which he develops are not essentially different from ours. Speaking about the Behring-Sea arbitration he says, that the American claim "failed to make full use of the notion of 'abuse of rights' as well as of the more general arguments that the freedom of the sea could not mean absence of any legal regulation whatsoever, and that it was inherent in the very idea of the common user of the produce of the sea that it implied reasonable limitations of its exercise".

We believe that the reasonable limitations in exercising the right of common user is exactly what we meant by the condition which should be fulfilled when making use of the high seas, namely that this should be done in such a way as not to cause unreasonable hindrance to another user.

If this assumption is right that Lauterpacht's version is the same as ours, then we feel that he used rather a strong wording in an additional paragraph 190 (III) in his edition of Oppenheim [42], p. 453,: "It is probable that the rigid application, in cases of this nature, of the principle of the freedom of the seas is open to criticism as failing to protect important interests of the littoral State against the unscrupulous exercise of a legal right", where he referred to the Behring-Sea arbitration and the arbitration of Asser in the case of the 4 American fishing boats seized by the Russians (Cape Horn Pigeon, James Hamilton Lewis, C. H. White and Kate and Anna).

The Memorandum on the Regime of the High Seas [22] gives a good idea of the high seas where it says (p. 1) that it has always "constituted both a means of communication and a source of wealth". It is, however, a great pity that the author of this admirable Memorandum has found fit to give the following appreciation of the notion "freedom of the high seas", where he says, p. 2: "The expression 'freedom of the high seas' is in reality a purely negative, worn-out concept, nothing more; it has no meaning for us, except as the antithesis of another, a positive concept, which has long since disappeared".

We said it is a pity, because we believe that it is not true and at any rate it is not necessary to put it this way.

All human utterances are only relatively true and nothing is absolute (not even this remark). If we use a notion like sovereignty, too many people are inclined to think that that word has an absolute meaning.

But how could any State claim absolute sovereignty where in a community of States the rights and powers of one State are necessarily limited for the single reason that that State is not the only one and other States have rights and powers which form ipso facto a limitation on the rights and powers of the first State.

We agree with the author of the Memorandum, *if* he meant to say that notions like the "freedom of the high seas" are apt to be interpreted in too wide a sense. The same is true where the notion sovereignty is concerned. But these notions are professional terms, and is it not going too far if we do away with professional terms, because laymen, or perhaps even lawyers who are not well worked in in these matters, give these terms the wrong connotation?

We do believe in principles, and we are strong supporters of keeping principles to lead the way in human relations, but we also believe that these principles should be applied with reason.

Freedom never means absolute freedom and therefore it was quite unnecessary to do away with the notion "freedom of the seas", obviously done by the author of the Memorandum because this principle would, if applied without reason, make either the freedom a harmful instrument or a barrier for new ways of using the high seas. The Memorandum, p. 3, reads: "Freedom of the seas was nothing more than a negation; it was the opposite of the idea of sovereignty of the high seas".

We have explained that we have a definite view on this question and we believe that Fulton [31] has convincingly proved that the so called sovereignty of the seas never existed in practice. Therefore the freedom of the seas might have been formulated as a contrast to the theory of sovereignty, but it was not an antithesis as the opposite practice did not exist.

Another attack, which may harm the principle of the freedom of the seas has been lanced by Reppy [157], who remarks first on p. 15, speaking about "de Jure Praedae" that "the arguments were marshalled and presented in the systematic form of a brief or a piece of special pleading, which in truth it was". There would be no objection to this statement if it had not been used to serve as a basis for the following accusation, on p. 17, Reppy in comparing "de Jure Praedae" with "de Jure Belli ac Pacis" writes: ".... that he (Grotius) was not intellectually honest when he offered his final work to the world as one uninfluenced by the events of his age, when in reality it was merely a revision of an earlier work, which was a direct reflection of the period in which he lived".

In other words Reppy finds, that a part of "de Jure Belli ac Pacis" was more or less a copy or improved copy of "de Jure Praedae" the latter being a piece of special pleading, hence Grotius should not have pretended, that his last work was developing the principles of International Law "out of the law of nature, on a plane of eternal principle, entirely disassociated, as he told us in the Prolegomena, from specific present or past incidents in world history ...." (loc. cit. p. 16).

It is true that Grotius [158] said in the Prolegomena, p. 29–30: "58. If any one thinks that I have had in view any controversies of our own times, either those that have arisen or those which can be foreseen as likely to arise, he will do me an injustice. With all truthfulness I aver that, just as mathematicians treat their figures as abstracted from bodies, so in treating law I have withdrawn my mind from every particular fact".

We must confess to be startled to hear that a piece of special pleading is not supposed to depict the truth and cannot be used to support a theory of law. We should have thought that this depends entirely on the mentality of the "advocate". At any rate we have no reason to doubt the fact that Grotius was convinced even in his "piece of special pleading" to have pleaded in support of a fact which actually existed long before his time, according to the authors he quotes, i.e. the freedom of the seas.

We limit ourselves to these few remarks and leave theoretical questions about the character of the high seas (res nullius, or res communis or res extra commercium). We believe that the high seas is a domain where States have common rights. It may become the nucleus of a new world of law. Maybe the first International Police force will be operating on the high seas, the only place where it could operate for the time being. We believe that this "common highway" should be kept free from interference of particular States and should be kept as large as possible.

SECTION 2. TERRITORIAL WATERS AND CONTIGUOUS ZONES

Just as in the Section on the high seas we are not going to give a systematic exposition of these subjects, but we will say a few words on the points bearing on the continental shelf.

We have seen in Chapter II that countries like Argentina have declared that the waters covering the continental shelf "are subject to the sovereign power of the nation" whereas countries like Chile and El Salvador enacted Decrees claiming a stretch of sea 200 miles wide along the coast. Both claims boil down to an extension of the territorial waters. Some claims are directly related to the continental shelf, some are not, but belong to the bombardment of claims, international lawyers had to register after the Truman Proclamations of 1945.

The question arises, can a country just extend its territorial waters indefinitely, without any limitation? At the Codification Conference in the Hague in 1930 no agreement was reached on the width of the territorial sea, but we can say that the minimum width claimed was 3 miles and the maximum was 12 miles. Do these recent claims indicate a tendency to go back to the times of the Treaty of Tordesillas?

There are several theories about the origin of the territorial seas.

First of all there is a theory that the marginal belt is a sort of residue from the larger claims of sovereignty. Fulton [31], p. 538, writes: "The sovereignty of the so-called territorial sea has sometimes been regarded as the direct remnant of a sovereignty which was previously asserted by particular nations over whole seas or large parts of them". Fulton believes this to be true in a general sense, and he mentions Norway, Sweden, Spain and Portugal in this connection. He tinks, however, that: "the territorial sea now held to pertain to Great Britain .... did not originate, in this way, by direct descent from the old claim to the dominion of the British seas. That claim simply died out and vanished in the lapse of time, without apparently leaving a single juridical or international right behind it. The British territorial waters .... were derived .... from the doctrine of Bijnkershoek. Even during the time when some nations were asserting a wide maritime dominion, and other nations were opposing such pretensions, there was a general recognition that every maritime State was entitled to exercise jurisdiction over some extent of the neighbouring sea".

We do not believe, that the territorial sea has anything to do with wider claims of some countries, exactly because only a few countries went in for these claims, but the majority did not. It would be difficult to explain how Holland and Belgium and France and Germany and the United States got their territorial sea because these countries never claimed extravagant parts of the ocean.

We have another reason not to believe in this theory, because the character of the territorial waters is different from the nature of the old claims, which even differed from one claim to another. The Portuguese and Spanish claims forbade navigation by foreigners, the English claim did not prohibit navigation, but demanded a salute, which was originally meant as a sign that no bad intensions existed, especially as the topsails had to be lowered. Fulton [31], p. 7, explains this custom. He says: "The king's officers had to ascertain .... whether (the ship) was a peaceful trader or a pirate". This sort of claim could be called dominium. The territorial waters, however, have more the character of sovereignty or at least of jurisdiction.

It is curious that in a time when the fleets were smaller, slower and less powerful in their artillery, claims concerning the sovereignty over vast sea-areas were made, whereas nowadays, no such claim exists in spite of the fact that to a far greater extent and over a far greater extension a possession of a part of the sea would be possible with warships and planes.

Ortolan [159] could ask with reason, p. 118: "Une nation quelconque peut-elle avoir la mer en sa puissance, en sa possession? Décuplez, centuplez toutes les flottes du monde réunies, mettront-elles la mer à la discrétion d'un peuple"? But since he wrote a larger control during a longer period over a larger area of the ocean has become possible. Besides, if the theory does give the genesis of the territorial waters correctly (quod non) it would still not explain why States would be free to reverse the evolution and indulge in atavism. It is true that a modern fleet could control an area more effectively than in former centuries, but then they are not the mightiest naval powers who have extended their territorial waters.

Is there any other point in the history which may enlighten us?

The cannon shot rule, first published by van Bijnkershoek, may of course be blamed.

Many times the increase of the range of guns has been invoked as a reason to increase the width of the marginal belt. Park [160], p. 2, says: "Shore batteries, while generally considered out of keeping with modern warfare, still have a range far in excess of three miles. What is three miles to a six hundred mile an hour airplane?". But before we proceed, how did the cannon range get mixed up with 3 miles? Azuni [161], p. 253–254, wrote in 1805: "Il serait donc raisonnable, à mon avis, que, sans examiner si la puissance, maîtresse du territoire, possède quelque tour ou une batterie construite en pleine mer, on déterminât fixement, et partout, que la juridiction sur la mer territoriale ne s'étendrait qu'à trois milles de distance de terre, ce qui est sans contredit la plus grande portée à laquelle la force de la poudre à canon puisse pousser un boulet ou une bombe". It was his compatriot Galiani, who told the world in 1782 that the maximum range of a gun was 3 miles [a]. Now this "trouvaille" of Galiani has echoed through Conference halls, has been repeated (we are tempted to say "parroted") and rewritten in monotonous repetition throughout the 19th century and we still hear it and read it in publications of our time. The quotations are not even always correct! Westlake [121], p. 188–189, wrote:

"The principle of a presumed limit to occupation was laid down by Bijnkershoek, who, taking into account only force exercisable from the shore, taught, first, as a general maxim, *imperium terrae finiri ubi finitur armorum potestas*, and secondly, as the application of that maxim to his own time, the range of cannon, then considered to be three sea-miles of sixty to a degree of latitude. Hence that distance, measured from low-water mark, became a commonplace among authors for the width of the littoral sea, and we may say that the agreement on it as a minimum is universal: no State claims less".

Let us first have Galiani examined by some experts on ballistics. As Galiani wrote in 1782 we will first of all take a range table of that time. The nearest we could find is a Dutch translation of 1839 by Gobius, of the second edition of a Treatise on Naval Gunnery by General Sir Howard Douglas, Bart [162]. On p. 309 in table XIX the results are given of trials with a gun of 12 pounds, with ordinary and oblong balls on the fortress Landguard in 1776.

The distance given for an ordinary ball of 11 pounds and 10 ounces, with an angle of elevation of 5 degrees, was 1789 yards (1 yard = 0,914 metre) or 1635 metres (1 mile = 1852 metres).

For the oblong projectile, with a weight of 23 pounds and 3 ounces, same angle of elevation, a distance of 1879 yards. is given, the equivalent in metres being 1715,4. Therefore both figures remained well under one mile. It is true that a bigger gun and a larger angle of elevation would have given a larger

---

[a] De' Doveri de' principi neutrali verso i principi guerreggianti e di questi verso :neutrali, Naples 1782.

distance, but on the other hand, it is not so much the maximum distance which interests us, as the distance at which a target could be hit with a certain amount of accuracy and be destroyed or at least severely damaged.

It should be kept in mind, also for the understanding of the figures which follow, that for accuracy the elevation should be the least possible (Sir Howard Douglas [163], p. 99, 167–168 and particularly p. 385: "No law of gunnery is more clearly demonstrated, and irrefutable than this — that elevation is, inversely, the exponent of accuracy".

Table VIII gives data of a steel ship-gun of 42 pounds, $9\frac{1}{2}$ ft. long, with an angle of elevation of 10 degrees, the maximum distance is 2900 yards or 2650,6 metres. This is not yet one mile and a half and the table is of a much later date, not long before 1839. With a smaller elevation (greater accuracy) for instance 3 degrees, the range is 1622 yards, or 1482,5 metres, much less than one mile [a].

Paixhans [164], p. 55, gives as the highest average range of a gun of 36 with massive ball, 3458 el (as the book is translated, the Dutch "el" is given, being 0,69 cm) which is 2386 metres or not yet $1\frac{1}{8}$ mile, during trials in Brest in 1824 (16 degrees elevation). With 3 degrees elevation the average range was 1611 el = 1111,6 metres, 200 metres more than $\frac{1}{2}$ mile (p. 55).

The original 5th edition of Sir Howard Douglas of 1860 [163] gives higher figures, which could be expected. On p. 172 (in the Section "Monster guns") the range of the "Pacha's" 130-pounder with an elevation of 20 degrees was for a shot of 128 pounds and a charge of 26 pounds, 4669 yards, that is 4267,4 metres or 2,3 miles. But with an elevation of 5 degrees the range was only 2151 yards = 1966 metres, about 100 metres more than one mile. This trial took place in 1842. Most of the figures given on p. 408–409 of Lafay [165], 1850, are lower except one, being a range of 4500 metres of a gun of 50 with an elevation

---

[a] The largest figure we could find was in table XVI, for a mortar of 13 inches, with a load of 20 pounds, an angle of elevation of 45 degrees, a distance of 4200 yards was recorded during trials in Woolwich in 1798. This would be 3704 metres or well over 2 miles. But then a mortar would not be used against ships, because the accuracy of these weapons is rather small and the probability of hitting a small target like a ship would be slight. They are used against targets with a big horizontal surface, whereas a ship is a vertical target.

of 30 degrees during trials in 1846–1847. But the range of the same gun at an elevation of 5 degrees is only 1737 metres; less than one mile!

Robertson [166], p. 75, gives the extreme range of 2500 paces, that is about one mile, of the heaviest gun used on board ships, the "culverin"- throwing a $17\frac{1}{3}$ lb. ball, according to a report drawn up in 1559.

Finally in a note on p. 205, Robertson says: "The sudden and extraordinary development of rifled ordnance which now took place had a revolutionary effect .... on land fortification. In '59 (1859) Sir William Armstrong giving evidence before a committee appointed by the War Secretary, stated that he could attain with a specially constructed gun a range of five miles.

The statement made a sensation for in the presence of such a gun most of the existing defences of our dockyards and depots were almost useless".

It is clear that the subject is worth a thorough research, but with the few data we have been able to find, it is more than likely that Galiani would fail for his exam in ballistics and that his statement is a rather uncritical attempt to combine 2 rules of completely different origin by asserting that they are one and the same. Walker [167] deserves the honour of having drawn the attention to this point, which should be investigated further.

We are inclined to agree with Walker's argumentation. Walker rightly says, p. 210, that van Bijnkershoek did not invent the cannon shot rule.

The Bijnkershoek rule had been in use for some time. According to Fulton [31], p. 156, on 6th May, 1610, a year after the Proclamation forbidding unlicensed fishing, by James I, the Dutch Ambassador had a formal conference with the English commissioners. On this occasion the Dutch delegation said: "…. by the law of nations the boundless and rolling sea was as common to all people as the air, 'which no prince could prohibit'. No prince, they said, could 'challenge further into the sea than he can command with a cannon, except gulfs within their land from one point to another' ", and this said Fulton was "the first occasion on which this principle for delimiting territorial waters, afterwards so celebrated, appears to have been advanced". Fulton further remarks that: "…. there are grounds for thinking that the idea may have originated in the fertile brain of Grotius".

Walker, p. 212, says (and this is very important) that the cannon shot rule "was not a doctrine of a maritime belt. It was a doctrine of port or fortress areas or zones within range of *actual* guns mounted on the shore" .... "The cannon shot rule was not a vague one (like the rule: 'within sight of land') — it did not depend on the hypothetical range of some imaginary gun.

It had no fixed distance, indeed, for it would vary with the calibre of particular cannon ....".

Walker goes on (p. 213): "With the 3 mile-limit as such Bijnkershoek had no direct contact. For him dominion depended, as with Grotius and others, on control, and by control in marginal seas was meant actual control by guns present on the spot". We therefore believe that Westlake [121], in the passage quoted above gave an incorrect picture of the genesis of the 3 mile-limit. Besides if Galiani was wrong in 1782 by giving guns a range which they would not even be able to reach in 1850 (with reasonable probability of hitting the target) how much further from the truth would van Bijnkershoek have been in 1702 *if* he had confused the two rules (quod non).

The 3 mile-rule, which should be called the league-rule, was born in the human brain, probably at several places, but at any rate in Scandinavia according to Walker, p. 216, 227 and 228, where he states that the Scandinavian league happens to be 4 miles, as against the three-mile league in general use elsewhere. (We believe that the "legua" used in Southern Europe amongst others by Columbus, was also 4 miles, may be Italian miles). (Concerning the 4-mile league see also Raestad in Acts [61], p. 143).

Walker says, p. 229: "It does not appear that Galiani is asserting that cannon range had been in the past or even then was identical with three miles. His language suggests that he is taking a convenient standard measure, the marine league, outside the range of then known cannon, but not so far as to be out of all relation to possible developments [a], and that he is proposing a maritime belt of that extent as a substitute for the

---

[a] We have heard these words before, in Chapter I, the 200 metre-isobath in connection with technical progress.

old cannon shot rule which required actual cannon in position".
We are happy about this milder criticism of Galiani, but we still
want to know what he knew about ballistics. But let us rather
try to unraffle what has been wrongly mixed [a]. We believe that
the one league rule was adopted as one of the many ways to
delimit a marginal belt. We mentioned the one league rule in a
British instruction for Ambassadors in 1673 (Chapter II, Section
6 above). Distance in general had been a criterion (the 100 miles
of Bartolus of Saxo-Ferrato, and the same width, according to
Azcárraga, p. 78, established by King don Jaime de Aragón
in a privilege given to the town of Cagliari; the 60 miles of Jean
Bodin, the 2 leagues of Valin (in narrow waters) etc.).

Was the choice of one league arbitrary? We believe not.
It was a humble desire, it is true, but in that "order" of distances,
it would have been arbitrary to choose 2 leagues, or half a league,
but just one is quite a logical thing to do, if one realizes that the
league was the unit of distance used in connection with the sea.
Gidel [169], p. 41, says: "La limite de trois milles .... eut une
force: celle de représenter l'unité nautique la plus connue; celle
de la lieue; c'est à celà qu'elle dut les acquiescements qui lui
vinrent". From all the criteria (cannon shot, depth, visibility
or "land kenning") the distance is the one most easily discernable
in practice if measured from a generally recognized base line,
but we are not going into that (pending the Anglo-Norwegian
Fisheries Case, where this base line is one of the points of friction).

In order to find an answer to our question whether a State
has the right to claim any width of territorial sea, we must
first state the fact that the cannon shot rule is a dead rule. A
rule based on imaginary power (guns which were not there)
should not have lived so long. The rule prolonged its sterile
existence under the cloak of the 3 mile-rule, to which it was
wrongly "coupled" by Galiani. It is curious that so long power
was respected where power was absent. And Grotius has shown
the right way. He did not say much in his Mare liberum [91], only
that he made an exception for bays. But he says very clearly
in de Jure belli ac pacis [158], p. 214: "It seems clear, moreover,
that sovereignty over a part of the sea is acquired in the same

[a] Borchard [148], p. 61, spoke about "the 1702-rule .... and its arbitrary identifi-
cation with three miles or one marine league ....".

way as sovereignty elsewhere, that is, as we have said above, through the instrumentality of persons and of territory. It is gained through the instrumentality of persons if, for example, a fleet, which is an army afloat, is stationed at some point of the sea; by means of territory, in so far as those who sail over the part of the sea along the coast may be constrained from the land no less than if they should be upon the land itself" (Lib. II, C.III, XIII, 2).

Cannon are not the only way to protect a coast. Not the British forts or coast guns withheld the Germans to invade Great Britain in World War II, but the British Navy and the R.A.F.. Grotius was right. Of course, coastal batteries are important in a defence scheme, but in an age of airplanes, they have lost some of their importance in that scheme. The remedy which is often suggested to extend the territorial waters, usually based on the greater range of coast guns, is outdated since the appearance of planes and V-weapons.

For defence an imaginary boundary line in the sea is not effective.

Sir Maurice Gwyer said in the Hague Conference in 1930 (Acts [61], p. 141–142): ".... if a State is threatened with attack from outside its territorial waters no country in the world would criticize that State for going beyond its territorial waters, in order to repel that attack. An additional area of territorial waters affords no additional safeguard".

For the duties involved for the signatories of the XIIIth Convention of the Hague of 1907, a wider zone would only mean a heavier burden.

For safety reasons, but for safety reasons only, we do not have to enlarge the marginal belt, but we may consider whether a contiguous zone (not too wide) would be the proper answer.

The cannon shot rule being dead (we will discuss an opinion to the contrary later) we have to see whether the 3 miles can stand as an absolute figure in International Law.

There may be three reasons why States, at least some States, are reluctant to extend the territorial waters. The *first*, given by Jessup [116], p. 18, note 52: "It may well be that England's resistance to any attempts looking toward an extension of territorial waters is motivated by her naval power.

At one stage of history that power was no doubt used to claim for its master the control of the high seas which all nations were ready to claim as their own. But when it became apparent that these claims were no longer tenable it would be logical for the nation which could by naval superiority command the open ocean, to favour a restriction of the zone which could be called territorial. The wider the ocean the wider is the power of him who controls it". This view may have been true 50 years before it was written. But in 1927 when Jessup wrote the British had already lost their naval superiority. And if we judge the statement now, there are even more reasons to doubt whether it still holds. If there are still neutrals in a future war, then they may claim a safety zone like the one claimed in the Declaration of Panama of 1939 and in the Treaty of Petropolis of 1947, a zone which is not a territorial sea, but only an area where no naval action will be tolerated, a zone much wider than the territorial sea.

It is true that the chance that there will be many neutrals in a future war is small, but nevertheless there is reason to believe that the width of territorial waters has lost a good deal of its significance in relation to naval supremacy.

A *second* reason which we believe is more generally applicable, is the one which was mentioned by Oppenheim [170] in his Report to the Institute of International Law in 1913 in Oxford, p. 405–406: "La raison pour laquelle je m'oppose à une extension de la limite de 3 milles est que la souveraineté reconnue d'un Etat riverain sur la mer territoriale ne comprend pas seulement des droits, mais implique aussi de lourdes responsabilités en temps de paix de même qu'en temps de guerre, et surtout pour les neutres en temps de guerre". This was not only the personal opinion of Oppenheim. In the "Bases of Discussion [126]", p. 162, in the reply of Great Britain we read exactly the same argument against a wider marginal belt: ".... the burden imposed on neutral States in time of war would be intolerable". Sir Maurice Gwyer used the same arguments during the Codification Conference in the Hague in 1930 (Acts [61], p. 140–142).

A *third* very important and realistic reason is that the State dreads either to be forced on a basis of reciprocity to accept the same extension of the territorial waters of the other contracting parties and losing that way important fishing grounds, or, in case of unilateral extension, to face the same result by way of retaliation from the side of another State.

There are countries particularly interested in the freedom of the high seas. For them a small territorial belt constitutes the least infringement on this freedom. Surie (Netherlands) said at the Hague Conference in 1930 (Acts [61], p. 20): "My Government has .... accepted a distance of 3 miles. This is regarded as sufficient and adequately safeguarding the freedom of the high seas, and p. 125 (The Netherlands) bases its decision (accepting the 3-mile limit), first, on the necessity of safeguarding the interests of commercial navigation on the high seas, and, secondly, on the consideration of not placing any too heavy obligations on the coastal State". (In the same sense the delegates from Japan and India, loc. cit. p. 21 and 25) and the British delegate Sir Maurice Gwyer, p. 123: ".... the three mile limit is the limit most in favour of freedom of navigation" and he gave as another reason for the support of the 3-mile limit that this rule was already adopted by maritime nations which possess nearly 80 per cent of the effective tonnage of the world [a].

On the other hand the desire to have a larger exclusive fishing area is the very cause that in many cases extension of the marginal belt is demanded. The 3-mile belt is not accepted by all the States. Therefore Westlake [121], p. 188–189, writes: "As a maximum the agreement is not universal, and it may be doubted whether is it so nearly such as to make it a rule of international law, while the increased range of cannon shot, as well as the increased need of protection for shore fisheries against trawl nets and other destructive devices, has made the reason for it quite obsolete and inadequate". We cannot accept his reasoning concerning the range of cannon shot, because, apart from mistakes made in the past, the cannon shot range, which should if consequently interpreted include the range of V-weapons and bomberplanes, would if adopted as a criterion for the width of territorial waters leave as little of the high seas as the depth criterion of Valin [23] would, if applied consequently, i.e. practically nothing, with the impossible situation that every country

[a] Compare with Table II of Boggs' [171] article published in April 1951, p. 204, giving 79,53 per cent, based on "Merchant Fleets of the World, June 30, 1950".

would claim enormous areas many times overlapped by the same claims of other countries.

We are not in particular devoted to the 3-mile limit, although we understand that this figure, or rather the league as unit was chosen, but it must be said that generally speaking a limited marginal belt, as we have known for about the last century, i.e. in the order of 3–12 miles, has worked rather satisfactorily in the past. Jessup [116] says about the 3-mile limit, p. 7: ".... once introduced it remained because the nations found it a convenient compromise between conflicting interests. When it ceases to be generally covenient it will probably be changed by general convention".

Riesenfeld [96], p. 279, put it this way: ".... an adjustment of the conflict between the reasonable, individual interests of coastal States and the interests of other nations in the international community was worked out in the concept of the marginal sea or territorial waters. This adjustment, as originally made, was based on the balance of interests as they stood during the eighteenth century.

Since questions of maritime warfare and neutrality rather than fisheries were the focal points of such interests at that time, the cannon shot principle was considered by many authors the sensible rule for this adjustment".

The answer on our question is taking shape. We may word it as follows: There may be weighty motives which would urge consideration of an extension of the territorial sea. Among such we would reckon either a reasonable extension of the area of exclusive fishery rights or a wider field for enforcement of protective legislation. We would not recognize motives of defence or neutrality, because the first is not relevant to the width of territorial waters as said above and for the second a contiguous zone may be sufficient, as we will explain presently.

The valid reasons, the examples concerning fishery rights, should be based on scientific data and approved by multilateral Convention.

Unilateral declarations or national legislation to this effect should be held as contrary to International Law, because a State by doing so takes without asking from the common domain. Besides for protection and conservation of fish, the method of areas covered by a multilateral Convention seems preferable.

At the moment we believe that the situation is such, that every country can claim whatever it wants, only it cannot enforce its claims against nationals of other States outside the limits recognized in International Law. Our suggestion will perhaps be difficult to realize, as long as pompous people, rather than wise and scientifically minded men, seem to have a finger in the pie in most countries or as Boggs [172], p. 241, describes the psychological phenomenon of the recent series of national actions as partly due "to a sort of 'cartographic chauvinism' and to a desire to 'keep up with the Joneses' ".

We are inclined to recognize claims to historic bays and historic waters as well as local peculiarities which may lead to differences in width of the marginal belt, as for instance on the Norwegian coast, or the coast of Portugal or of Chile.

We would welcome an International Conference where the "bases of discussion" would be prepared on scientific reports as far as possible and which would start from the principle that within reasonable limits, differences in the width of territorial waters and recognition of some historic waters etc. would not be excluded. We agree with van der Lee [173], who discussing the breadth of territorial waters wrote, p. 8: ".... it is not believed that uniformity should be aimed at".

We still have to discuss the question of safety for neutral maritime States. To start with it is not immediately apparent why States should need a marginal belt for safety or security any more than a safety or security zone along their land frontiers.

However, there are two questions. One is damage ashore caused by a naval action outside the territorial waters and the second one is damage to shipping inside the territorial waters by the same cause. Extending the territorial waters in the second case would not help and a contiguous zone is indicated. Of course such a zone would be a solution for the first case as well. We find this idea in Art. 4 of the rules adopted by the Institute of International Law in 1894: "In case of war a neutral littoral State has the right to fix, by declaration of neutrality or by special notification its neutral zone beyond six miles up to the range of coast artillery" (Scott [92], p. 114 or Annuaire [93], Vol. 13, p. 328).

The Declaration of Panama of 1939 [174] went much further [a].
We quote the following passages, p. 19:

"The Governments of the American Republics meeting at
Panama, have solemnly ratified their neutral status in the
conflict which is disrupting the peace of Europe, but the present
war may lead to unexpected results which may affect the funda-
mental interest of America and there can be no justification
for the interests of the belligerents to prevail over the rights
of neutrals causing disturbances and suffering to nations which
by their neutrality in the conflict and their distance from the
scene of events, should not be burdened with its fatal and painful
consequences", and p. 20:

"Resolve and hereby declare:

1. As a measure of continental self-protection, the American
Republics, so long as they maintain their neutrality, are as
of inherent right entitled to have those waters adjacent to the
American continent, which they regard as of primary concern
and direct utility in their relations, free from the commission
of any hostile act by any non-American belligerent nation,
whether such hostile act be attempted or made from land, sea
or air. Such waters shall be defined as follows . . . .".

Here follows a description of the boundary line from which it
appears that the area is about 300 miles wide. Then the Declaration continues:

"3. The Governments of the American Republics further
declare that whenever they consider it necessary they will
consult together to determine upon the measures which they may
individually or collectively undertake in order to secure the
observance of the provisions of this Declaration.

4. The American Republics, during the existence of a State
of war in which they themselves are not involved, may undertake
whenever they may determine that the need therefore exists,
to patrol, either individually or collectively, as may be agreed
upon by common consent, and in so far as the means and resources
of each may permit, the waters adjacent to their coasts within
the area above defined.

(Approved October 3, 1939)".

---

[a] A suggestion in this direction is reported by Moore [175], p. 703: "The President
(Mr. Jefferson, in an informal conversation) mentioned a late act of hostility committed by a French privateer near Charleston, S.C., and said we ought to assume,
as a principle that the neutrality of our territory should extend to the Gulf Stream,
which was a natural boundary, and within which we ought not to suffer any hostility
to be committed".

The Brazilian Government issued the following Declaration on Continental Waters (loc. cit. p. 21):

"The sea outside territorial water, only three miles from our coast, from our cities and even from our capitals, not only is not ours, but in it we are at the mercy of any action contrary to the free and peaceful expansion of our sovereignty, of our continental relations and even of the maritime communications between ports of the same country.
To the defence of the continental territorial integrity, we must add, therefore, as an inseparable part of an American political whole, the security of continental waters".

France, Great Britain and Germany protested against the Panama Declaration. We give as an example the British Admiralty Statement on the Panama Declaration [176], p. 69:

".... In olden times many extravagant claims were put forward by the various nations as to the limit of their territorial waters, but since those days such claims have been drastically modified and it is now generally recognized that no country can properly claim jurisdiction over large areas of ocean nor the right to control or exclude the movements of foreign ships on the high seas; this applies equally to belligerent operations though a belligerent can of course restrict his operations of his own free will if he so wishes. .... Great Britain in common with many other countries has long refused to recognize claims to a territorial belt of great width".

(New York Times, October 14, 1939) [a]

Hyde [177] made the following remark, p. 466: "With a common appreciation of the scope of the existing right of the coastal State to prevent certain forms of activities on the high sea that mark no assertation of dominion thereon, there is proportionally lessened the sense of need of a wider marginal sea, or of a zone of control adjacent to it".
The idea of the Panama Declaration has in spite of the protests been laid down in Art. 4 of the Inter-American Treaty of Reciprocal Assistance, signed at Petropolis, September 2, 1947 [178], but not in the case of neutrality but as defence zone for armed attack.

---

[a] For the text of the French and the German notes see Anglo-Norwegian Fisheries Case, British Reply [84], Annex 37, p. 23–25 (or Hackworth Digest of International Law, Vol. VII, p. 704–708).

We quote from Art. 3 the following passages:

Art. 3. "1. The High Contracting Parties agree that an armed attack by any State against an American State shall be considered as an attack against all the American States ....

3. The provisions of this Article shall be applied in case of any armed attack which takes place within the region described, in Article 4 or within the territory of an American State".

Art. 4 gives a description of the region to which this Treaty refers.

This Treaty reflects the modern strategical conception that a frontline is not the national frontier itself, but some line outside, which if passed by an aggressor, endangers the security of the State. This idea is expressed in General Mac Arthur's address before Congress on April 19, 1951 [179].

We quote the following passage, p. 2: "Prior thereto the Western strategic frontier of the United States lay on the littoral line of the Americas, with an exposed island salient extending out through Hawaii, Midway and Guam to the Philippines. That salient proved not an output of strength but an avenue of weakness along which the enemy could and did attack. The Pacific was a potential area of advance for any predatory force intent upon striking at the bordering land areas.

All this was changed by our Pacific victory. Our strategic frontier then shifted to embrace the entire Pacific Ocean, which became a vast moat to protect us as long as we held it. Indeed, it acts as a protective shield for all of the Americas and all free lands of the Pacific Ocean area. We control it to the shores of Asia by a chain of islands extending in an arc from the Aleutians to the Mariannas, held by us and our free allies.

From this island chain we can dominate with sea and air power every Asiatic port from Vladivostok to Singapore — with sea and air power, every port, as I said, from Vladivostok to Singapore — and prevent any hostile movement into the Pacific".

It means a tendency of extension of spheres of interest, not to be mixed up with the area under the sway of sovereignty. We mentioned it only in order to underline the distinction which should be kept in mind.

We now turn to the actual decrees by which territorial waters were extended recently.

Two States of the United States extended their territorial sea into the waters covering the continental shelf in the Gulf of Mexico, namely Louisiana by the Act No. 55, to declare the sovereignty of Louisiana along its sea-coast and to fix its present sea-coast boundary and ownership, 30 June, 1938. (Acts passed by the Legislature of the State of Louisiana, 1938, p. 169) [153], p. 114–116, and Texas by the Act declaring the sovereignty of Texas along its sea-coast, 16 May, 1941, as amended by Act of 23 May, 1947. "General and Special Laws of the State of Texas", 47th Legislature (1941), c. 286, p. 454; 50th Legislature (1947), c. 253, p. 451 [153], p. 41–43.

First then an extract of the Louisiana Act: "Whereas dominion, with its consequent use, ownership and jurisdiction, over its marginal waters by a State has found support because it is the duty of a State to protect its citizens whose livelihood depends on fishing, or taking from said marginal waters the natural products they are capable of yielding; also has found support in that sufficient security must exist for the lives and property of the citizens of the State;

Whereas, according to the ancient principles of international law it was generally recognized by the nations of the world that the boundary of each sovereign State along the seacoast was located three marine miles distant in the sea, from low-water mark along its coast on the open sea; ....

Whereas, the said three-mile limit was so recognized as the seaward boundary of each sovereign State, because at the time it became so fixed three marine miles was the distance of a cannon shot and was considered the distance at which a State could make its authority effective on the sea by the use of artillery located on the shore;

Whereas, since the said three-mile limit was so established as the seaward boundary of each sovereign State, modern cannon have been improved to such an extent that now many cannon shoot twenty-seven marine miles and more and by the use of artillery located on its shore a State can now make its authority effective at least twenty-seven marine miles out to sea from low-water mark;

Whereas, by the Act of Congress of February 20th, 1811, by which the State of Louisiana was admitted to the United States as a State, the southern boundary of Louisiana was fixed as follows: 'thence bounded by the said gulf to the place of beginning, including all islands within three leagues of the coast';

Whereas, therefore the gulfward boundary of Louisiana is

already located in the Gulf of Mexico three leagues distant
from the shore, a width of marginal area made greater, by the
above Act and agreement, than the well-accepted and inherent
three-mile limit;

Whereas, a State can define its limits on the sea; ....

Section 1. Be it enacted by the Legislature of Louisiana, that
the gulfward boundary of the State of Louisiana, is hereby
fixed and declared to be a line located in the Gulf of Mexico
parallel to the three-mile limit as determined according to said
ancient principles of international law, which gulfward boundary
is located twenty-four marine miles further out in the Gulf of
Mexico than the said three-mile limit.

Section 2. That, subject to the right of the Government of
the United States to regulate foreign and interstate commerce
under Section 8 of Article 1 of the Constitution of the United
States, and to the power of the Government of the United
States over cases of admiralty and maritime jurisdiction under
Section 2 of Article 3 of the Constitution of the United States,
the State of Louisiana has full sovereignty over all of the waters
of the Gulf of Mexico and of the arms of the Gulf of Mexico
within the boundaries of Louisiana, as herein fixed, and over
the beds and shores of the Gulf of Mexico and all arms of the
said Gulf within the boundaries of Louisiana, as herein fixed.

Section 3. That the State of Louisiana owns in full and complete
ownership the waters of the Gulf of Mexico and of the arms of
the said Gulf and the beds and shores of the Gulf of Mexico and
the arms of the Gulf of Mexico, including all lands that are
covered by the waters of the said Gulf and its arms either at
low tide or high tide, within the boundaries of Louisiana, as
herein fixed. " ....

"Note. In a suit by the United States against the State of
Louisiana, the United States Government alleged that the
United States was and is 'the owner in fee simple of, or possessed
of paramount rights in, and full dominion and power over, the
lands, minerals, and other things underlying the Gulf of Mexico,
lying seaward of the ordinary low-water mark on the coast of
Louisiana and outside of the inland waters, extending seaward
twenty-seven marine miles and bounded on the east and west,
respectively, by the eastern and western boundaries of the
State of Louisiana'.

The Supreme Court of the United States applied in this case
its decision in United States v. California (332 U.S. 19); it
pointed out that 'There is one difference, however, between
Louisiana's claim and California's. The latter claimed rights
in the three-mile belt. Louisiana claims rights twenty-four
miles seaward of the three-mile belt. We need note only briefly
this difference. We intimate no opinion on the power of a State

to extend, define, or establish its external territorial limits or on the consequences of any such extension vis à vis persons other than the United States or those acting on behalf of or pursuant to its authority. The matter of state boundaries has no bearing on the present problem. If, as we held in California's case, the three-mile belt is in the domain of the nation rather than that of the separate States, it follows *a fortiori* that the ocean beyond that limit also is. The ocean seaward of the marginal belt is perhaps even more directly related to the national defense, the conduct of foreign affairs, and world commerce than is the marginal sea. Certainly it is not less so. So far as the issues presented here are concerned, Louisiana's enlargement of her boundary emphasizes the strength of the claim of the United States to this part of the ocean and the resources of the soil under that area including oil'.

U.S. Supreme Court, 5 June, 1950, Official Reports of the Supreme Court, Vol. 339, p. 699, 701, 705".

The Texas Act is very similar. We therefore reproduce only the parts which are different from the Louisiana Act: "....

Whereas, the first Congress of the Republic of Texas passed an Act (1 Gammel's Laws, 1193) defining the boundaries of the Republic of Texas and declaring that its boundaries began at the mouth of the Sabine River and ran West along the Gulf of Mexico three (3) leagues from land to the mouth of the Rio Grande, then up to the principal stream of said river to its source; and the Congress of the United States (5 U.S. Statutes at Large, 797) proposed to the Republic of Texas that it be admitted into the Union, and that Texas should retain all vacant and unappropriated land lying within its limits; and the Congress of the Republic of Texas thereafter passed a Joint Resolution accepting the terms of annexation proposed by the United States (2 Gammel's Laws, 1200), and such action of the Congress of the Republic of Texas was ratified by popular vote of the people of Texas, and Texas was admitted to the Union by virtue of a Resolution of Congress passed December 29, 1845, under which the State of Texas retained all of its public lands (9 U.S. Statutes at Large, 108); and the first Legislature of the State of Texas declared: 'That the exclusive right to the jurisdiction over the soil included in the limits of the late Republic of Texas was acquired by the valor of the people thereof, and was by them vested in the Government of said Republic; that such exclusive right is now vested in and belongs to the State (Acts, First Legislature, 1846, p. 155); and under the Treaty of Guadalupe Hidalgo, the boundary line between the Republic of Mexico and the United States was defined as commencing in the Gulf of Mexico, three (3) leagues from land, opposite the mouth of the Rio Grande; it is clear that the Republic of Texas and the State of Texas have from the earliest days asserted title to the ownership of that portion of the Gulf of Mexico, and the soil at the bottom thereof, out to the limit of three (3) marine leagues from shore' ....

Whereas, the State of Texas owns the waters of the sea and the waters of the arms of the sea, and the sea-shore and the shores of all arms of the sea as far inland as the high-water mark within the territory of the State of Texas; ....

Be it enacted by the Legislature of the State of Texas:

Section 1. That the gulfward boundary of the State of Texas is hereby fixed and declared to be a line beginning in the Gulf of Mexico at the mouth of the Sabine River; thence on a grid bearing S. 35 degrees 55 minutes and 22 seconds E. to the farthermost edge of the continental shelf from the Gulf Shore line; thence in a westerly and southerly direction with the edge of the continental shelf to a point opposite the mouth of the Rio Grande River; thence to the mouth of the Rio Grande River ....;

Section 3. .... and that all of said lands are set apart and granted to the Permanent Public Free School Fund of the State, and shall be held for the benefit of the Public Free School Fund of this State according to the provisions of law governing the same.

Section 4. That this Act shall never be construed as containing a relinquishment by the State of Texas of any dominion sovereignty, territory, property or rights that the State of Texas already had before the passage of this Act."

"Note. The original text of Section 1 of the 1941 Act read as follows:

'Section 1. That the gulfward boundary of the State of Texas is hereby fixed and declared to be a line located in the Gulf of Mexico parallel to the three (3) mile limit, as determined according to said ancient principles of international law, which gulfward boundary is located twenty-four (24) marine miles further out in the Gulf of Mexico than the said three (3) mile limit'. ...

The Supreme Court has made only the following short comment on the Texas Acts of 1941 and 1947: 'Texas in 1941 sought to extend its boundary to a line in the Gulf of Mexico twenty-four marine miles beyond the three-mile limit and asserted ownership of the bed within that area. And in 1947 she put the extended boundary to the outer edge of the continental shelf. The irrelevancy of these acts to the issue before us has been adequately demonstrated in United States v. Louisiana'. Official Reports of the Supreme Court, Vol. 339, p. 707, 709, 720".

According to the Columbia Law Review [180], p. 317: "The primary purpose of this statute is to assert the state's right to any oil wells that may be drilled in the Gulf of Mexico beyond the present states boundary". As to the right of the State of Louisiana the article continues, p. 318–319: "To the federal government was delegated by the Constitution the powers of admiralty jurisdiction and commerce regulation but this delegation cannot be construed as a cession of the waters over which they are to be exercised. The ownership of the waters remains in the state and aside from the above mentioned powers the state also has jurisdiction over them . . . ."

However, the article then takes a turn in reasoning which, we believe, is incorrect, where it says: "When it became plain that the great fishery banks were being depleted by modern fishing methods, various conventions recognized property rights in free swimming fish [26]". Note 26 referred to reads: "North Seas Fisheries Convention of May 6, 1882. Cf. Manchester v. Massachusetts, 139 U.S. 240 (1891)". It must be said that neither in the Convention of 1882 nor in the case Manchester v. Massachusetts the word property is used. Article 2 of the 1882 Convention [153], p. 179, merely reserves the exclusive rights to fish to the fishermen of the coastal State, within the three-mile limit, whereas in Manchester v. Massachusetts [181] the Supreme Court decided the question whether a state of the United States had the right to enact and enforce conservation regulations within its own boundaries.

The following passages are of interest, p. 164: " . . . . if Massachusetts had continued to be an independant nation, her boundaries on the sea, as defined by her statutes, would unquestionably be acknowledged by all foreign nations, and her right to control the fisheries within those boundaries would be conceded",

and p. 165: "The Court also held that the Act (Act of the Legislature of Massachusetts, approved May 6, 1886 . . . . for the Protection of the Fisheries in Buzzard's Bay) was not repugnant to the clause of the Constitution which conferred upon Congress the power to regulate commerce . . . .",

and p. 166: "The extent of the territorial jurisdiction of Massachusetts over the sea adjacent to its coasts is that of an independant nation; and except so far as any right of control over this territory has been granted to the United States, this control remains with the State. Within what are generally recognized as the territorial limits of States by the law of nations, a State can define its boundaries on the sea and the boundaries of its countries ....",

and finally, p. 167: ".... the right of control exists in the State in the absence of the affirmative action of Congress taking such control, the fact that Congress has never assumed the control of such fisheries is persuasive evidence that the right to control them still remains in the State.

Judgment affirmed. Decided March 16, 1891".

We therefore believe that the argument is incorrect on this point and we would have preferred to say: a Convention, i.e. the Convention of 1882 recognized the exclusive right to fish in the territorial sea by nationals of the littoral State [a]. But this does not overthrow the argument, that the United States have always adhered to the three-mile limit ever since Thomas Jefferson, Secretary of State, wrote a note to the British Ambassador in 1793 informing him of the provisional acceptance of this limit by the United States. Therefore the article states rightly, on p. 322, that implicit in the present acceptance by the United States of the three-mile limit is the assurance that none of its states will attempt to occupy any further part of the sea. "The statute which Louisiana has passed is opposed to the international policy of the United States", and p. 323–324: "It is anomalous that in spite of the United States' adherence to the three-mile limit, the boundary of Louisiana was set at nine miles from the coast by the Congressional Act which admitted that state to the Union .... Any attempted enforcement of the present nine-mile boundary of Louisiana could with justice be contested by other nations", and p. 325: "since annexation of new territory is likely to have international significance, it might more wisely be held that such state legislation as the present Louisiana statute is void for all purposes".

It is curious to note that this cannon shot rule, dead since many years, has been exhumed by Louisiana and Texas for this purpose. (We feel to be in good company concerning our views on this rule by the statement of Schücking during the

---

[a] *Article II* [153], *p. 179.*

Codification Conference at the Hague in 1930 (Acts [61], p. 13), who said: "…. we are all aware, gunshot range cannot serve as a criterion to-day").

Apart from these two Decrees there have been issued recently some national laws, extending the territorial waters. Although not linked with the continental shelf, we mention the following Decrees because they also seem to fit in the recent avalanche of extension of sovereignty rights:

a. *Saudi Arabia*, No. 6/4/5/3711, May 28, 1949 [182] [a], and

b. *Egypt*, of 15 January, 1951 (Official Gazette, Vol. 78, No. 6 (18 January, 1951), Arabic edition, p. 4) (Translation by the Secretariat of the United Nations) [153], p. 307; both concerning the territorial waters, and extending these waters by adopting a rather unusual policy concerning baselines. By doing so the marginal belt may be as far as 18 miles, and sometimes even further from the shore. We mention these Decrees together because they are very similar.

The United States Government presented a note to the Saudi Arabian Government on December 19, 1949. We quote the following passage (Selak [65], p. 675): "…. The United States finds itself compelled to take exception to certain provisions thereof, deeming such provisions to be unsupported by accepted principles of international law, and to reserve all its rights and the rights of its nationals with respect thereto, namely:

1. All provisions to the effect that the inland waters of the Kingdom include waters outside of ports, harbors, bays, and other inclosed arms of the sea along its coast; and

2. All provisions to the effect that the coastal sea, i.e. the marginal sea, of the Kingdom extends seaward of a belt of three nautical miles along its coast or around its islands". A similar note was sent to the Government of Egypt on June 4, 1951 (Appendix 13 [81], p. 27–28).

The Government of the United Kingdom sent a note to the Government of Egypt on May 28, 1951 (Appendix 7 [81], p. 9–10).

We also mention in this connection the Act concerning the coastal waters of the Federal Peoples's Republic of Yugoslavia, of 1 December 1948 [153], p. 132–135, and the note from the Govern-

---

[a] See also U.N. Legislative Series [153], p. 89.

ment of the United Kingdom to the Yugoslavian Government of 5 May, 1949 (Appendix 5[81], p. 5 and further notes p. 5 and 6).

The picture we derive from all this is, that a certain rule has developed from vague principles, that a State has sovereignty rights in the adjacent sea to a distance of at least three miles from low-water mark. Some publications on the subject excel in uncritical copying, inaccuracy and incorrectness [a]. On the other hand they actually show what little convincing evidence there exists of a firm rule on the subject. Amongst writers exist vastly different opinions about the character of the rights of the coastal State, but here at least has evolved a rather dominant State practice that the territorial waters belong to the territory of the littoral State. The only place where this is to be found in written International Law is Article 2 of the Convention on International Civil Aviation of Chicago of 1944 [101]:

"For the purposes of this Convention the territory of a State shall be deemed to be the land areas and territorial waters adjacent thereto under the sovereignty, suzerainty, protection or mandate of such State". (See also Acts [61], p. 209).

We may also say, that it is a generally accepted rule of International Law that the sea-bed and subsoil under the territorial waters belong to the coastal State.

The sovereignty over the territorial waters is conditioned [b] by the right of innocent passage; and certain restrictions on criminal and civil jurisdiction concerning ships passing through the territorial sea (except after leaving inland waters) (Articles 8 and 9 of the Appendix 1, Acts [61], p. 215).

The point, however, which interests us most at the moment

---

[a] An example of incorrect writing is an article by Brittin [188], who quotes on p. 1541 a quotation by Walker [187]. Walker quoted (p. 210) Scott's Introduction of the translation of van Bijnkershoek's De Dominio Maris: "In the day of Bijnkershoek, a cannon carried approximately three miles; hence, the statement that a nation may occupy and exercise ownership over water three miles within low-water mark. This was the solution proposed by the young publicist; this was the solution accepted by the nations; this is the solution still obtaining, unless modified by express consent". Walker rightly criticized the quoted passage. But Brittin uses the quotation for quite another purpose. He writes: "That quotation was written in 1702 (sic.!); suffice it to say that it is the generally recognized rule to-day", thus distorting history and suggesting that Scott wrote in 1702 and Bijnkershoek lived, who knows how many years (centuries) before that date!

[b] *or rather: qualified.*

is, can we say that at least International Law limits the right of sovereignty of the coastal State to a margin between 3 miles and 12 miles (the latter being the maximum mentioned during the Codification Conference of 1930). In an attempt to answer this question one has to be extremely careful not to indulge in wishful thinking. Although we would defend the general interest rather than the particular interest of a State and are therefore inclined to limit the territorial waters and would be reluctant to recognize contiguous zones but for very limited purposes and a strictly limited distance we must confess that the law on this subject is uncertain. We rather agree with Scelle [184], who is himself a staunch adherent of the freedom of the seas, but who also admits, p. 425: "En réalité, il n'existe aucune règle coutumièrement établie, mais seulement des règles fixées par les Etats, soit de façon unilatérale, soit plus rarement, de façon conventionnelle, et dont ils imposent le respect dans la limite où ils en ont le pouvoir .... En somme, c'est l'anarchie".

We feel that this is, in view of the recent developments, an extremely unsatisfactory situation. On one side it is difficult to see why the situation has changed so much since 1930, that such *enormous* deviations from the 3–12 mile order should be necessary, on the other hand it is understandable that States look for means to defend their rights in a world where technical development has increased the danger of their interests being harmed in time of peace as well as in time of war. It also has to be realized that Governments are constantly under pressure of people demanding protection of their private interests, via their representatives in the Parliaments, people who are only thinking of their own business or at the most, that of their country.

It is rather disturbing to notice how many people are still motivated by thoughts of national profit or gain without any consideration as to other nation's needs and rights and without feelings of reasonable sharing or at least co-operation.

In the field of claims on stretches of water, either demanded for the State for fishery purposes or other commercial reasons, we find the following mentioned by Jessup [151], p. 129:

"On. May 5, 1938, the Senate passed S. 3744 which had been introduced by Senator Copeland; no action was taken upon the bill by the House prior to adjournment. The Copeland bill recites that 'the shallow depths of the Behring-Sea must be regarded as a slightly submerged margin of the American Continent' and that 'geologists have concluded that this part of the Behring-Sea does not partake of the qualities of a true ocean basin and that the so-called 'continental shelf' is no more or less than another of the several old Alaska beach deposits'. The bill further recites the need for protecting mineral deposits, fisheries and animal-life in this area, and then proceeds to provide:

that jurisdiction of the United States is hereby declared to extend to all the waters and submerged land adjacent to the coast of Alaska lying east of the international boundary in the Behring-Sea between the United States and the Union of Soviet Socialist Republics, as defined in the treaty between Russia and the United States, concluded at Washington on May 30, 1867, whereby Alaska was ceded to the United States, and lying within the limits of the continental shelf, the edge of such continental shelf having a depth of water of one hundred fathoms, more or less".

We mentioned already the bill introduced by Mr. Dimond in the Preliminary Conclusion of Chapter II.

We quote in this connection section 3 of that bill (loc. cit. p. 130): "Section 3 then follows the lead of the Anti-Smuggling Act of 1935 by authorizing the President to proclaim a 'salmon-fishery law-enforcement' area in any place in those adjacent waters where he finds that a vessel or vessels are hovering or being kept for the purpose of catching 'Alaska salmon which are en route to the lakes, rivers, or other inland waters of Alaska to spawn ....'. No such salmon-fishery law-enforcement area 'shall include any waters more than one hundred nautical miles from the place or immediate area where the President declared such vessel or vessels are hovering or being kept ....'. Nor shall they 'include any waters more than one hundred fathom deep, or any waters lying west of the .... boundary between the United States and the Union of Soviet Socialist Republics in the Behring-Sea' ".

It is curious to note that we find here an early predecessor of the Proclamation of 1945. It is still in a crude state and seems to include thoughts, which later appeared again in both Proclamations of 1945, showing that originally the "law-enforcement areas" (later called "conservation zones") were linked with the continental shelf. It is true that the bill never became law, but the strong protests of the United States to Chile and Peru, make us think of Goethe who said when he saw a prison-wagon pass: "Poor people what you have done, I had in mind".

We cannot strongly enough urge the necessity of an International Conference to be convened at an early date in order to attempt to check this most unfortunate development. We call it unfortunate because this development is bound to be a fertile ground for conflicts, which will certainly arise as soon as these claims will be enforced against the nationals of other States.

Let us hope that the work of the International Law Commission (the results on this item will be discussed presently) and the decision of the International Court of Justice in the Anglo-Norwegian Fisheries Case will provide the Governments with a sound foundation for the Conference referred to above.

Section 3. Navigation

Analysing the physical obstruction caused to navigation by an installation built in the sea, we arrive at the following picture.

An installation, let us say 60 by 100 metres in surface, is built on poles, at a distance of 50 miles off shore in waters of a depth of 30 metres. This installation will surely be marked on the charts, as all obstructions to shipping are being marked on the charts as soon as the information contained in the "Notice to mariners" comes in. This installation will also undoubtedly be illuminated, for the time being at least according to local regulations. As examples of such regulations we quote the second sentence of Article 6 of the Treaty between the United Kingdom and Venezuela of 1942 [110] [a]:

"In particular passage or navigation shall not be closed or be impeded by any works or installations which may be erected, which shall be of such a nature and shall be so constructed, placed, marked, buoyed and lighted, as not to constitute a danger or obstruction to shipping" [b];

---

[a] *See also* [153], *p. 45.*

[b] *This provision is also to be found in the Submarine Areas of the Gulf of Paria (Annexation) order in Council, 6 August 1942, Art. 5 (1)* [153], *p. 47.*

Article 25 of the Petroleum Act 1945 of the Bahamas [139] [a]:

"A licensee or lessee shall, if called upon so to do by the Board, illuminate between the hours of sunset and sunrise with respect to a submarine area in a manner satisfactory to the Pilotage Board for New Providence .... all derricks, piers, survey marks or any other installations erected in any submarine area .... included in a licence or lease";

Article 15 (*b*) of the Oil Mining Regulations of 2 September, 1949, of British Honduras [153], p. 32, is very similar to the quoted Article 25 of the Regulations of the Bahamas;

and Article 20 of the Submarine (Oil Mining) Regulations of 22 May, 1945, of Trinidad and Tobago [153], p. 38, as revised by Government Notice No. 99 issued 1 June, 1945:

"The licensee shall not carry on any operation authorized by the licence in such a manner as to effect the closing of the marine areas specified in Section 2 of the Submarine Areas of the Gulf of Paria (Annexation) Order, 1942, or any of them, and any works or installations erected by the licensee shall be of such nature and shall be so constructed, placed, marked and buoyed as not to constitute a danger or obstruction to shipping, and the licensee if required by the Governor to do so shall illuminate between the hours of sunset and sunrise, in a manner satisfactory to the Harbour Master, all derricks, piers, survey marks or any other installations erected in the area included in a licence. The means of illumination shall be such as is approved or required by the Harbour Master";

and Rule Q of the Rules and Regulations governing drilling etc. of Texas [141]:

"In order to provide ample safeguards against collisions proper signal lights or buoys shall be installed at, or in immediate vicinity of well and structures erected in the water. Signal lights shall be kept burning except during daylight hours" [b].

---

[a] *See also* [153], *p. 31.*

[b] Even the small exploration devices, constituting a possible obstacle for shipping, have to be marked, according to Article III of the Rules and Regulations governing, geophysical .... etc. explorations .... of Texas [140]: "All pipes, buoys and all other markers used in connection with the seismic work shall be distinctly marked giving the name of the party using same and if not constituted of easily destructible materials so they will cause no substantial damage to any boat which may collide therewith shall be properly flagged in the daytime and properly lighted at night. All buoys and other markers shall be anchored by sash weights or similar weights which cannot cause injury to the nets of commercial fishermen, except those buoys placed in the water in accordance with rules and regulations of the Department of the Army or the United States Coast Guard".

We believe that it would be preferable to establish eventually international rules concerning marking of such installations.

However, the question now to be answered is: is such an installation, marked on the charts, illuminated, and we may add supplied with a proper fog warning device, an obstacle to shipping? Our answer is theoretically yes, practically no.

In pure theory and "ad absurdum" we could say that pure freedom of navigation would only exist if but one ship sailed the oceans. As soon as a second one appears, the first one *might* be hindered in its movements. But now more realistic: a ship may anchor in or outside the territorial waters. As such it constitutes an obstacle. Hence the rule laying down that it shall warn shipping of its presence by bell signals in fog and by showing anchor-lights at night. What is the difference between a ship at anchor or an installation as described? The latter does not sway round its anchor on the tide or changing wind, like the ship may do, but is fixed on its poles. Moreover the installation is marked on the chart, the ship is not! If anchoring is free and within the common use of the high seas, which it is, why should it not be permitted to erect an installation? Of course they must be equipped with warning devices which have to be operated. Both have in common that they can in case of fog be detected by radar.

Much depends of course on the place where the installation is erected. This factor plays a more serious role if a number of these installations are going to be built in the same area. For a "cluster" of these installations the answer to the question, whether they constitute a hindrance for shipping largely depends on the location, and of course the distances between them. If located somewhere in a bay or gulf, not frequented by shipping we can safely say that no practical obstacle exists. We believe that this is at the moment applicable to the situation in the Gulf of Mexico. It is, however, quite clear that if oil companies were left free in putting their constructions wherever they wanted, irrespective of other interests, a serious obstruction of shipping may arise, for instance in narrow passages, straits frequented by shipping or approaches to harbours, in short in shipping routes. For the sake of the argument the picture may become clear at once, if we imagine that oil geologists discover the most fabulous oil reserves under the Strait of Dover. Of course, building installations just there would be an intolerable obstruction of a highly frequented shipping passage.

Where approaches to harbours are concerned, we can leave it safely to the coastal State to limit concessions which would entail obstruction of traffic to and from the harbours of that State.

Where frequented shipping lanes are concerned, not of direct interest to the coastal State adjacent to the shelf, interest of oil-companies and shipping may clash. This is one of the reasons which we adduce for the idea of creating an international body before which both interests could be submitted and weighed, and which should be given the power to decide, which of the two interests in a given case should prevail.

But a crazy example as given above will not easily occur and we believe that we can say, that in view of the remarks just made, the anxiety for obstruction is exaggerated.

Now as to the right of a State to build installations in the sea, we have to distinguish between the territorial seas and the high seas. A State having sovereignty over its territorial waters, can we object to this State allowing installations to be built in these waters, i.e. in his own "territory"? On the face of it perhaps not. The sovereignty of the coastal State is, however, limited by the right of innocent passage which all ships have in these waters. Would not this right be forestalled if the coastal State allowed the territorial waters to be crowded with constructions?

This is not so fantastic as it may sound.

The territorial sea being in most cases only 3 miles wide, may leave, because of natural obstacles, i.e. shallow water banks, rocks, or coral reefs, only a relatively narrow passage, which could easily be blocked by a few installations. The question arises, what prevails, the right of the State to build on its "territory" or the right of inoffensive passage through that "territory". Personally we should think that the greater right, namely the right of the greater community, i.e. all the sea-faring nations of the world, would prevail above the smaller right, i.e. the right of the smaller community, the coastal State. We have to go into this question somewhat deeper. That all merchant ships have the right of innocent passage through territorial waters is a

generally accepted rule of International Law. In the Bases of
Discussion [126], p. 65–71, the answers of the Governments on
point IX were unanimously in the affirmative. State practice
has proved this rule to be a generally accepted one. Moreover
in Article 2 of the Statute on Freedom of Transit of the Convention of Barcelona of 20 April, 1921 [185], p. 27, relating to transit
of persons, goods and vessels, it was agreed that "Contracting
States will allow transit in accordance with the customary
conditions and reserves across their territorial waters". This
freedom and the freedom of navigation on the high seas are
both parts of the general and well recognized principle, referred
to in 1949 by the International Court of Justice in the Corfu
case [186], p. 22, as the principle of the freedom of maritime
communication.

The coastal State may, however, enact and enforce certain rules
with which ships should comply during their passage. Article 6
and the observation thereon of the Appendix 1 of the Report
adopted by the Committee on April 10th, 1930: The Legal
Status of the Territorial Sea (Acts, 1930 [61]), p. 214, reads:

"Foreign vessels exercising the right of passage shall comply
with the laws and regulations enacted in conformity with international usage by the coastal State, and, in particular, as regards:
(a)   The safety of traffic and the protection of channels and
      buoys;
(b)   The protection of the waters of the coastal State against
      pollution of any kind caused by vessels;
(c)   The protection of the products of the territorial sea;
(d)   The rights of fishing, shooting and analogous rights belonging
      to the coastal State.
The coastal State may not, however, apply these rules or
regulations in such a manner as to discriminate between foreign
vessels of different nationalities, or, save in matters relating
to fishing and shooting, between national vessels and foreign
vessels".

"*Observations.*

International law has long recognized the right of the coastal
State to enact, in the general interest of navigation, special
regulations applicable to vessels exercising the right of passage
through the territorial sea. The principal powers which international law has hitherto recognized as belonging to the coastal
State for this purpose are defined in this article.

It has not been considered desirable to include any special provision extending the right of innocent passage to persons and merchandise on board vessels. It need hardly be said that there is no intention to limit the right of passage to the vessels alone, and that persons and property on board are also included ....".

Moreover the coastal State has a right to judge whether the passage is innocent.

Article 5 (and the comment) of the Appendix 1 mentioned above reads:

"The right of passage does not prevent a coastal State from taking all necessary steps to protect itself in the territorial sea against any act prejudicial to the security, public policy or fiscal interests of the State, and, in the case of vessels proceeding to inland waters, against any breach of the conditions to which the admission of those vessels to those waters is subject".

"Observations.

The article gives the coastal State the right to verify, if necessary, the innocent character of the passage of a vessel and to take the steps necessary to protect itself against any act prejudicial to its security, public policy, or fiscal interests. At the same time, in order to avoid unnecessary hindrances to navigation the coastal State is bound to act with great discretion in exercising this right. Its powers are wide if a vessel's intention to touch at a port is known, and include, inter alia, the right to satisfy itself that the conditions of admission to the port are complied with".

The steps the coastal State may take to protect itself are varied. Sojourn, not necessary for nautical reasons (anchorage during fog) or in case of distress, is usually not allowed. Von Liszt [187], p. 144, mentions as acts objected to, for instance, manoeuvres or hydrographical survey. On the other hand as the French reply states, Bases of Discussion [126], p. 67: "In time of peace also, free access may be abolished in order to allow military or naval manoeuvres to be carried on, or for the security of the shore". In the English reply it was said in the last paragraph (loc. cit. p. 68): "...., and the vessels cannot claim to transport through territorial waters persons or goods whose presence there is prejudicial to the safety, good order or revenues of the State".

Smith [188], p. 32–33, writes: "The international right of passage in no way diminishes the inherent right of every state to take such measures in its own territory, whether land or water, as it may judge to be necessary for the protection of its own interests and a voyage ceases to be 'innocent' if its purpose involves any violation of thóse interests". A recent practical example of closing parts of the territorial sea is the following:

During the conflict between the Netherlands and the Indonesian Republic, a conflict which could not be called a war, but a revolt, or insurrection, some parts of the territorial waters were closed in order to prevent the import of weapons and landing of rebellious forces or propagandists, and to prevent export of stocks belonging to private firms, which had fallen into the hands of the insurgents.

We should distinguish theoretically between mere passage through territorial waters and passage to and from a port. However, it is clear that in practice this distinction is extremely difficult to maintain, especially if one has to cope, as in this case, with small native sailing craft.

It was not a question of "policing of territorial waters to intercept access to insurgents" or "self blockade", as Oppenheim [189], p. 629, calls it. It was not a closing of harbours held by the insurgents either. It was a proper closing of certain stretches of the territorial waters, because the passage was presumed not to be innocent, or based on the principle of self-defence.

It was based on the same principle as laid down in the Italian Law, No. 612, of 16 June, 1912 [153], p. 81. Article 1 reads:

"The passage and sojourn of national or foreign merchantmen may be prohibited, at any time whatever and in any determined place whatever, within or without the seas of the State, when it is recognized as necessary in the interest of the national defence".

(See also Report to the Council of the League of Nations [52], p. 43) [a].

---

[a] The relevant executive orders in the year 1947 were based on Art. 8 of the "Territoriale Zee- en Maritieme Kringen Ordonnantie 1939" (Netherlands East-Indies Statute Book 1939 No. 442) and the "Ordinance No. 171/DvO/VII A. 3 van het Militair Gezag, of 21st February, 1942" (Javasche Courant 1942, No. 17a).

Nothing is said, however, about our question: is the coastal State entitled to build constructions in its territorial waters which would entail obstruction to innocent passage.

The innocent passage should, according to de Bustamante [190], not be understood as reducing the rights of the littoral State (p. 205): "on ne peut limiter le sens de cette expression (passage inoffensif) à l'absence d'intention de causer du dommage, ni entendre qu'elle réduit d'une manière quelconque les droits et les pouvoirs de l'Etat riverain .... Il serait absurde qu'ils puissent .... passer entre les lieux qui ont besoin d'isolement pour être consacrés à la propagation ou la conservation des éponges ou des perles, ou lancer des substances qui causent la destruction ou l'éloignement des poissons ....".

With a "clair voyant" eye he has foreseen our problem when he says: "Il n'est pas non plus possible qu'ils s'opposent en vertu de ce droit de transit, à la construction par l'Etat riverain de stations pour l'amérissement des aéroplanes ....". They could not object, he says, to the building of a railway from the continent over a number of islands to the furthest island, as happened in Florida (to Key West), closing the passage completely and on p. 206: ".... ce droit international de transit.... ne s'oppose pas et ne doit pas s'opposer à l'exercice passé, actuel ou futur, d'aucun des droits de l'Etat riverain .... et il faut laisser la voie ouverte à toutes les nouvelles inventions et à tous les incessants progrès de notre merveilleuse civilisation".

During the Codification Conference in the Hague in 1930, the American delegate Miller proposed (Acts 1930 [61], p. 58) the insertion at the beginning of the Basis No. 19 reading: "A coastal State is bound to allow foreign merchant ships a right of innocent passage through its territorial waters" the words: "subject to the rights of the coastal State to the use of the territorial waters or the subsoil for its national purposes" and he explained this proposal by saying: "Our desire is to emphasize what we consider in this connection to be the superior right of the coastal State. Reference was made here the other day by Dr. Schücking to the question of mines in the subsoil. He spoke, if I remember correctly, of petrol. It might well be necessary to erect a structure which would require ships passing through territorial waters to make a detour round that installation of a coastal State. I believe it is also quite a common practice for the coastal State to close a portion of the territorial waters for a particular time. While, therefore, there is no desire to limit the ordinary right of innocent passage for merchant ships, an expression should be included in the Basis which would indicate that, in principle, the right of the coastal State, when it is necessary for national purposes to use a portion of the territorial sea either on the surface or the subsoil, is superior to the right of innocent passage".

Of the opposite opinion is van der Lee [173], p. 11: "An argument to be invoked against the sovereignty of the coastal State over the bed and the subsoil of the territorial sea could be that the right of innocent passage might be endangered by acts of sovereignty of the coastal State in the bed and the subsoil of the territorial sea. In such a case, e.g. the construction of installations, an established rule of international law would be violated if innocent passage became impossible. The coastal State has a perfect right to erect, e.g. oil derricks within the territorial zone, provided sufficient measures are taken to safeguard innocent passage of foreign vessels".

Fauchille [70], p. 18, writes: "Les seuls travaux et les seules installations qu'un Etat peut être admis à réaliser au fond de la mer sont ceux qui ne doivent pas nuire à l'usage de la mer par les autres Etats, c'est à dire gêner leur libre navigation à la surface ou troubler d'une manière quelconque leurs droits communs". Article 4 of the Appendix of the Report adopted by the Committee in 1930, Acts 1930 [61], p. 212, reads: "A coastal State may put no obstacles in the way of the innocent passage of foreign vessels in the territorial sea ....".

We believe that it is a general principle of law, that in a community the individual cannot exercise rights which would endanger or constitute a hindrance to the community to which he belongs. Private property is not an absolute right, and cannot be an absolute right in a community. The principle of rather unlimited freedom and powers of the proprietor as codified in the Civil Codes of France, Belgium and the Netherlands, says Belinfante [191] [a], and the freedom of making contracts

---

[a] He refers to Art. 625 of the Netherlands Civil Code and Art. 537 of the French "Code Civil", the first sentence reading: "Les particuliers ont la libre disposition des biens qui leur appartiennent sous les modifications établies par les lois".

are no longer the basis on which the legislator has built his subsequent regulations. The individual can only exercise his rights in so far as in doing so they do not interfere with public regulations, guaranteeing public peace and order. The domain of private law decreases, whereas the domain of public law extends.

More and more public interests are recognized to prevail above private ones. If a landowner's real property is in the way of a projected highway, a part of his property will be expropriated for the benefit of the public. The State has "eminent domain" i.e. it retains a right over the estates of individuals to resume them for public use.

In spite of the reaction after the war, which shows an increase of nationalism, a renewed tendency to fence with the dangerous notion "sovereignty", we believe that the general tendency of International Law being the mirror of the public legal and moral conscience is an increasing feeling of dependency on each other, of solidarity. A State cannot act as freely nowadays as it used to do. The objection will be made, that this can only be done by a voluntary sacrifice of sovereign rights. We think, however, that the word "voluntary" is a cloak to save face. The transition of relations in the world is psychologically speaking difficult to digest for those who have known "better" times. We admit that a certain amount of wishful thinking is partly responsible for these utterances.

We even admit, that if such a tendency is living, it is certainly hampered in its realization by the conditions we find in the world of to-day. This, however, is no proof of the non-existence of the tendency. We should look further than just this period in which human relations are badly "indisposed", let us hope for not too long a period. The big lines are drawn by the undeniable intensifying of human intercourse all over the world through technical developments. This will lead to developments which are parallel to the ones we have witnessed in the relation of the individual to the State where he lives. The increasing power of public interest in the State will be followed by the increasing power of public interest in the world. The common interest of humanity will prevail above the individual interest of the States.

One of the first and oldest public rights is that of freedom of traffic and it is not surprising that this freedom has settled itself so early in International Law. This freedom should certainly be recognized as belonging to a higher order than the individual interests of the State even in its "own" territorial waters.

We therefore conclude by saying, that a State is free in building constructions in its territorial waters, but that the free passage should not be barred completely. No objection can reasonably be made if a ship is forced to diverge from her course, or, more conforming the practice, has to follow a course steering well free of the obstacle, and probably somewhat longer than the one she would have chosen if no obstacle had been in her way. The only thing we can demand is that proper passage is ensured.

We should think that if this principle can be accepted it applies a fortiori to the high seas. The freedom of the seas, we have seen, means that everybody is free to do in that domain as he thinks fit, always on the condition that other people who are as free as he is will not be unreasonably deprived of their freedom. Tolerance in a common domain is one of the first qualities necessary.

We cannot see why such an upheaval is made by the sponsors of the principle of the freedom of the seas about a use of that freedom which, admittedly, happens to be a new kind of use but which is nevertheless a use of that freedom. It is rather remarkable that the advocates "par excellence" of the principle of the freedom of the high seas, object vividly to the erection of platforms, perfectly marked on the charts and well equipped with visual, auditive and may be electronic signal devices, whereas they remain silent, when hundreds of thousands of mines are layed in war time in the high seas [a], the areas in which danger exists being notified, it is true, to neutral and after the war to all States, but forming a dangerous obstacle to shipping for

---

[a] An attempt to limit mine laying to the territorial waters was made by the Institute of International Law in 1910 and 1911 (Scott [92], p. 166 et seq.) and in 1913, at Oxford a Manual concerning the Laws of Naval War governing the relations between belligerents, was adopted (Scott, p. 174 et seq.). Aricle 20 reads: "It is forbidden to lay automatic contact mines, anchored or not, in the open sea". No diplomatic or "Peace" Conference being held on the subject, the rules of the VIIIth Hague Convention of 1907, are still in force. Both rules are out of date since the acoustic, magnetic and pressure mines were introduced.

years after the cessation of hostilities. Only narrow channels have been swept so far, for instance along the Dutch coast only one mile wide, which is in view of prevailing weather conditions, flatness of the sea-bed so that soundings do not give much help, and tidal currents, a dangerously narrow lane. Moreover the mines are often swept off their anchors, the devices to make them harmless when floating are not always working and statistics prove a heavy toll of ships lost by mine explosions.

We maintain that building constructions in the high seas is using the freedom of the seas just as much as navigating on these seas, or fishing in these seas or laying telegraph cables on the bottom of these seas. If Grotius had known about telegraph cables and oil derricks, he would have included them in the kind of use one can make of the high seas. It would be ridiculous if a free man in the street who can walk wherever he wants, would object to a tramline being made across that street, which will force him in future to stop when a tramcar passes.

The only thing we can reasonably ask is to avoid building these installations just there where congestion of shipping exists. In order to achieve a reasonable application of different ways of using the high seas, it would be recommendable to create an international body, judging plans submitted to it and giving binding decisions.

The other obstacle to shipping, not a physical one, may be the fact that either the installations will be surrounded by a "security zone" which will be forbidden for shipping, or one in which at least some prohibitive regulations will be enforced, concerning the use of open fire for instance.

It may well be that if more installations are built in a certain area the "security zones" will overlap or nearly touch each other (much will depend of course on the width of these zones) with the result that practically an area of considerable dimensions will be closed for shipping or at least that the ships themselves will rather avoid "clusters" of installations for their own safety and will thus be forced to prolong their course.

On top of this, it is more than likely that in wartime whole areas where installations are built will be closed for shipping, either in the case of the coastal State remaining neutral and not wanting any hostilities to be committed in the neighbourhood of these installations with possible catastrophic results, or in the case of belligerents, from the view point of defence of the assets of their war economy.

All these points should be taken into consideration when the subject will be discussed at an International Conference, which we hope will be convened in order to regulate the continental shelf-exploitation. We envisage that such a Conference would lay down in a Convention rules with the object to keep shipping lanes free, for instance on the lines of the second sentence of Article 6 of the Treaty between the United Kingdom and Venezuela of 1942 [110] reading:

"In particular passage or navigation shall not be closed or be impeded by any works or installations which may be erected, which shall be of such a nature and shall be so constructed, placed, marked, buoyed and lighted, as not to constitute a danger or obstruction to shipping".

It may also delegate the power to make such rules to an international body. The international community would then act as was done in the national sphere when weirs (the "sero's" we discussed in the section on Sedentary Fisheries in Chapter II) were forbidden to be in the way of shipping in Article 36 of the Magna Charta in 1215. Rowland [192], p. 55, gives the following translation of Article 36: "All weirs for the time to come shall be demolished in the rivers Thames and Medway, and throughout all England except upon the sea-coast [a]".

SECTION 4. ARTIFICIAL ISLANDS

When investigating the question what legal status a drilling platform should have, a comparison with so-called artificial islands is to be expected. As a matter of fact, the subject was

---

[a] The numbering of the articles differs in the various publications. For instance in Taswell Langmead [193], p. 93, it is Article 33. He gives also another name for weirs i.e. "kydelli", and says: "the object was .... to remove from rivers all obstacles likely to interfere with navigation". Stubbs [194] gives on p. 292 in "John" A.D. 1215, Articles of the Barons, Article 23: "Ut omnes kidelli de cetero penitus deponantur de Tamisia et Medewaye et per totam Angliam", and on p. 300: "John" A.D. 1215, Great Charter of Liberties, Magna Charta, Article 33 as above with the addition: "nisi per costeram maris".

broached by Mr. Frank. R. Newton in a letter of August 20, 1918, to the State Department, according to Hackworth [195], p. 679–680.

In the answer of the State Department printed in Hackworth, it is said that the letter has been received, "concerning the alleged discovery of a large oil pool in the Gulf of Mexico about 40 miles from land on a reef where the water is less than 100 feet deep. .... You suggest that an artificial island might be erected and inquire whether such island could be brought under the jurisdiction of the United States and whether it could protect your rights by giving you a lease to the property .... In reply the Department informs you that the United States has no jurisdiction over the ocean-bottom of the Gulf of Mexico beyond the territorial waters adjacent to the coast. Therefore, it does not appear possible for the United States to grant to you the leasehold or other property rights in the ocean-bottom which you desire. The Department further informs you that, unless the erection of an artificial island interfered with rights of the United States or of its citizens, or formed the subject of a complaint made upon apparently good grounds, by a foreign government, it is not likely that this Government would object to the erection by American citizens of such an island as you suggest.... It may also be observed, although the Department can give no assurances on the subject, that it would seem possible that, if an island were constructed 40 miles from the coast of the United States by the efforts of American citizens and inhabited and controlled by them in the name of the United States, this Government would assume some sort of control over the island".

The last paragraph does resemble the provisions concerning Guano Islands in the United States Code, Title 48, Chapter 8, Section 1411 [196], p. 5431–32:

"Whenever any citizen of the United States discovers a deposit of guano on any island, rock or key, not within the lawful jurisdiction of any other Government, and not occupied by the citizens of any other Government, and takes peaceable possession thereof, and occupies the same, such island, rock or key may, at the discretion of the President, be considered as appertaining to the United States (R.S. par. 5570). (Derivation Act, August 18, 1856, ch. 164, par. 1, 11 Stat. 119)",

and par. 1417: "All acts done, and offenses or crimes committed, on any island, rock or key mentioned in section 1411
of this title, by persons who may land thereon or in the waters
adjacent thereto, shall be deemed committed on the high seas,
on board a merchant ship or vessel belonging to the United
States; and shall be punished according to the laws of the United
States relating to such ships or vessels and offenses on the high
seas, which laws for the purpose aforesaid are extended over
such islands, rocks and keys (R.S. par. 5576) (ch. 164, par. 6,
11 Stat. 120)",
and finally par. 1418 "Employment of land and naval forces
in protection of rights.
The President is authorized at his discretion to employ the
land and naval forces of the United States to protect the rights
of the discoverer or of his widow, heir, executor, administrator
or assigns. (R.S. par. 5577). (Derivation Act, August 18, 1856,
ch. 164, par. 5, 11 Stat. 120)".

After the Proclamation of President Truman with respect
to the natural resources of the continental shelf, proclaiming
that the United States regards the natural resources of the
continental shelf contiguous to its coasts as appertaining to
the United States and subject to its jurisdiction and control,
it is hardly conceivable, that the United States would consider
an installation built by permission of the United States and
with the object to extract the resources of the continental
shelf appertaining to the United States, as anything else but
likewise appertaining to the United States. Similar provisions
as those relating to the guano islands will probably be made
concerning criminal jurisdiction and employment of (land or)
naval forces for protection of rights. This supposition can be
made with even stronger reason, for those countries which claim
the sovereignty over the continental shelf.

The problem is therefore more a theoretical one, and perhaps
only important as far as it may enlighten us about the status
of the waters around the installation. What we mean is of course,
whether these waters are high seas or whether around the
installation a marginal belt is to be recognized.

A few words may be said about the theoretical problem.
These installations can of course only be compared with constructions on rocks permanently covered by the sea. Jessup [116]
refers to a case where such an installation was involved, p. 68:

"The British ship Frances E was seized by the coastguard cutter
Sankee 16 miles off the Florida coast with a cargo of liquor,
seven hundred miles off her declared route. The point where the
arrest occurred was 12 miles from Sea Horse Reef, a beacon
built on a shallow reef, the beacon projected above the water
but the reef was wholly submerged. In arguing for condemnation
of the vessel under the treaty with Great Britain, the district
attorney wished to consider this reef as 'the coast of the United
States, its territories or possessions'. This contention was rejected
by the court, which held that the treaty in using those terms
'had no reference to maritime structures erected in the water
and having no coast' (the decision was reversed on other grounds
by the Circuit Court of Appeals for the Fifth Circuit on May 12,
1926, 13F (2d) 74. But there was no disagreement with the point
here discussed)". The reason for this decision seems to be that
a beacon is not inhabitable, which is often taken as a criterion
whether to consider a thing projecting permanently above the
surface of the sea as an island (artificial island). For instance
Gidel [169], p. 684, gives the following definition: "Une île est une
élévation naturelle du sol maritime qui, entourée par l'eau,
se trouve d'une manière permanente au-dessus de la marée haute
et dont les conditions naturelles permettent la résidence stable
de groupes humains organisés. Sont assimilées aux îles naturelles
les îles artificielles satisfaisant aux mêmes conditions et dont
la formation par l'action de phénomènes naturels a été provoquée
ou accélérée au moyen de travaux". It would appear that an
oil-drilling platform built on poles does not come under this
definition.

The literature about artificial islands very often uses as its
main example the lighthouse. Hence the monotonous repetitions
of the casual remarks about the "lighthouse built upon a rock
or upon piles driven into the bed of the sea" etc. by Sir Charles
Russell being an answer on a question put by Mr. Phelps during
the Behring-Sea Arbitration (Moore [197], p. 900), words which
he probably never intended to be more than just a part of his
argument, but not an axiom which should be a mainstay for
future students of International Law and chewed and rechewed
in numerous books and articles. At any rate we believe that
oil-drilling installations cannot be compared with lighthouses,

because they are built for quite a different purpose, and it cannot be said (for the time being) that oil-installations are built for the general interest of mankind, in spite of some assurances that they are (a point which we will discuss later). Of course an oil-installation is inhabitable and would therefore conform to the criterion. But we cannot without further reasoning take over what has been said or better, what has been the practice about lighthouses.

For a lighthouse it is generally recognized that the State building it, gets a territorial right. It does not follow that this rule would apply to oil-installations, unless the practice of States would show general recognition to this effect, or better, such rights would be agreed upon in an universal Convention regulating the whole problem of oil-exploitation of the continental shelf.

The other important question of course is, should an artificial island, or an oil-installation be surrounded by a marginal belt? Again the examples in the doctrine are mostly referring to lighthouses. Oppenheim [170], p. 410, did not think that lighthouses should have a marginal belt: ".... il vaudrait mieux traiter les phares sur le même pied que les bateaux-phares amarrés .... Pour cette raison je proposerais d'ajouter à l'article 1er de l'avant projet .... Il n'y a pas de droit de souveraineté sur une zone de la mer qui baigne les phares". This opinion is also expressed in Oppenheim's book [42], p. 454, i.e. that there is no difference with a lightship. Indeed, both are marked on the charts, but a marginal belt round a ship is even more unacceptable than round a lighthouse [a].

Lindley [128] based the same conclusion on the following argument, p. 67: ".... if a rock or barren island is not occupied for its own sake, but merely to facilitate fishing and navigation in

---

[a] In this connection it may be remembered what Fulton [81], p. 262, wrote about the proposals of Pennington to the King and to the Admiralty: ".... any foreign ship attacked by another foreigner in the narrow seas might put herself under the protection of any of the King's ships by coming under its lee 'in the same manner as under a castle on shore' .... 'whether his Majesty's ffloatinge ffortes shall not have ye same privilege in succoringe and defendinge them as ffortes a Land hath'. Ships of war have long been regarded as being essentially an extension of the territory of the State to which they belong; but no writer ever suggested that the water around them on the high sea should be looked upon as partaking of the same character. The sea round a King's ship, within the range of the guns on board, was to be sanctuary like the waters of the King's Chambers, a sort of territorial girdle which it carried about with it like an aureole round the head of a saint".

the surrounding ocean, it does not appear that this would be a sufficient justification for extending the sovereignty of the occupying State over those waters". Westlake [121], p. 190, brings in the criterion whether "a mere rock and building" can be "so armed as really to control the neighbouring sea", and because that is not the case with lighthouses, he denies them a marginal belt. Of course the control of a sea-area is not dependent on the presence of a few guns on a stable platform, but rather on the fact whether the navy or airforce of the coastal State has supremacy in the sea-area concerned or in the air.

The oil-installation is in a different position. It offers more place for armanent, but apart from that it is a more plausible object of attack being an asset in the war economy of the coastal country.

Speaking about artificial islands we do not include contraptions like those meant in the Report to the Council of the League of Nations 1927 [52], p. 41, i.e. "islands" which are, as the Report says, "artificially created by anchorage to the bed of the sea, and which have no solid connection with the bed of the sea, but which are employed for the establishment of a firm foundation, e.g. for enterprises designed to facilitate aerial navigation ....". We believe that such contraptions should not be called artificial islands, because they float and in fog they should sound the bell like a ship. Even if they are anchored by two anchors so that they cannot sway on the tide or current or wind, they have all the characteristics of a ship and nothing in common with an island. In the Report the same opinion is expressed (p. 41): "Such fictitious islands must be assimilated to vessels voyaging on the high seas".

We should not have spent so many words on these devices, which for the purpose described, became outdated before they were born because planes have learned to make the "jump" without refueling. But we will meet in Chapter V on offshore drilling, devices which combine both qualities, a sort of amphibious contraption: pontoons, which are towed to the spot and then filled with water, sink and settle on the bottom, to be lifted again after the well is made successfully or not successfully.

It seems to deserve a strong support to pigeonhole these "pontoons" in the category "ships" with the floating aerodromes because of their relatively short "sedentary habitat" and to deny them like the floating aerodromes a marginal belt.

In the answers of several countries about the definition of an island (Point VI) to the Preparatory Committee, reference was made to artificial islands. Bases of Discussion [126], p. 52–53:

Germany: ".... It should therefore be laid down that artificial islands (artificial constructions) should be assimilated to natural islands, provided that they rest on the sea-bottom and have human inhabitants".

Denmark: .... "Where the territorial waters of two States are in contact, neither of them would be entitled to modify the existing delimitation to the prejudice of the other, by the construction of artificial islands, lighthouses", etc.; p. 53:

Great Britain: "An island is a piece of territory surrounded by water and in normal circumstances permanently above high water. It does not include a piece of territory not capable of effective occupation and use. His Majesty's Government considers that there is no ground for claiming that a belt of territorial waters exists round rocks and banks not constituting islands as defined above, and would view with favour an international agreement to this effect in order that there may be no doubt as to the status of the waters round such rocks and banks and round artificial structures raised upon them".

Netherlands: "For the purpose of the decisions on Points IV and V, an island should be understood to be any natural or artificial elevation of the sea-bottom above the surface of the sea at low tide".

These quotations may strengthen the feeling, that in those years the problem was not considered to be very important.

Gidel [112] (Vol. I, p. 503) makes the establishment of such structures rightly dependent on the acceptance by other States: "C'est donc seulement dans la mesure où les installations dont il s'agit seront considérées par les autres Etats comme compatibles avec le principe de la liberté de la haute mer qu'elles pourront être créées. Mais la question se pose de savoir jusqu'où peut aller le droit d'opposition des Etats tiers et quels effets il convient de lui reconnaître".

He then gives as an example the construction of works in the Dover Strait in connection with the tunnel project and says that there seems to be no doubt that other States could veto such constructions if liable to obstruct shipping. He refers in particular to Chapter III of an English Report on the subject of 28 February, 1930, in which other means of communication, apart from a tunnel in the subsoil, were discussed, such as a bridge or tubes resting on the sea-bed with two ventilation towers or tubes resting on poles driven into the sea-bed. In both cases the Report ends in saying that it is hardly conceivable to arrive at an international agreement, allowing to establish these obstructions to navigation, an agreement which would be indispensable, (p. 504–505).

Gidel [169], p. 682 et seq., makes a distinction between two kinds of artificial islands. One is an artificial "alluvion", the other one a completely human made construction. The first kind has not appeared on the continental shelf so far, but it can well be imagined that such an "island" is brought into being with human help, if conditions so allow, for the purpose of exploiting the subsoil. Besides it remains questionable whether any real difference can be made between such an "island" and one which is not built on poles, but made by dumping sand and stones and clay into the sea, in shallow waters, in order to create an island which does not differ in nature from a real one, but for the fact that it has been made by men. Such an island could, if big enough, perhaps be used to erect derricks or even to sink mineshafts in order to exploit other minerals than oil, like uranium.

Gidel allows to the first kind the character of real islands, because they are the result of a simple acceleration in the natural accretion of the land. In other words it would have happened anyhow, but without human help only slower. The modification of the territorial waters outer-limit will be slight, another reason for Gidel to recognize them as islands. But our man-made island could, as far as we can judge, also only be built in shallow water and therefore will probably be not too far from the coast. The first reason to allow the "aluvion" island territorial waters (or better to take it into consideration when mapping out the baseline for the territorial waters) does not exist in our case, but the second reason does.

The question is, what will be the State practice, whenever such an island consisting of dumped material is built?

We would not be surprised if the remaining rather subtle difference with the "Gidel-type-island" would not deter the coastal State from considering it as a real island. Once that is done, and acquiesced in by other States, the next step may be to consider as a real island our "dumped" one which is too far off the coast as to cause a bulge in the outer-limit of the territorial waters, thus giving it its own marginal belt.

Our type of artificial islands have not yet been made, as far as we know and we believe certainly not for the purpose of extracting minerals from the subsoil, but we consider it not altogether impossible and we thought fit to draw attention to this possibility. But then, what is the real difference between an island made of sand and clay and one made of iron poles? It seems that the kind of material does not make any difference in law. The only real difference is the permanent character of the first.

Gidel may even have had this kind in mind, although he refers to constructions envisaged in connection with the tunnel in the Strait of Dover, discussed in the first Volume [112], p. 502. He refers to them as (third Volume [169], p. 683): ".... l'édification de toutes pièces, si l'on peut dire, d'émergences ....". On the question if *these* "artificial islands" should have territorial waters he is, we must say, salva reverentia, not very clear. He says, p. 683: "Pour donner à la question une réponse définitive, il faudrait être en possession d'un grand nombre de données techniques relevant d'hydrographie et de l'art de l'ingénieur", but then as a provisional opinion he says: "Il ne paraît pas douteux que l'on doive attribuer à ces îles artificielles une mer territoriale, lorsqu'elles sont édifiées à une telle distance des côtes que cette attribution ne réagît pas sur le tracé d'autre zones voisines de mer territoriale". He then refers to the answer of Denmark in the Bases of Discussion which we have quoted above. But in his final conclusion (p. 684), which we also quoted, he refers only to the kind of artificial islands of the aluvion type, which have been accelerated by men.

At any rate, like Johnson [198] rightly remarks, p. 214: "There are strong grounds for arguing that an artificial island should also be a permanent rather than a temporary installation, because it could hardly be expected that every time a structure is erected on piles driven into the sea-bed, even if only for a temporary purpose, it should carry territorial waters with it".

This is particularly applicable to oil-installations. We can distinguish 3 types. One is the drilling platform, where the result of the work is a "dry hole". This will certainly be pulled down. The second is the platform where a producing well is made. Here it depends. If the well needs constant survey, the platform will be maintained. If the well does not need survey, a pipe-line will be made to a collecting platform and the original installation will disappear. The third type we mentioned already is the collecting platform. The problem exists actually only for the latter type and the ones, where the well needs constant or at least periodical survey and maintenance. However, a producing well unfortunately does not go on producing eternally. Therefore we could really limit the problem to the sand and clay islands. They are, once made, probably permanent, and are the only ones for which a marginal belt may be considered, if ever this type of islands will be made.

The Committee on rights to the sea-bed and its subsoil submitted the following proposal to the Conference of the International Law Association at Copenhagen 1950 [24], p. 15–16, II (4) and (5):

"(4) The coastal State which is erecting or has erected any installation of the description referred to in I (5) above, being an installation which reaches above sea-level, should be entitled to exercise over a limited portion of the waters above the continental shelf such control and jurisdiction as is required for the protection of such installation, but no such installation should of itself be considered as an 'island' or an 'elevation of the sea-bed' within the meaning of international law. Such limited portions of the high seas above the continental shelf should be referred to as 'safety zones'.

(5) Each safety zone should normally be defined by a circle with a radius of 500 metres around the installation in question. This suggestion is made in view of the fact that, according to the legislation of various countries, the safety zone around an oil-well (within which smoking and the lighting of fires is prohibited) is defined in this way".

As an example of such a rule we mention the following: In the State-wide Order governing the drilling for and producing of oil and gas in the State of Louisiana [199], Section VIII under B we read:

"No boiler. open fire, or electric generator shall be operated within 100 feet of any producing oil- or gas-well, or oil-tank".

The proposals of the International Law Commission will be discussed in Section 7.

SECTION 5. ISLANDS ON THE CONTINENTAL SHELF

There exists a tendency to consider islands situated on the continental shelf adjacent to a State as pertaining for that reason to that State.

Lakhtine [200] gives the text of a Russian Declaration of (29) September, 1916, stating the discovery of some islands north of Siberia. The Declaration then goes on, p. 43: "Le Gouvernement Impérial de Russie a l'honneur de notifier par la présente aux Gouvernements des Puissances alliées et amies l'incorporation de ces terres dans le territoire de l'Empire de Russie.

Le Gouvernement Impérial profite de cette occasion pour faire ressortir qu'il considère aussi comme faisant partie intégrant de l'Empire les îles Henriette, Jeannette, Bennett, Hérald et Ouyédinénié, qui forment avec les îles Nouvelle Sibérie, Wrangel et autres situées près de la côte asiatique de l'Empire une extension vers le nord de la plateforme continentale de la Sibérie". This Declaration was confirmed by the Government of the U.S.S.R. The islands were again mentioned and it was said that they formed (op. cit. p. 44): "la continuation septentrionale du plateau continental sibérien" (a slightly different wording, which, we believe, does not change the meaning). According to François (First Report [10], p. 34), this note was dated 4 November, 1924. In a third note reference is made to an Act claiming as territory of the U.S.S.R. lands and islands situated in the Arctic Sea. The note then reads, that in conformity with this Act: "sont declarées territoire de l'Union toutes terres et îles découvertes ou qui pourraient être découvertes à l'avenir, qui ne sont point reconnues par le Gouvernement de l'Union territoire d'un Etat étranger à la date de la promulgation de l'Arrêté et qui se trouvent dans l'Océan Glacial dans les limites du secteur, formé par les méridiens encadrants les côtes septentrionales de l'Union. Ces méridiens sont precisés dans l'Arrêté, dont une copie est annexée .... ".

François remarks in his First Report [10], p. 34: "The term 'continental platform' is clearly not used in the same sense as that employed today: it does not refer to an under-water plateau. The rights claimed by the Soviet Union in polar waters should, in the view of the Rapporteur, be considered in relation to the 'theory of sectors' .... ". We agree that the third note refers to this "theory of sectors", but the first and second notes do not. Besides François starts to say (same page, just under the heading "The Continental Shelf"): "As early as 1916 the theory of the 'continental shelf' .... made its appearance in two different places, in Spain and in Russia", which suggests that he, writing this, considered the Russian claim as connected with the continental shelf. So do we, because otherwise the term "plateforme continentale" or (second note) "plateau continental" have to be interpreted as referring to the territorial plane of Siberia, which would be an unusual interpretation. Besides the idea is the same. The plane continues under water and emerges again here and there, thus forming the island referred to, which therefore belongs to the plane.

Argentina claims the Falkland Islands. Not long after Argentina's notification to the International Telegraph Bureau in 1925 concerning the call-sign of the wireless station on Laurie Island indicating that she claimed all the South Orkney Islands, a formal declaration was made through the International Postal Bureau "that Argentine jurisdiction extended de jure over the Falkland Islands .... " (Waldock [201], p. 332). Waldock continues: "In 1939 the Argentine Press began to formulate Argentina's claim to a sector, but it does not seem to have been until 1942 that Argentina officially laid claim to the whole sector between longitudes 25° W. and 68°34′ W. and south of latitude 60° S [a]". If this is correct the Falklands lying north of

---

[a] The Declaration of the Argentine Delegation at the meeting of the Ministers of Foreign Affairs of the American Republics in Panama, 1939 [174], p. 21, reads: "The Argentine Delegation declares that in waters adjacent to the South-American Continent, in that territorial extent of coasts which, in the zone defined as free from any hostile act, corresponds to the Argentine Republic, it does not recognize the existence of colonies or possessions of European countries, and adds that it specific-

the parallel of 60 degrees south would be outside the sector.

The claim could then not be based on any theory of sectors. The claim rests mainly according to Fenwick [69], p. 347: "Upon the long abandonment of the Islands by Great Britain between 1774 and 1810 ....". Azcárraga [32], p. 64, seems to suggest another possible ground for the claim: "The well-known Islas Malvinas (Falkland Islands), which are claimed by the Argentine, lie on the South-American continental shelf and, just like our Canary Islands or, for example, Japan, they are the high summits of submerged mountain ranges which emerge from the sea many miles from the Argentine coast". He makes this remark after having discussed the Argentine Declaration proclaiming sovereignty over the epicontinental sea and the continental shelf, which we have discussed partly in Chapter II.

Young [73], p. 853, goes even further in this connection, where he says: "The extent of the shelf and sea claimed by Argentina is nowhere defined in the instrument. This studied lack of precision may be of practical importance in connection with Argentina's Antarctic claims, which comprise a sector bounded by the 25th and 74th meridians west of Greenwich. In support of these, Argentina has laid considerable stress on the argument of geographical propinquity; and though the waters between South-America and Antarctica considerably exceed 100 fathoms in depth, there are connecting geological structures between the two land masses which might be brought within an expanded continental shelf-doctrine", thus actually suggesting, that the claim of Argentina on a part (sector) of Antarctica is based on the fact that this "island" is more or less situated on the continental shelf of Argentina.

If this were indeed the idea at the base of the claim, the qualification "far-fetched" would be a mild one. Antarctica is just as much a continent as South-America and the continental shelf as usually understood, does not connect the two continents. But whether Young's suggestion can stand, is questionable. Azcárraga [32] on p. 64, infers on the contrary that "this lack

---

ally reserves and maintains intact the legitimate titles and rights of the Argentine Republic to islands such as the Malvinas, as well as to any other Argentine territory located within or beyond the said zone".

of precision may redound to the prejudice of the limit of the Argentine claims to the Antarctic sector ....".

However, to base a claim on an island on the fact, that it is situated upon the continental shelf adjacent to the claiming State, seems to us unacceptable. The absurdity of this idea becomes manifest at once when we imagine the consequences which it would entail. France, Belgium and Holland, for instance, could claim the British isles, because they are situated on the European continental shelf, adjacent to these three countries.

Section 6. Cables and Pipelines

The oil-exploitation of the continental shelf has yet given rise to another "use" of the high seas, very much akin to the laying of telegraph cables: the laying of undersea pipelines. The pipelines will either be used to connect producing wells with a collecting platform, or to connect this platform (or the producing wells directly) with the shore.

The principle of the freedom of the sea involves the freedom to lay submarine telegraph and telephone cables. The protection of these cables is regulated in the Convention of Paris of 14 March, 1884 [202] [a].

The subject of laying pipelines is so similar to that of telegraph cables, that the International Law Commission decided that the principle applying to cables should also apply to pipelines (S.R. 65 [130], p. 11).

In his Second Report [26], François gives a detailed history of the protection of telegraph cables, p. 30–36. Amongst other resolutions he mentions those of the London Conference of 1913 (p. 32) and of the Institute of International Law in 1879, 1902 and 1927. He then (p. 34) proposes as a basis of discussion 7 Articles, the first reading: "All States may lay telegraphic cables and pipelines".

The subject has been discussed in the 114th, 115th, the 124th and 125th meeting (S.R. 114 [109], p. 12–20, S.R. 115 [148], p. 5–7, S.R. 124 [203], p. 22–24 and S.R. 125 [204], p. 3–9). The pipelines of the type we mentioned are not communicating two different countries, but are connecting an oil-installation of the coastal

---

[a] *Also,* [153]*, p. 251 et seq.*

State on the continental shelf with the coast of that State. The type envisaged by the Commission for a moment, connecting two countries, does not exist. It is the case where a pipeline would be laid on the continental shelf of another country. Long pipelines would need pumping stations between the beginning and the end of the pipeline, which would entail the building of installations on the sea-bed (or maybe on an island). At any rate the last type is not likely to come into being soon and, besides, is not the type typical of the exploitation of the continental shelf. In the 115th meeting (S.R. 115 [148], p. 5) the Commission decided not to mention pipelines in an Article, but to mention them in the comment.

The Article and the comment proposed by the Commission in its Report covering the Third Session [36], p. 57–58, read:

ARTICLE 5

"Subject to the right of a coastal State to take reasonable measures for the exploration of the continental shelf and the exploitation of its natural resources, the exercise by such coastal State of control and jurisdiction over the continental shelf may not exclude the establishment or maintenance of submarine cables.

1. It must be recognized that in exercising control and jurisdiction under Article 2, a coastal State may adopt measures reasonably connected with the exploration and exploitation of the subsoil, but it may not exclude the laying of submarine cables by non-nationals.
2. The Commission considered whether this provision should be extended to pipelines. If it were decided to lay pipelines on the continental shelf of another country, the question would be complicated by the fact that pumping stations would have to be installed at certain points, and these might hamper the exploitation of the subsoil more than cables. Since the question does not appear to have any practical importance at the present time, and there is no certainty that it will ever arise, it was not thought necessary to insert a special provision to this effect".

We hope that the Commission will discuss this matter again at its next session, because the pipelines connecting the collecting platforms with the shore are lying on the continental shelf under the *high seas*, are liable to be damaged by anchoring ships or fishing vessels and will, if damaged, cause pollution of sea-water and will give rise to conflicts of an international character. Some provisions have to be made for protection. We are even thinking of a "safety zone" of let us say 250 metres width, following the pipeline on both sides, where it should be forbidden to anchor. Pipelines should be marked on the map. This proposal will probably meet strong opposition because of strategical considerations.

It should be kept in mind that damaging a pipeline has a more serious character than damaging a telegraph cable, because of the pollution of the sea-water which would be the result. Not only the private interest of the owner, but the general interest is involved, perhaps even more than in the case of cables. Such provisions should either be laid down in a Convention or in the Convention an international body should be given the power to make regulations for each case, following, if this is more acceptable, draft rules ("règlement-type" or "projet de règlement) added as an Annex to the Convention.

We believe that the subject deserves further consideration. We will deal with the special aspect of pipelines (in connection with the resolution of the Institute of International Law of 1902) in war-time in the last Chapter.

SECTION 7. DISCUSSIONS INTERNATIONAL LAW COMMISSION AND PROVISIONAL CONCLUSIONS

We will leave out what has been said about an international body, because that subject will be dealt with separately in the last Chapter.

François mentions the problem of obstruction by installations in his First Report [10], p. 32, as follows: ".... To a greater or lesser degree all these installations restrict the possibility of using the high seas and their erection must therefore be subject to the express or tacit agreement of the other States". He added to this statement (S.R. 66 [7], p. 21) that he "felt that a violation of the principle of freedom of the high seas would occur, whenever such waters had to be passed through in order to reach the sub-soil".

Kerno pointed out (S.R. 69 [67], p. 5): ".... that the continental shelf should be taken to mean the sea-bed and the subsoil, and to exclude the waters covering them". In the same meeting (loc. cit. p. 12) el Khoury cited the solution adopted in Moslem law: "A well dug in the desert enjoyed an easement. It was forbidden to dig another well within a radius of 300 metres". Hudson (p. 11) thought that installations were temporary and should not have territorial waters. The Commission decided in principle against territorial waters round an installation and for a safety zone (p. 12).

The Commission laid down in its Report covering the Second Session [14] the very important principle (p. 22):

"There could be no question of such right of control and juris-diction over the waters covering those parts of the sea-bed. Those waters remained under the regime of the high seas. The exercise in them of navigation and fishing rights might be impaired only in so far as was strictly necessary for the exploitation of the sea-bed and subsoil. For works and installations established in the waters of the high seas for working the sea-bed and subsoil, special security zones might be set up, but they could not be classed as territorial waters".

The Second Report of François [26], goes on on this basis. Point 6 of his proposals on p. 69 reads (as far as relevant to our Chapter): "The exploration and exploitation of the sea-bed and subsoil of the continental shelf outside the territorial waters is permissible only in so far as it does not substantially interfere with shipping and fisheries, e.g. in so far as it does not constitute an obstruction of traffic routes ....". In point 7 another con-dition is introduced (p. 70):

"*a*) that interested parties (e.g. governments, shipping and fishing interests, airlines, etc.) must be duly notified in advance of the intended construction of such installations, and

b) that such installations must be equipped with efficient warning apparatus (lights, sound signals, radar, buoys, etc.)"

Finally (same page) a safety zone with a radius of 500 metres around the installation is proposed.

Hudson remarked in the 115th meeting (S.R. 115 [148], p. 9): ".... But in the case of established traffic routes, the coastal State would be required, when exercising its right of control and jurisdiction as laid down for the purposes of exploiting the continental shelf, to avoid interfering with such traffic routes ....". The further discussions being mainly concerned with drafting can be omitted. The results of these discussions are the following.

The International Law Commission made 3 statements about the waters covering the continental shelf. In the Report covering the third session [36], *a.* p. 55, note 9 on Article 1:

"The continental shelf referred to in this article is limited to the submarine areas outside territorial waters. Submarine areas beneath territorial waters are, like the waters above them, subject to the sovereignty of the coastal State";
b. p. 57, Article 3: "The exercise by a coastal State of control and jurisdiction over the continental shelf does not affect the legal status of the superjacent waters as high seas";
c. p. 58, Article 6 (2): "Such installations shall not have the status of islands for the purpose of delimiting territorial waters, but to reasonable distances safety zones may be established around such installations, where the measures necessary for their protection may be taken".

In spite of these proposals, which we do not object to, we are afraid that coastal States will not remain uninterested about their installations on the shelf in case of war. We admit immediately that if ever a war would break out again, there will not be many neutral States. But if any maritime States with a shelf remain neutral we expect that they will proclaim a security zone like in the Declaration of Panama of 1939, the border-line running well outside the farthest installations, with the object to forbid any hostilities to be committed within the area thus encircled. This of course has nothing to do with the extent of the territorial waters.

Further Article 6 (1), p. 58, reading: "The exploration of the continental shelf and the exploitation of its natural resources must not result in substantial interference with navigation or fishing. Due notice must be given of any installations constructed and due means of warning of the presence of such installations must be maintained", with the following comment:

"1. It is evident that navigation and fishing on the high seas may be hampered to some extent by the presence of installations required for the exploration and exploitation of the subsoil. The possibility of interference with navigation and fishing on the high seas could only be entirely avoided if the subsoil could be exploited by means of installations situated on the coast or in territorial waters; in most cases, however, such exploitation would not be practicable. Navigation and fishing must be considered as primary interests, so that the exploitation of the subsoil could not be permitted if it resulted in substantial interference with them. For example, in narrow channels essential for navigation, the claims of navigation should have priority over those of exploitation.

2. Interested parties, i.e. not only governments but also groups interested in navigation and fishing, should be duly notified of the construction of installations, so that these may be marked on charts. Wherever possible, notification should be given in advance. In any case, the installations should be equipped with warning devices (lights, audible signals, rader, buoys, etc.).

3. The responsibility for giving notification and warning, referred to in the last sentence of paragraph 1 of this article, is not restricted to installations set up on regular sea-lanes. It is a general duty devolving on States regardless of the place where such installations are situated.

4. While an installation could not be regarded as an island or elevation of the sea-bed with territorial waters of its own, the coastal State might establish narrow safety zones encircling it. The Commission felt that a radius of 500 metres would generally be sufficient, though it was not considered advisable to specify any definite figure".

That navigation and fishing are considered as primary interests, must, we believe, be explained in this way, that they represent the general interest, whereas the oil-exploitation is, at least for the time being, more particularly the private interest of the coastal State. Of course oil is a generally needed commodity, but we feel that as long as the world-production fulfils the needs, and as long as the activities of a particular producer are not strictly necessary to satisfy the total demand, these activities are limited to the order of commercial competition rather than belonging to the order of the general interest.

Contiguous zones touch our subject only slightly. Therefore, we will not follow the discussions of the Commission on this item. The Commission proposes Article 4 (Report Third Session [36], p. 62): "On the high seas adjacent to its territorial waters, a coastal State may exercise the control necessary to prevent teh infringement, within its territory or territorial waters, of its customs, fiscal or sanitary regulations. Such control may not be exercised more than twelve miles from the coast". In connection with what we have said above explanatory note 4 is of interest:

"4. The proposed contiguous zones are not intended for purposes of security or of exclusive fishing rights. In 1930 the Preparatory Committee of the Codification Conference found that the replies from governments offered no prospect of reaching agreement to extend beyond territorial waters the exclusive rights of coastal States in the matter of fishing. The Commission considers that in that respect the position has not changed".

The main conclusions we arrived at in this Chapter are:

1. Building installations in the high seas is as much a legal use of the high seas as navigation and fishing, on the understanding that the general interest prevails above the private interest.
2. Building installations in the territorial waters should be done in such a way as not to bar free passage completely.
3. Freedom of navigation is subject to the freedom to use the high seas by other people and in other ways.
4. Use of the high seas should be made in such a way as not to cause unreasonable hindrance to another user.
5. Oil-installations are not to be considered as (artificial) "islands", because of their temporary character.
6. A Conference should be convened to check the undesirable tendency of extension of territorial waters, on the basis that uniformity should not be aimed at.
7. A Conference should be convened to draft a Convention covering all items connected with the exploitation of the resources of the continental shelf in connection with shipping and fishery interests.

CHAPTER IV

MINERAL RESOURCES

SECTION 1. PROCLAMATIONS, DECLARATIONS AND DECREES

Mention will be made of those parts or provisions, which have not been quoted in previous Chapters.

1. The first instrument was the Treaty between the United Kingdom and Venezuela of 1942 [110] [a], relating to the submarine areas of the Gulf of Paria. The following articles or parts thereof will be quoted:

Article 1. "In this Treaty the term 'submarine areas of the Gulf of Paria' denotes the sea-bed and subsoil outside of the territorial waters of the High Contracting Parties to one or the other side of the lines A–B, B–Y and Y–X.

Article 2 (1). His Majesty The King declares that he for his part will not assert any claim to sovereignty or control over those parts of the submarine areas of the Gulf of Paria which lie westerly of the line A–B, or southerly of the lines B–Y and Y–X respectively described in Article 3 of the present Treaty, and that he will recognize any rights of sovereignty or control which have been or may hereafter be lawfully acquired by the United States of Venezuela over the said parts of the submarine areas of the Gulf of Paria.

(2) The President of the United States of Venezuela declares that he for his part will not assert . . . . "

(follows a similar wording, but relating to the other side of the dividing line). Article 3 gives a description of the dividing line.

Article 4 "(1) The High Contracting Parties shall, as soon as practicable after the coming into force of this Treaty, appoint

---

[a] *See also* [153], *p. 44.*

a mixed Commission to take all necessary steps to demarcate the lines A–B, B–Y and Y–X by means of buoys or other visible methods on the surface of the sea or on the land as the case may be. Any buoys or other means employed shall, however, conform in all respects to the provisions of Article 6 of this Treaty" . . . .

2. Then follows in chronological order *Argentina*:

Decree No. 1,386, concerning mineral reserves, 24 January, 1944. "Boletín Oficial de la República Argentina", Vol. 52, No. 14,853 (17 March, 1944), p. 6. Translation by the Secretariat of the United Nations [153], p. 3.

Article 2. "Pending the enactment of special legislation, the zones at the international frontiers of the national territories and the zones on the ocean coasts, as well as the zones of the epicontinental sea of Argentina, shall be deemed to be temporary zones of mineral reserves; . . . .".

3. *United States.* Proclamation by the President with respect to the natural resources of the subsoil and sea-bed of the continental shelf.

September 28, 1945 [62] *a*

"Whereas the Government of the United States of America, aware of the long range world-wide need for new sources of petroleum and other minerals, holds the view that efforts to discover and make available new supplies of these resources should be encouraged; and

Whereas its competent experts are of the opinion that such resources underlie many parts of the continental shelf off the coasts of the United States of America, and that with modern technological progress their utilization is already practicable or will become so at an early date; and

Whereas recognized jurisdiction over these resources is required in the interest of their conservation and prudent utilization when and as development is undertaken; and

Whereas it is the view of the Government of the United States that the exercise of jurisdiction over the natural resources of the subsoil and sea-bed of the continental shelf by the contiguous nation is reasonable and just, since the effectiveness of measures to utilize or conserve these resources would be contingent upon coöperation and protection from the shore, since the continental shelf may be regarded as an extension of the land-mass of the coastal nation and thus naturally appurte-

a See also [153], *p. 38.*

nant to it, since these resources frequently form a seaward extension of a pool or deposit lying within the territory, and since self-protection compels the coastal nation to keep close watch over activities off its shores which are of the nature necessary for utilization of these resources;

Now, therefore, I, HARRY S. TRUMAN, President of the United States of America, do hereby proclaim the following policy of the United States of America with respect to the natural resources of the subsoil and sea-bed of the continental shelf.

Having concern for the urgency of conserving and prudently utilizing its natural resources, the Government of the United States regards the natural resources of the subsoil and sea-bed of the continental shelf beneath the high seas but contiguous to the coasts of the United States as appertaining to the United States, subject to its jurisdiction and control. In cases where the continental shelf extends to the shores of another State, or is shared with an adjacent State, the boundary shall be determined by the United States and the State concerned in accordance with equitable principles. The character as high seas of the waters above the continental shelf and the right to their free and unimpeded navigation are in no way thus affected.

In witness whereof, I have hereunto set my hand and caused the seal of the United States of America to be affixed".

United States. Executive Order of the President reserving and placing certain resources of the continental shelf under the control and jurisdiction of the Secretary of the Interior.

September 28, 1945 ^a

"By virtue of and pursuant to the authority vested in me as President of the United States, it is ordered that the natural resources of the subsoil and sea-bed of the continental shelf beneath the high seas but contiguous to the coasts of the United States declared this day by proclamation to appertain to the United States and to be subject to its jurisdiction and control, be and they are hereby reserved, set aside, and placed under the jurisdiction and control of the Secretary of the Interior for administrative purposes, pending the enactment of legislation in regard thereto. Neither this Order nor the aforesaid proclamation shall be deemed to affect the determination by legislation or judicial decree of any issues between the United States and the several states, relating to the ownership or control of the subsoil and sea-bed of the continental shelf within or outside of the three-mile limit".

---

^a *See also,* [153] *p. 41.*

4. *Mexico.* Presidential Declaration, 29 October, 1945 [72] [a].

The Declaration starts to describe the continental shelf .... "delimited by a two-hundred metre-isobath" and after having indicated the presence of minerals, continues:

"For these reasons, the Government of the Republic claims the whole continental shelf adjacent to its coasts and all and every one of the natural riches, known or still to be discovered, which are found in it ...." [b].

5. *Argentina.* Declaration proclaiming sovereignty over the epicontinental sea and the continental shelf, October 9th, 1946 [74] [c].

"The Executive Power, in Article 2 of Decree No. 1,386, dated January 24, 1944, issued a categorical proclamation of sovereignty over the 'Argentine Continental Shelf' and the 'Argentine Epicontinental Sea', declaring them to be 'transitory zones of mineral reserves';

The State, through the medium of the Yacimientos Petrolíferos Fiscales (Public Petroleum Deposits Administration) is exploiting the petroleum deposits discovered along the 'Argentine Continental Shelf', thereby confirming the Argentine nation's right of ownership over all deposits situated in the aforesaid continental shelf";

Referring to the paragraph concerning the right of every nation to consider as national territory the entire extent of .... the adjacent continental shelf (quoted in Chapter II, Section 3 under Chile) the Preamble continues:

---

[a] *See also* [153] *p. 13* (*Slightly different translation*).

[b] *The Presidential Decree*, incorporating in the property of "Petróleos Mexicanos" the subsoil of the lands covered by the territorial waters of the Gulf of Mexico and other lands specified therein, 25 February, 1949, Diario Oficial, Vol. 173, No. 10 (11 March, 1949), p. 4. Translation by the Secretariat of the United Nations [153], p. 14, *although* in the Preamble *referring* to the claim of the "submarine shelf" in the Declaration of 29 October, 1945 and *stating* under III "that, in the case of hydrocarbons, the jurisdiction of the Government of the Republic is indisputable, under the express terms of Article 27, paragraph 4, of the Federal Constitution, whereby the Nation is empowered to undertake the exploitation of the petroleum resources of the subsoil of the continental shelf, through the intermediary of the Public Petroleum Institution known as Petróleos Mexicanos ...." nevertheless Art. 1 includes in the property of Petróleos Mexicanos the subsoil of the lands covered by the territorial waters of the Gulf of Mexico .... *only* to a distance of 5 kilometres from the low-water mark, that is less than 3 miles.

(It seems to appear from the Preamble that Art. 27 of the Constitution *has* been amended, as a result of the Declaration of 29 October, 1945, although the United Nations, Legislative Series, Vol. I [153], does not contain that amendment).

[c] *cf. Decree No. 14,708, concerning national sovereignty over Epicontinental Sea and the Argentine Continental Shelf, 11 October 1946.* "*Boletín Oficial de la República Argentina*" Vol. 54, No. 15,641 (5 December 1946) p. 2, [153] p. 4.

"The doctrine in question, aside from the fact that it is implicitly accepted in modern international law, is now deriving support from the realm of science in the form of serious and valuable contributions, according to the testimony offered by numerous national and foreign publications and even by official educational programs; and

The manifest validity of the thesis invoked above, as well as the determination of the Argentine Government to perfect and preserve all the attributes inherently bound up with the exercise of national sovereignty, make it advisable to formulate the declaration pertinent to this matter, thereby amplifying the effects of the aforesaid Decree No. 1,386".

(For Art. 1 of the operative part of the Declaration, declaring the continental shelf to be "subject to the sovereign power of the Nation", see Chapter II, Section 3).

6. *Chile*. Presidential Declaration of 25th June, 1947 [72] p. 188 [a].

The Preamble mentions "the fact that the exploitation of resources contained in the continental shelf which are essential to the national life is already under way, as is the case with the coal-mines, which are being worked and will continue to expand into the territory which is covered by water ...."

Section 1 of the operative part of the Declaration reads: "The Government of Chile confirms and proclaims the national sovereignty over the whole continental shelf adjacent to the continental and insular coasts of the national territory whatever its depth may be, claiming, consequently, all the natural riches which exist on, in, or under said shelf, known or to be discovered".

7. *Peru*. a. Presidential Decree of 1 August, 1947 [72], p. 190 [b].

The Preamble contains amongst others the following paragraphs: "That natural wealth exists in said platform, and it is indispensable to proclaim that this wealth forms part of the national patrimony; ....

That Article 37 of the Constitution of the State lays down that the mines, lands, forests, and in general all natural sources of wealth pertain to the State, except where others have legitimately acquired rights";

Section 1 of the operative part reads: "It is hereby declared that the national sovereignty and jurisdiction extend to the submarine platform or continental and insular shelf adjacent to the continental and island coasts of the national territory, whatever may be the depth and the extent of said shelf".

---

[a] *See also* [153] *p. 6. (translation slightly different).*
[b] *See also* [153] *p. 16. (translation slightly different).*

b. Supreme Resolution No. 121, granting to the State Petroleum Enterprise oil-reserves in Tumbes and Piura, 27 April, 1948, "El Peruano: Diario Oficial", Vol. 108, No. 2,200 (8 May, 1948), p. 1, translation by the Secretariat of the United Nations [153], p. 18.

Section 2 reads: "Likewise to grant to the State Petroleum Enterprise a concession in respect of the submerged oil-fields off the north coast within the limits specified in the Supreme Decree of 1 August, 1947; it being understood that the southern limit of the submerged area so conceded will be Latitude South four degrees fifteen minutes (4°15′) in the proximity of Cape Blanco, and that it will extend from that point as far as the frontier with Ecuador".

8. *Costa Rica.* a. Decree-Law No. 803 of 2 November, 1949 [153], p. 9. Article 1, Decree Law No. 116 of 27 July 1948 shall read as follows:

Article 1. "National Sovereignty is confirmed and proclaimed in the whole submarine platform or continental and insular shelf adjacent to the continental and insular coasts of the national territory, at whatever depth it is found, and the inalienable right of the Nation to all the natural wealth which exists in the said shelf or platform is reaffirmed"....

b. Political Constitution, 7 November, 1949, La Gaceta: Diario Oficial, Vol. 71, No. 251 (7 November, 1949), p. 2069. Translation by the Secretariat of the United Nations [153], p. 300.

.... Article 6. "The State exercises complete and exclusive sovereignty in respect of the air space above its territory and in respect of its territorial waters and continental shelf, in accordance with the principles of international law and the treaties in force"

9. *Honduras.* (See for 4 Decrees Chapter II, Section 3).

10. *Panama.* Constitution, 1 March, 1946, "Gaceta Oficial", Vol. 43, No. 9,938 (4 March, 1946), p. 18. Translation by the Secretariat of the United Nations [153], p. 15.

Article 209. "The following belong to the State and are of public use and, in consequence, cannot be the object of private appropriation: .... (4) The aereal space and the submarine continental shelf which appertain to the national territory ....".

11. *El Salvador.* (See Chapter II, Section 3).

12. *Ecuador.* (See Chapter II, Section 3).

13. *Guatemala.* Petroleum Law, enacted by Legislative Decree No. 649, 30 August, 1949. "Diario de Centro America", Vol. 56, No. 46 (27 September, 1949), p. 505. Translation by the Secretariat of the United Nations [153], p. 10.

Article 1. "All deposits or natural reserves of petroleum within the land or sea boundaries of the Republic, up to the extremity of the continental shelf or platform of the Republic, shall, whether they lie on or under the earth, lakes, rivers, or seas, be the property of the nation. The direct dominium over them is inalienable and imprescriptible".
.... Article 29. "The Executive Power may grant prospecting concessions in any area included within the boundaries of the national territory, territorial waters and continental shelf or platform of Guatemala ...."
.... Article 36. "The Executive Power may grant operating concessions in any areas included within the boundaries of the national territory, territorial waters and continental shelf or platform of Guatemala ....".

14. *Nicaragua.* Political Constitution of Nicaragua, 1 November, 1950. "La Gaceta", Vol. 54, No. 235 (6 November, 1950), p. 2209, Translation by the Secretariat of the United Nations [153], p. 15.

.... Article 5. "The national territory extends between the Atlantic and the Pacific Oceans and the Republics of Honduras and Costa Rica. It also comprises: the adjacent islands, the subsoil, the territorial waters, the continental shelf, the submerged foundations (zócalos submarinos), the air space and the stratosphere.
Such frontiers as may not yet be determined shall be fixed by treaties and by law".

15. *Brazil.* a. Decree No. 28,840 integrating into national territory the adjoining part of the continental shelf, 8 November, 1950. Diario Oficial, Vol. 89, No. 264 (18 November, 1950), p. 16,617. Translation by the Secretariat of the United Nations [153], p. 299.

Article 1. "It is formally proclaimed that part of the continental shelf which adjoins (correspondente) the continental and insular territory of Brazil is integrated into that territory, under the exclusive jurisdiction and dominion of the Federal Union".

Article 2. "The utilization and exploration of products or natural resources of that part of the national territory shall be subject in all cases to federal authorization or concession" ....

16. *Philippines*. a. Petroleum Act of 1949, enacted by Republic Act No. 387, 18 June, 1949. "Official Gazette", Vol. 45 (1949), p. 3192 [153], p. 19.

Article 3. "State Ownership. All natural deposits or occurrences of petroleum or natural gas in public and/or private lands in the Philippines, whether found in, on or under the surface of dry lands, creeks, rivers, lakes or other submerged lands within the territorial waters or on the continental shelf, or its analogue in an archipelago, seaward from the shores of the Philippines which are not within the territories of other countries, belong to the State, inalienably and imprescriptibly".

17. *Saudi Arabia*. Royal Pronouncement concerning the Policy of the Kingdom of Saudi Arabia with respect to the subsoil and sea-bed of areas in the Persian Gulf contiguous to the coasts of the Kingdom of Saudi Arabia, May 29, 1949 [205], p. 156 [a]. From the Preamble we quote:

.... "Deeming that the exercise of jurisdiction over such resources by the contiguous nation is reasonable and just, since the effectiveness of measures to utilize or conserve these resources would be contingent upon coöperation and protection from the shore and since self-protection compels the coastal nation to keep close watch over activities off its shores which are of a nature necessary for the utilization of these resources; ....".

(The operative text is quoted in Chapter I, Section 1).

18. *Iran*. Bill relating to Persian Gulf subsea resources approved by the Council of Ministers and submitted to Majlis, May 19, 1949 [206], p. 347. (For the text see Chapter I, Section 1).

19. *United Kingdom. Bahrain*. Proclamation No. 37/1368, June 5, 1949 [207], p. 185 [b]. From the Preamble we quote:

.... "Whereas the right of any coastal government to exercise its sovereignty over the natural resources of the sea-bed and the

---

[a] *Also in* [153], *p. 22.*
[b] *Also in* [153], *p. 24. (slightly different translation).*

subsoil in the vicinity of its shores has been established by international practice through the action taken by other governments . . . ."

From the operative text: "We, Salman Ibn Hamad Al Khalifah Ruler of Bahrain, hereby declare that the sea-bed and the subsoil of the high seas of the Persian Gulf bordering on the territorial waters of Bahrain and extending seaward as far as limits that we, after consultation with the neighbouring governments, shall determine more accurately in accordance with the principles of justice, when the occasion so requires, belong to the country of Bahrain and are subject to its absolute authority and jurisdiction.

There is nothing in this proclamation that may be interpreted as affecting dominion over the islands or the status of the sea-bed and the subsoil underlying any territorial waters.

There is nothing in this proclamation that may be interpreted as affecting the character of the high seas in the waters of the Persian Gulf overlying the sea-bed and beyond the limits of the territorial waters, or the status of the air space above the waters of the Persian Gulf beyond the territorial waters, or fishing, or the traditional rights of pearling in these waters".

Abu Dhabi	Proclamation,	10 June, 1949 [153], p. 23	
Ajman	,,	20 June, 1949 [153], p. 23	
Dubai	,,	14 June, 1949 [153], p. 25	
Kuwait	,,	12 June, 1949 [153], p. 26	
Qatar	,,	8 June, 1949 [153], p. 27	
Ras Al Khaimah	,,	17 June, 1949 [153], p. 27	
Sharjah	,,	16 June, 1949 [153], p. 28	
Umm Al Qaiwain	,,	20 June, 1949 [153], p. 29	

The texts of these Proclamations are similar to the Proclamation of Bahrain.

The Bahamas. a. (Alteration of Boundaries) Order in Council 1948, 26 November, 1948, No. 2574[208], p. 250[a].

"Whereas it is desirable to extend the boundaries of the Colony of the Bahamas so as to include the continental shelf contiguous to the coasts of the Colony:

Now, therefore His Majesty, in pursuance of the powers conferred upon Him by the Colonial Boundaries Act, 1895, and of all other powers enabling Him in that behalf, is pleased, by and with the advice of his Privy Council, to order, and it is hereby ordered as follows:

----

[a] *Also in* [153], *p. 31.*

1. This order may be cited as the Bahamas (Alteration of Boundaries) Order in Council, 1948.
2. The boundaries of the Colony of the Bahamas are hereby extended to include the area of the continental shelf which lies beneath the sea contiguous to the coasts of the Bahamas.
3. Nothing in this Order shall be deemed to affect the character as high seas of any waters above the continental shelf and outside the limits of territorial waters".

b. Petroleum Act, 3 April, 1945 [139] *a*.
(The relevant articles have been quoted in Chapter II, Section 12 and Chapter III, Section 3).

British Honduras. *a.* (Alteration of Boundaries) Order in Council, 1950, No. 1649, 9 October, 1950 [209], p. 210 *b*.
(The text is similar to the Order in Council of the Bahamas).

b. Oil-mining Regulations, 2 September, 1949, British Honduras, Statutory Instruments, No. 56 of 1949 [153], p. 32.
(The relevant articles have been mentioned in Chapter II, Section 12 and Chapter III, Section 3).

Jamaica. (Alteration of Boundaries) Order in Council, 26 November, 1948, No. 2575 [210], p. 1664 *c*.
(The text is similar to the Order in Council of the Bahamas).

Falkland Islands. (Continental Shelf) Order in Council, 1950, 21 December, 1950, No. 2100 [209], p. 682 *d*.
(The text is similar to the Order in Council of the Bahamas, except for a detailed description of the boundary-line of the continental shelf).

Trinidad and Tobago. *a.* Submarine areas of the Gulf of Paria (Annexation) Order in Council, 6 August, 1942 "United Kingdom, Statutory Rules and Orders, 1942", Vol. I, p. 919 [153], p. 46.
"Whereas the Gulf of Paria and the adjacent waters are bounded by the coasts of Venezuela and the Island of Trinidad respectively:
And whereas the Government of the Republic of Venezuela have annexed to Venezuela certain parts of the submarine areas of the Gulf of Paria:
And whereas it is expedient that the rest of the submarine areas of the Gulf of Paria should be annexed to and form part of His Majesty's dominions and should be attached to the Colony of Trinidad and Tobago for administrative purposes:

---

a Also in [153], *p. 30.* *b Also in* [153], *p. 304.* *c Also in* [153], *p 33.* *d Also in* [153] *p. 305.*

Now, therefore, His Majesty is pleased, by and with the advice of His Privy Council, to order, and it is hereby ordered, as follows:

1. This Order may be cited as the Submarine Areas of the Gulf of Paria (Annexation) Order, 1942.

2. In this Order the expression 'submarine areas of the Gulf of Paria' means the sea-bed and subsoil situated beneath the waters, excluding territorial waters, bounded as follows: .... (follows description)

shall be annexed to and form part of His Majesty's dominions and shall be attached to the Colony of Trinidad and Tobago for administrative purposes, and the said submarine areas are annexed and attached accordingly.

4. Nothing in this Order shall:

(a) Affect, or imply any claim to, any territory above the surface of the sea or any part of the high seas, or

(b) Prejudice any rights of passage or navigation on the surface of the sea.

5. The Governor of the said Colony shall, as soon as may be after the date of this Order, make regulations to ensure;

(1) That the marine areas within the limits specified in section 2 of this Order shall not be closed to navigation, and that any works or installations which may be erected shall be of such nature and shall be so constructed, placed, marked, buoyed and lighted as not to constitute a danger or obstruction to shipping" ....

b. Submarine (Oil Mining) Regulations, 22 May, 1945, Government Notice No. 87: Proclamations, Orders, Regulations, etc., 1945, p. 101 [153], p. 33.

(The relevant articles have been mentioned in Chapter II, Section 12 and Chapter III, Section 3).

20. *Pakistan.*

(a) Declaration by the Governor-General, 9 March, 1950. The Gazette of Pakistan, Extraordinary, 14 March, 1950, p. 123[153] p. 303.

"I, Khwaja Nazimuddin, Governor-General of Pakistan, hereby declare in pursuance of clause (bb) of sub-section (1) of section 5 of the Government of India Act, 1935, that the sea-bed along the coasts of Pakistan extending to the one hundred fathom contour into the open sea shall, with effect from the date of this declaration, be included in the territories of Pakistan".

SECTION 2. WHAT IS THE VALUE OF PROCLAMATIONS DECLARATIONS AND NATIONAL DECREES IN INTERNATIONAL LAW?

As the Proclamations by President Truman have been followed, without much delay, by other Declarations and by national Decrees, more or less on the same subject, in a rather rapid sequence, we may assume that the American Proclamations have given an impetus to the latter ones.

Therefore it seems useful to investigate what the attitude of the United States towards such Declarations and national Decrees has been in the past, and is in these days.

The following note quoted by Moore [175] is instructive in this respect and also in connection with the question we endeavoured to answer in Chapter III, i.e. whether a State can extend its territorial waters indefinitely. Moore, p. 706: "The undersigned, Secretary of State of the United States, having taken the instructions of the President, will perform the duty of answering the note which was addressed to the undersigned on the 8th of October by His Excellency Senor Gabriel G. Tassara, minister plenipotentiary of Her Catholic Majesty the Queen of Spain.

In that paper Mr. Tassara informs the undersigned that Her Catholic Majesty's Government is surprised that a United States naval officer cruising in the waters of Cuba has fallen into the error of claiming that the jurisdictional belt of the island of Cuba does not extend beyond three miles, whereas the Government has fixed the limit as six miles on the open sea. Mr. Tassara proceeds under his instructions to say that in fixing that limit Her Catholic Majesty's Government has conformed to all the rules of the law of nations. Mr. Tassara next observes that the principle which is generally recognized is that maritime jurisdiction extends to the range of a cannon ball, and that even abiding by this principle, which every nation has modified at its will, the belt fixed by Spain goes no farther than the modern improvements in artillery. Mr. Tassara, pursuing the subject, remarks that no international compact is required for the determination or recognition of a jurisdiction which is not at all excessive, ..." and p. 707: "Impressed by these general views, the United States are not prepared to admit that Spain, without a formal concurrence of other nations, can exercise exclusive sovereignty upon the open sea beyond a line of three miles from the coast, so as to deprive them of the rights common to all nations upon the open sea ...." and p. 708: "Spain presented substantially the same claim to this Government in the case of the El Dorado in 1856, and Mr. William L. Marcy, then Secretary of State, by direction of the President, announced that the United States could not concede the extension of Spanish sovereignty beyond three miles in the seas which surround the island of Cuba.

Upon the grounds which have been set forth, the President feels himself obliged to decline to give to the naval commanders of the United States the instructions proposed to him by Her Catholic Majesty's Government ...."

(Mr. Seward Sec. of State, to Mr. Tassara, Span. min. Dec. 16, 1862; MS. Notes to Spain, VII, 331) and p. 709:

"Mr. Tassara is nòt understood to deny this proposition. But he insists that this principle has its exceptions, and that some states, and among them the United States, habitually claim and exercise a wider jurisdiction. While this fact is cheerfully admitted, it does not seem to the undersigned conclusive in favor of the claim of Spain. The exceptions are so few and so special that they do not disturb or impair the general principle that three miles is the legal boundary of external maritime jurisdiction. Mr. Tassara seems to assume, however, that as there are some existing and acknowledged exceptions to that principle, so there are also other existing exceptions which ought to be acknowledged, and that the jurisdiction for which he is now contending is such an exception. He very truly assumes that wherever such an exception actually exists, evidence of it will be found in the statutes or decrees of the maritime power which asserts it. As such evidence, he quotes several royal decrees of Spain, some ancient and others modern, which declare that the jurisdiction of Spain in the waters which surround her coasts extends to the limit of six miles.

Nevertheless it *cannot be admitted*, nor indeed is Mr. Tassara understood to claim, that *the mere assertion of a sovereign, by an act of legislation, however solemn, can have the effect to establish and fix its external maritime jurisdiction*. His right to a jurisdiction of three miles is derived not from his own decree but from the law of nations, and exists even though he may never have proclaimed or asserted it by any decree or declaration whatsoever. He cannot, by a mere decree, extend the limit and fix it at six miles, because, if he could, he could in the same manner, and upon motives of interest, ambition, or even upon caprice, fix it at ten, or twenty, or fifty miles, without the consent or acquiescense of other powers which have a common right with himself in the freedom of all the oceans. Such a pretension could never be successfully or rightfully maintained. The statutes which Mr. Tassara has recited are therefore regarded as showing what certainly is by no means unimportant, that Spain at an early day asserted, and has on different occasions since that time reasserted, in her domestic legislation, a claim to an exceptional jurisdiction of three miles in addition to the three miles of jurisdiction conceded by the law of nations ...." and p. 710:

"Nations do not equally study each other's statute books, and are not chargeable with notice of national pretensions resting upon foreign legislation .... The case of the El Dorado seems to be the only one in which this claim of Spain has been brought to the notice of this Government ...." and finally p. 711–712: "In view of the considerations and facts which have been thus presented, the undersigned is obliged to state that the Government of the United States is not prepared to admit that the jurisdiction of Spain in the waters which surround the island of Cuba lawfully and rightly extend beyond the customary limit of three miles". (Mr. Seward, Sec. of State, to Mr. Tassara, Spanish Min., Aug. 10, 1863, MS. notes to Span. Leg. VII, 407). (underlined by us). We may add, that the Spanish Decree dates from 1760 according to the Annuaire [93], p. 126: ".... l'Espagne, de son côté, exerce sa puissance, depuis la real cedula du 17 décembre, 1760, sur une zône de six milles autour de ses côtes, ce qui d'après Riquelme, n'a jamais donné lieu à aucune protestation ou réclamation de la part des autres puissances. (Elementa de derecho publico internacional I, p. 23)".

If an American Secretary of State, surely expressing the opinion of the American Government, could write that about a national law, which had at that time been in force for over 100 years and which had not escaped the attention of the American Government in view of the admittance in regard to the "El Dorado" case, what would then be the attitude towards a unilateral proclamation by a foreign (foreign from the American view-point) Power *now*, as to the validity of such Proclamation in International Law? If International Law has any meaning,

States should act under that law exactly as they expect other States to behave towards them in accordance with that law. This is the simple application of Kant's "categorical imperative": act in such a way that the rule of conduct of your will can serve at the same time as a rule for a general law. In other words when a State acts, it should test the act by considering the general result which would follow if all States acted in the same way. Of course we are not discussing power politics, but law. Neither has this remark anything to do with the contents of proclamations or decrees, but only with the method of action.

It was on similar lines that Nielsen [211] warned against an extension of jurisdiction in 1923. He said, p. 36:

"I think it has been suggested that, with a view to preventing a nefarious form of smuggling with which the authorities of our Government have lately had to deal, the seizure of vessels on the high seas might be justified under the so-called right of self-preservation, which Halleck calls 'imperfect war' (Halleck, International Law, Vol. I, p. 113). From an examination of the instances in which this principle of self-preservation has been invoked, it would appear that they have been generally limited to efforts to thwart acts of a military character ...." and p. 37:

"I am not disposed to think that the United States would undertake such a radical extension of the right of self-preservation, which has given rise to so much controversy in the past.

I have no interest in taking any time in a endeavor to point out how, by expedients and subtle lines of reasoning, the United States might escape the full consequences of a rule which it has insisted firmly on imposing on other nations".

We take for our discussions the Truman Proclamations, because they are the "mildest" in character. What we are able to say about them, will be applicable to the other ones "a fortiori".

There are several angles of approach to this problem. One way of reasoning would be to say that if no law existed on the point, a unilateral proclamation, declaration or a national decree (we will call these for convenience unilateral acts) can initiate a rule of law. A certain nucleus of a rule of law has certainly been created. "Condensation" around such a nucleus may take place if conditions are favourable, i.e. if no disturbances either physically or spiritually, in the form of protests, occur during a certain lapse of time. But we would not be able to say that the unilateral act creates at once a rule of law.

The proclamation constitutes a precedent. This has been appreciated by the other States who enacted decrees or issued declarations, in as far as they referred to the preceding ones, (one has only to compare in chronological order the preambles of the relative instruments of some South- and Central American Governments). Some wishful thinking has been at work where it was said in the Chilian Declaration [72] that "an international consensus recognizes that each country has the right to consider as national territory the whole extent .... of the continental shelf", or as in the Proclamation of Bahrain [207], that "the right of any coastal government to exercise its sovereignty over the natural resources of the sea-bed and subsoil .... has been established by international practice".

However, customary law is based on precedents. A custom is based on some single act in the past. Today such single acts take place, and will grow into customs. But in all our considerations we cannot get round the fact that time is necessary. Customary law is the result of growth, and growth takes time. The time required certainly depends on the amount of precedents. We should think that if all the States of the world would accept a certain line of behaviour, one could sooner recognize a rule of International Law, than if only one or a few would adopt such practice. In our case we must admit that a considerable number of States have at least adopted the attitude, that a coastal State has the right of control and jurisdiction over the resources of the continental shelf adjacent to its coast. If we count the United Kingdom with all its colonies and protectorates as one (which is reasonable, because these dependencies will naturally follow the same policy) there are 19 countries which have so far adopted that attitude. A proclamation alone is, we believe, not sufficient to create a custom. Some action is needed. Oppenheim [42], for instance, says that we can speak of a custom (p. 25): "when a clear and continuous *habit* of *doing certain actions* has grown up under the aegis of the conviction that these actions are, according to International Law, obligatory or right". As such action we would consider the acts performed by the coastal

State as a consequence of the proclamation or decree. Farming out "blocks" on the continental shelf for exploration or exploitation to applicants of their own choice, in particular to their nationals only, making regulations concerning exploration and exploitation, would be acts of the government in this sense.

But can a coastal State lease submarine lands, without considering itself to have a title on these lands? Surely according to the old adage "nemo plus iuris ad alium transferre potest, quam ipse habet", the coastal State considers itself to have a right, i.e. the right to take and consume minerals found outside its territory, from which the right to farm out exploitation of these minerals follows. Of course the South-American States go further by claiming not only the resources but the whole shelf. Two lines of thought are possible. The coastal State appropriates "territory" which is not under the sovereignty of any other State and acquires sovereignty over that territory, i.e. occupation. Or, the State takes possession of that "territory", even if this act could not be considered to be strictly legal, and acquires a title after a lapse of time, i.e. prescription.

The "territory" in this case is, where the Truman Proclamation is concerned, only a part of what may be called "territory" i.e. the part called "minerals". In the case of the South-American Republics it is the continental shelf and in both cases the question arises, may we use this term for something which is not normally understood under the notion "territory"? This demonstrates clearly the reason why in preambles the term "submerged lands" is used, i.e. to smooth the path in answering our question in the affirmative.

The only leg we can stand on if we want to consider the continental shelf as "territory", is the fact that sea-bed and subsoil have been considered in the past as property of the Crown, especially in England, as we have seen in Chapter II. The sea-bed and subsoil were thought of in terms of land, over which the Crown had sovereignty. Moreover the subsoil has been and still is exploited through mines. But even if we could use the term "territory" in this connection are we allowed to apply the theory of occupation? Can a State occupy the sea-bed and subsoil? We will deal with this question more particularly in the next Section, but we want to suggest right here, that a distinction has to be made between the sea-bed and the subsoil, for the simple reason, that in the case of the subsoil encroachment on rights of other States is not very likely to occur, but in the case of the sea-bed the rights of other States are involved, as we will see later. For the reasons we will mention in the next Section we think that if occupation is a notion to be applicable in our case, it can only be used for the subsoil, and not for the sea-bed.

The conditions necessary to recognize occupation are the "animus possidendi" or rather the "animus occupandi" and the "possessio corpore". Should occupation be notified to other States? The Truman Proclamation could then be interpreted as such notification because it announces the will to assume rights on the minerals (part of the "territory"). At the same time the resources (at least a part of them) are taken possession of in reality. The other unilateral acts, however, only announce the "animus occupandi" but the notification is not accompanied by actual taking of possession (except in the Persian Gulf where drilling on the shelf is taking place). However, notification is not necessary. In the Clipperton-case (Briggs [212], p. 173) it was said that "the precise obligation to make such notification is contained in Article 34 of the Act of Berlin (1885) .... which .... is not applicable to the present case". The unilateral acts would appear to be superfluous and irrelevant for the validity of the occupation. In the North-American case "administration" is established. We believe that only in that case (and may be in the Persian Gulf) it would be possible to recognize occupation of the resources of the subsoil. The application of the notion occupation seems somewhat strained. The fact that in the Truman Proclamation (in the first paragraph of the Preamble) the Government of the United States "holds the view that efforts to discover and make available new supplies of these resources should be encouraged", constitutes another difficulty in adopting the notion "occupation". Clearly the rights claimed concern resources which have been discovered, but also those which have not been discovered. The presence of the last category is merely assumed. On "terra firma" occupation of an undiscovered land seems difficult to accept.

Discovery of the presence of minerals as a result of exploration, or better of actual drilling, would only give an inchoate title. Effective occupation would only follow when possession is taken of the resources, which will follow, however, in the case of drilling, often immediately afterwards. In normal cases, if no taking of possession follows, the inchoate title would perish. Here, however, the right is established before the discovery in many cases, and therefore discovery would be irrelevant.

Can we apply the notion "prescription"? The main difficulty here would be the lapse of time necessary to recognize a title. Granted the right exercised (possession or control and jurisdiction) is continuous and undisturbed, how long should this situation be maintained in order to recognize the acquisition of "territory"?

Here the authors differ in opinion. Oppenheim [42], p. 527, says: "No general rule can be laid down". Scelle [184], p. 160: "le Droit International ne comporte pas de délais déterminés des prescriptions, et il est un peu naïf de la part de certains auteurs d'avoir essayé de les chiffrer (quarante ou cinquante ans) ....". Fauchille [70], p. 762: "Il nous semble .... difficile d'abandonner aux circonstances de chaque espèce la fixation de la durée de la prescription, parce qu'un pareil système serait nécessairement une source de conflits, les Etats intéressés devant résoudre la question en sens opposé, suivant leur propre intérêt: .... Un délai fixe nous paraît donc nécessaire. Mais lequel? Les facilités qu'ont atteintes de nos jours les communications entre les Etats ne doivent pas, à notre sens, imposer un délai par trop long; il importe, au surplus, dans l'intérêt de la paix internationale, que les situations incertaines soient le plus rapidement réglées dès lors qu'un temps suffisant a été laissé, pour protester, à l'Etat dépouillé d'une partie de son territoire et au peuple soumis à une nouvelle domination. Pour ce double motif, nous estimons qu'on pourrait étendre au droit international le terme de 30 ans admis par le droit privé".

State-practice gives a few examples. Fulton [31], p. 425, discussing the fisheries struggles between England and Holland writes, that Cromwell said that the Dutch "could not establish their right by prescription, for by the civil law it required a hundred years for a just prescription, and the States had not existed so long as an independent nation". In the Treaty between Great Britain and the United States of Venezuela, respecting the settlement of the boundary between the colony of British Guiana and the United States of Venezuela, Washington, February 2, 1897 [213], p. 57, the Parties agreed that the Arbitrators should be governed by Rules, to be taken as applicable to the case. Rule (a), p. 60, reads: "Adverse holding or prescription during a period of fifty years shall make a good title. The Arbitrators may deem exclusive political control of a district, as well as actual settlement thereof, sufficient to constitute adverse holding or to make title by prescription".

In our case, it will be difficult to find a lead as to the period of time which would be necessary. One factor should, however, be taken into consideration. Hurst [214], p. 155, writes: "It is strange that the issue of the President's Proclamation, now more than three years ago, attracted so little attention outside the United States. I think this can only be explained by the fact that in September 1945, the Allies were just finishing the struggle against Nazi Germany. The ideas of every Allied national were then directed entirely to the coming of peace".

We would be inclined not to count for the calculation of whatever period may be taken as necessary, the first 3 years after the cessation of hostilities in World War II, because the Governments of most countries had quite different things to think about than the unilateral acts we are discussing.

The third possible approach to our problem is to assume that law existed on this point, in other words that the coastal State has "ipso iure" sovereignty over the resources or over the continental shelf. In that train of thought the unilateral acts would be declaratory of existing rights.

This theory seems to us quite unacceptable. The continental shelf is a relatively new conception. How could rights concerning a geographical formation exist before people were aware of its existence? Or do we have to assume that the rights were born at the same time that the soundings of scientific deep-sea research expeditions, like those of the Challenger, the Valdivia, the Siboga, the Meteor and so on, started to draw the 100 fathom-isobath on the charts? And who had heard of resources in the continental shelf 30 years ago?

Quite another thing is the wish that International Law will develop in that direction. The last two alternatives are given in the Copenhagen Report of the International Law Association [24]. On p. 13 under I, is a set of proposals under the heading: "The Conference affirms that the following principles *are* existing international law"; and on p. 15, under II: "The Conference, in order to provide a reasonable and equitable solution of the problems connected with the sea-bed and its subsoil, suggests that international law *should be developed* in the following directions: ....".

We can only agree with the last alternative. If no Convention will be convened on the subject the development will be that after a certain lapse of time the rights of the coastal States will be recognized in International Law. But which rights, those claimed by the United States or the more substantial ones of the other States? Of course this development would only take place if undisturbed. We noted already certain protests. The situation would be very uncertain and would give rise to conflicts if this subject would not be discussed internationally. Therefore we hope that an International Conference will take place and a Convention will be signed to lay down the principles on which this important subject should be founded.

The often mentioned continental shelftheory or doctrine, according to Feith's Report to the Copenhagen Conference (see Introductory Report Copenhagen [24], p. 12) fills a gap in International Law and considers the continental shelf outside territorial waters as "belonging to the coastal state, even without any proclamation and/or effective occupation by that state". The State practice so far, is according to "the adherers of this theory — a recognition by many important nations of that principle of the continental shelf doctrine". We find this statement not quite logical. If the doctrine fills a gap, it is recognized that no rule of International Law exists on the subject. If the doctrine purports to bring a new rule, we must confess that this method of creating International Law is quite a novel one.

What is the role of the International Law Commission in this respect? This Commission, according to Article 1 of its Statute [215]: "shall have for its object the promotion of the progressive development of international law and its codification".

In our opinion the Commission is fulfilling on this topic, the first of the two tasks mentioned in Article 1 of its Statute. The uncertainty of the situation comes well to the fore in the Award of the Umpire Lord Asquith of Bishopstone in the Arbitration between Petroleum Development (Trucial Coast) Ltd. and the Ruler of Abu Dhabi [216]. On 11 January, 1939, the Ruler of Abu Dhabi entered into a written contract with the Petroleum Development (Trucial Coast) Ltd., whereby the Sheikh purported to transfer to that Company the exclusive right to drill for and win mineral oil within a certain area of Abu Dhabi. The agreement contained an arbitration clause. A dispute arose.

The subject matter of the Award was (p. 2): "The arbitration is to determine what are the rights of the Company with respect to all underwater areas over which the Ruler has or may have sovereignty, jurisdiction, control or mineral oil rights. The Company claims that the area covered by the Agreement of the 11th January, 1939, ... includes in addition to the mainland and islands:

(1) All the sea-bed and subsoil under the Ruler's Territorial Waters (including the Territorial Waters of his islands), and (2) All the sea-bed and subsoil contiguous thereto over which either the Ruler's sovereignty, jurisdiction or control extends or may hereafter extend, or which now or hereafter may form part of the area over which he has or may have mineral oil rights". The issues as far as relevant to our problem appear from the following questions referred to Arbitration (p. 2): (iii) "At the time of the Agreement did he (the Sheikh) own (or as the result of a Proclamation of 1949 did he acquire) the right to win mineral oil from the subsoil of any, and, if so, what submarine area lying outside territorial waters? (iv) If yes, was the effect of the Agreement to transfer such original or acquired rights to the Claimant Company? (The Sheikh in 1949 — ten years after this agreement — purported to transfer these last rights to an American Company — the 'Superior Corporation': which the Petroleum Development Company claim he could not do, since he had already ten years earlier parted with these same rights to themselves)".

Article 2 (a) of the Contract, in the translation relied upon by the Respondent, the Sheikh, and accepted by the Claimants, reads (p. 4): "The area included in this Agreement is the whole of the lands which belong to the rule of the Ruler of Abu Dhabi and its dependencies and all the islands and the sea-waters which belong to that area. And if in the future the lands which belong to Abu Dhabi are defined by agreement with other States, then the limits of the area shall coincide with the limits specified in this definition".

The question arose, which law was applicable in interpreting the agreement. In order to answer this question the Umpire put himself another question: What is the doctrine of the continental shelf, is it an established rule of International Law or not? In the relevant part of the Award the Umpire, referring to the draft Articles 1, 2 and 3 contained in the Report on the third session, of the International Law Commission [36], says (p. 19):

"These draft Articles have been prayed in aid by the Claimants with the implication that they are, or are intended to be the expression of principles which are already part of international law, not merely of principles which ought to, or might with advantage form part of, that law in future. If this is indeed the contention of the Claimants, I am of opinion that it is ill-founded"

He comes to the same conclusion as we did, and relied in particular on paragraph 6 of the comments on Article 2 (which we will quote presently) and in which it is stated that the International Law Commission is equally of opinion that it can hardly be said that the unilateral action has already established a new customary law. The Umpire then continues:

"I therefore cannot accept these Articles as recording, or even purporting to record, established rules: and if they do not, if they are mere recommendations as to what such rules might with advantage be, if adopted by International Convention, they clearly cannot affect the construction of the contract of 1939. (f) Pausing here before dealing with the last question, viz.: the effect, if any, of the *negotiations* on the meaning of the contract; and considering only the possible effect on the construction of the contract of the doctrine of the Shelf; I would summarise as follows the Claimant's argument and my conclusions about it: The Claimant's primary contention is (1) that the doctrine of the Shelf is settled Law, (2) that it always was so, and therefore that it was so in 1939; ergo, the meaning which some of the expressions in the contract would or might otherwise have borne is enlarged by the inclusion therein of the Shelf. For instance, in Article 2 either the expression 'the whole of the lands which belong to the rule of the Ruler of Abu Dhabi' or the expression 'and the sea-waters which belong to that area', are so enlarged by the inclusion of an area in this case measuring over 10,000 square miles of *extra*-territorial marine subsoil. The argument falls to the ground if I am right in rejecting the premiss on which it rests, namely, that the doctrine of the Shelf has become, and, indeed, was already in 1939, part of the *corpus* of international law.

Again, if I am right in rejecting that premiss, the second way in which they put their case also fails; here they rely on the proviso to Article 2 which says that 'If in future the lands which belong to Abu Dhabi are defined by agreement with other States, then the limits of the area' (of the Concession) 'shall coincide with the limits specified in this definition'. The argument is that the Concession is by these words expressly to extend to any after-acquired area of Abu Dhabi, and that the effect of the proclamations of 1949, if not retrospective, cannot be less than to add the Shelf to the area originally covered as from the date when the proclamations were promulgated. This argument also fails if I am right in thinking that the premiss on which it rests is invalid; but I think it would fail independently of that since there has been no definition of anything 'by agreement with other States', and I should have thought in any case that the definition referred to was limited to one affecting dry land, whether epirot or insular".

. . . .

"6. *Conclusions and Award.*

It follows, if I am right, that the Claimants succeed as to the subsoil of the territorial waters (including the territorial waters of islands) and that the Sheikh succeeds as to the subsoil of the Shelf; by which I mean in this connexion the submarine area contiguous with Abu Dhabi outside the territorial zone; viz.: the former is included in the Concession, and the latter is not; and I award and declare to that effect.

I would only add in conclusion a word about the Qatar Arbitration over which Lord Radcliffe presided. I have reached a result in this case which happens closely to correspond with that reached by Lord Radcliffe in that case, on other facts and a different agreement. There is in fact little connection between the two Arbitrations if only because in the Qatar Agreement there was no allusion in the contract to 'sea-waters' at all. If Lord Radcliffe instead of merely recording his conclusions had expounded the principles on which he had reached them, I should have derived invaluable and authoritative guidance from such an exposition; but as he took the course he did, I am to that extent *inops consilii*, and have only departed from his (perhaps more prudent) method and gone into general principles at the express invitation of the parties: to whose legal representatives I would wish to express my deep indebtedness.

(*signed*) ASQUITH OF BISHOPSTONE" [a]

In the literature we find the following opinions. Andrassy [217], p. 84, speaking about the unilateral acts, says: "Ces actes ne peuvent pas modifier le droit actuel ni créer des règles nouvelles. Mais on a prétendu, notamment aussi au sein de l'International Law Association, que le grand nombre d'actes unilatéraux peut produire un effet que les actes isolés sont impuissants à produire. Ce point de vue ne peut pas être accepté", and p. 86: "Le droit existant ne peut pas contenir une règle comme celle invoquée par la théorie du plateau continental, parce qu'il n'y a là-dessus aucune disposition conventionnelle, et une règle coutumière n'a pas eu le temps de se former. Il est certain qu'elle n'existait pas au moment de la proclamation du Président Truman", and p. 87: "Les espaces sous-marins n'ont pas été accessibles à l'activité humaine jusqu'aux temps tout récents. Au moment où ils le sont devenus, des règles spéciales régissant leur acquisition ne pouvaient encore exister".

---

[a] "The proceedings were held at 5, Rue Le Tasse, Paris, France, from Tuesday 21rst August, 1951, to Tuesday, 28th August, 1951.

Sir Walter Monckton, K.C.M.G., K.C.V.O., M.C., K.C. with him Professor H. Lauterpacht, K.C., Mr. G. R. F. Morris, and Mr. R. Dunn (instructed by Messrs. Bischoff & Company, Solicitors, London), appeared on behalf of Petroleum Development (Trucial Coast) Limited.

Mr. N. R. Fox-Andrews, K.C., with him Professor C. H. M. Waldock, K.C., Mr. Stephen Chapman, and Mr. J. F. E. Stephenson, (instructed by Messrs. Holmes, Son & Pott, Solicitors, London) appeared on behalf of His Excellency, the Ruler of Abu Dhabi".

Referring to the proclamations of the Rulers of some territories on the Persian Gulf, where, as we have seen in the Bahrain instrument, it was said that a right has been established in international practice by the action of other States, Lauterpacht [218] says, p. 393: "Four years in international relations is on the face of it too brief a period to make possible the creation of a new rule of customary law .... A 'consistent and uniform usage practised by the States in question' — to use the language of the International Court of Justice in the Asylum case — can be packed within a short space of years. The 'evidence of a general practice as law' — in the words of Article 38 of the Statute — need not be spread over decades". He thinks, however, that the period of time is not important in this matter, p. 394: ".... assuming that we are confronted here with the creation of new international law by custom, what matters is not so much the number of states participating in its creation and the length of the period within which that change takes place, as the relative importance, in any particular sphere, of states inaugurating the change. In a matter closely related to the principle of the freedom of the seas the conduct of the two principal maritime Powers — such as Great Britain and the United States — is of special importance".

Yepes made a speech in the 67th meeting of the International Law Commission, which struck us as quite an exceptional statement concerning customary law. He said, S.R. 67[25], p. 6: "It could be seen then that the Proclamation of 28 September, 1945, might rightly be regarded as one of the most important documents of our epoch and that it constituted a veritable customary law to which the Commission should give recognition by incorporating it in its code of international law".

We will give the opinion of the International Law Commission after discussion of a few other aspects of the subject.

We come to the provisional conclusion that the value of the unilateral acts is an initiative impulse to a new development in International Law, and at the same time, from a national point of view an act of economic self-defence, a sort of conservatory seizure or attachment to prevent other States from exploiting the resources of the shelf.

SECTION 3. THE MEANING OF CONTROL AND JURISDICTION AND
OTHER RIGHTS CLAIMED

Two statements concerning the resources of the subsoil and
sea-bed of the continental shelf are made in the Proclamation
of President Truman:

1. the United States regards them as appertaining to the United
   States, and
2. regards them subject to its control and jurisdiction.

The first statement, we believe, cannot mean anything
else than a property claim, and we should think that it makes
ipso facto the second statement superfluous.

In the Executive Order of the President, these resources
are "Reserved, set aside, and placed under the jurisdiction and
control of the Secretary of the Interior for administrative purposes".

It may be important to know what interpretation the Secretary
of the Interior has given to this wording. In his Letter of transmittal, the Secretary of the Interior, Harold L. Ickes, in his
Annual Report, 1945 [48], p. X, referring to this Proclamation as
well as to the Proclamation by the President with respect to
Coastal Fisheries, says:

"Two Presidential proclamations assert our sovereignty over
the mineral resources of this ground, and our jurisdiction over
the fishery resources of the high seas contiguous to our lands.
The food and mineral resources of these areas are worth billions
of dollars ...." and .... "the cost of the survey may run to
several millions of dollars if we include the cost of ships and
equipment that have served their war purpose for the Navy
and which are still in the Navy's possession. Even if we did count
the cost of these essentials in the cost of the survey, which would
be doubtful bookkeeping, the shelf would still be cheap. Alaska
cost us $ 7,200,000; the Danish West Indies $ 25,000,000; and
the Louisiana Purchase amounted to $ 27,000,000. The Conti-
nental Shelf cost only the forethought that was required to
assert our *sovereignty* over it" (underlined by us). This interpretation brings the resources as well as the continental shelf,
in which they are contained, under the sovereignty of the
United States. We should think that this interpretation, by the

Secretary of the Interior himself, who has been charged by the President in the Executive Order to administer the jurisdiction and control over these resources, is of the utmost importance. Similar interpretations have been given, for instance by Hurst [214], who starts to say; p. 159 that "the text of the Proclamation does not purport to effect any extension of the sovereignty of the United States. In fact the language used points the other way; the operative part of the Proclamation describes the Continental Shelf as 'beneath the high seas' and again the last sentence of the operative part says 'The character as high seas of the waters above the Continental Shelf and the right to their free and unimpeded navigation are in no way effected' ".

We may interrupt Hurst for a moment to give our opinion that this reasoning does not seem to hold water, in so far as it peruses the words "the right to their free and unimpeded navigation are in no way effected" as a proof that no extension of sovereignty is claimed. First of all because this right of free and unimpeded navigation might have been meant as a favour, as a sacrifice of one of the rights of sovereignty, the rest, however, being claimed to the full extent. Secondly it may be argued that free and unimpeded navigation is also recognized to exist in the territorial sea, albeit qualified by the adjective "innocent". The territorial sea is under the sovereignty of the coastal State, although this sovereignty is qualified by the right of innocent passage. But the subsoil under the territorial sea is generally considered to be under the absolute sovereignty of the coastal State, without qualifications or conditions. Therefore we believe that this part of the argument of Hurst cannot stand. Hurst goes on ....: "The international law world would hesitate to admit that if the waters above the Continental Shelf are high seas, they are, or can be subject to the sovereignty of the State. The Proclamation must therefore be read as not affecting any immediate extension of American sovereignty over the Continental Shelf".

Here again we beg to disagree with Hurst. Hurst is discussing only the American Proclamation relating to the Continental Shelf. He is right in saying that if the waters above the Continental Shelf are high seas, these *waters* cannot be subject to the sovereignty of the State. But that is not the point. He is discussing the Continental Shelf and jumps to the conclusion that therefore the Proclamation must be read as not affecting any immediate (this adjective is not clear in this context) extension of American sovereignty over the Continental Shelf. We fail to see the logic in this reasoning. Why could not the sovereignty be claimed over the sea-bed and subsoil alone, without the waters above.

He feels the possible criticism when he writes on p. 164: ".... Hitherto it has, I believe, been generally assumed that the limit of a State's sovereignty is a vertical straight line stretching upwards and downwards ad infinitum from the starting point. Was the Continental Shelf policy intended to introduce a new system? A system under which the limits of a State's sovereignty would be a line which made a gigantic zigzag ....".

We answer that this system is not new. In the first place we mentioned already the difference in "quality" of the rights of sovereignty going vertically from the subsoil underneath the territorial sea into that sea. The sovereignty changes from an absolute into a relative one. But there are stronger examples.

We refer to the note related to the last sentence of Section 2 of Chapter I, where we mentioned a borderline between coal deposits owned by Germany and the Netherlands, independent of the frontier between the two countries. (Treaty of 17 May, 1939, Netherlands Statute book No. 30). We meet here in State practice the "zigzag", referred to by Hurst.

It seems that there is strong evidence that control and jurisdiction is the same right, or nearly the same right as sovereignty. Several authors are of this opinion and in spite of the initial thoughts, quoted above, Hurst comes to the conclusion, p. 162: "One cannot read this Proclamation without feeling that within the area of its Continental Shelf, the United States is claiming rights which are as large as sovereignty; ....".

Brierly said (S.R. 68[49], p. 8): "If the littoral State had exclusive rights of control and jurisdiction over the subsoil, it could be regarded as enjoying sovereignty". Waldock [219] argues, relying on the wording of the Executive Order, p. 32: ".... the Proclamation (Truman) looks very like an act of appropriation", which would, we think come very near a right of sovereignty. Vallat [220],

p. 336, writes: " .... It is difficult to see what distinction there is between control over the 'natural resources' and control over the subsoil and sea-bed themselves. Anything of value might equally be regarded as a use of or interference with their 'natural resources'. Therefore, it does not seem that the use of this expression imports any real limitation, and the claim may be taken as relating to the subsoil and sea-bed themselves. Indeed, the contemporaneous press release spoke simply of 'jurisdiction over the continental shelf'. Moreover, 'jurisdiction and control' are tantamount to sovereignty. Thus, notwithstanding the restrained language of the Proclamation, it does appear to amount to a declaration that the Government of the United States regards the sovereignty over the continental shelf as belonging to the United States. The failure to state this expressly may have been due to the realization that a bare unilateral declaration could not, in international law, vest any legal right to sovereignty in the United States. The mere statement that the United States regards the continental shelf as subject to its jurisdiction and control is not equivalent to actual occupation or control which is necessary to found a legal claim to sovereignty according to accepted views of International Law".

Cohn [221], p. 30, speaks of the "sea-bed of the continental shelf, which was summarily put under the sovereignty of the United States ....". (Undoubtedly he meant sea-bed and sub-soil).

In the Memorandum on the Regime of the High Seas [22], however, it is said, p. 81: "It would appear wrong to identify with sovereign rights the rights to the United States' continental shelf claimed under the President's Proclamation. The Proclamation makes no claim, in respect of the United States' continental shelf, to all the national rights exercised in the conditions recognized by international law; it merely lays claim to certain limited and specialized rights in respect of certain areas beyond the outer limits of United States' territorial waters". According to this vision the United States waived rights. It could have made claim to *all* the national rights exercised in the conditions recognized by International Law.

In our opinion the author of the Memorandum can only have had in mind the possibility to claim sovereignty rights over a part of the subsoil, like the countries exploiting mines under the sea-bed or like the rights which England and France might have claimed over a part of the tunnel under the Strait of Dover.

No other rights are recognized in International Law, unless the author is of opinion that the further claims made by the South-American Republics for instance, i.e. the sovereignty over the continental shelf, are already recognized in International Law, a thesis which we do not accept.

But in claiming that the resources of the subsoil appertain to the United States, this State appropriates a part of the subsoil, and in fact by exploiting these resources acts as owner, at least as being entitled to dispose over these resources. It is true that the material or substance of the continental shelf subsoil, not falling under the category of minerals, are not claimed, but the liberty is taken to work that material, to drill holes in it for instance.

Hurst [214], p. 162, calls it an "exclusive" control, although the word "exclusive" does not appear in the Proclamation. But in fact it is exclusive. Without any doubt the United States would object if Mexico drove a tunnel in the subsoil in order to extract minerals in such a direction that it would enter the continental shelf adjacent to the United States, but outside the three-mile limit. In other words the United States exclude other States. The South-American Republics refer to the claim of sovereignty rights in the Proclamation of President Truman. This may be a wrong interpretation, inspired by wishful thinking, or looking for support to justify their own claims. It may, however also be that the Governments of these Republics could not see any real difference between control and jurisdiction on one side and sovereignty on the other.

All other States except the Philippines claim sovereignty over the sea-bed and subsoil of the continental shelf, although different expressions are used, like "belongs to" or "is integrated into the territory, under the exclusive jurisdiction and dominion of the Federal Union" (Brazil). The Philippines Petroleum Act speaks of "All natural deposits or occurrences of petroleum or natural gas .... found in .... the continental shelf .... belong to the State". Strictly speaking this could mean the shelf as a whole, because it consists of deposits or sediments, but we assume that only mineral deposits are meant, if the words "deposits" and "occurences" both relate to petroleum.

The United Kingdom annexes the continental shelf in the colonies. The boundaries are "extended" to include the area of the continental shelf.

It seems that the extension of sovereignty is pretty well general. The International Law Commission proposes in Article 2 of the Report of its Third Session [36], a form, which seems to be in between the United States' claim on the resources only and the claims of the other States on the continental shelf, by saying that the continental shelf is subject to the exercise by the coastal State of control and jurisdiction *for the purpose* of exploration and exploitation of its natural resources. Whether this limitation means anything in practice is questionable. It is difficult to imagine that in practice control and jurisdiction over the continental shelf could be exercised for any other purpose than exploration and exploitation of the natural resources of the sea-bed and subsoil. What we regret in particular is, that the Commission speaks of the continental shelf in Article 2 and does not mention in this Article the sea-bed and subsoil. These are mentioned in the comment, which will be quoted later.

SECTION 4. SEA-BED AND SUBSOIL

In all the documents we met so far these two notions are mentioned together in one breath.

The resources of the sea-bed and the subsoil belong to the coastal State or are under the control and jurisdiction of that State. What are the resources of the sea-bed? The sea-bed is nothing else than the surface dividing the sea from the subsoil. But it is an evasive thing. If we dig into it, if we dredge sand or shells out of it, the material once forming part of that dividing surface is taken away, but the sea-bed is still there. The bottom of the pit we have made forms the new sea-bed. It is true that the sea-bed is the roof of the subsoil, but it is infinitely thin. Even if we take only a tiny quantity of material from it, this material actually belongs to the subsoil. One could argue that in practice the upper layer is called sea-bed and the deeper layers subsoil. We would ask, how deep is this upper layer, and nobody could answer this question. If we dump sand or clay on the sea-bottom, the sea-bottom rises. In short the sea-bed is indestructible. How then can we extract resources from the sea-bed? If we dredge mud from the sea-bottom in order to extract tin-ore as is done in the drowned river valleys in the Singkep-tin concession in Indonesia [222], we do not take anything from the sea-bed, we only displace the sea-bed to a lower level. The material, the mud with the tin-ore we get to the surface, we appropriate, is in fact subsoil and nothing else. It is a generally accepted fact, that the sea-bed and subsoil under the territorial waters belong to the coastal State. It is a generally accepted fact that the subsoil, even outside the territorial waters of the coastal State can be appropriated or occupied by that State, as is proved by the acquiescence of the States in the exploitation of submarine mines (although we believe that none of them actually pass the three-mile limit).

But it remains to be seen whether in International Law it can be said that the sea-bed can be occupied. If the Sinkep-tin concession would have reached the three-mile limit and tin-ore is still available outside that limit the Company will go on dredging outside that limit and we doubt if any State would object. But that is not a matter of occupation of the sea-bed, it is only a penetration necessary to get hold of the subsoil. It is a form of *use*. Let us compare for a moment the use made of the sea-bed with the use made of the high seas. We believe that the status or regime of the sea-bed is very akin to that of the high seas. The use made of the high seas is in similar ways made of the sea-bed. Where navigation is concerned the sea-bed is used to anchor, is used for soundings, (with the sounding lead as well as for echo soundings) and can be used by a submarine to rest on in comparatively shallow waters. Where fishing is concerned it is the abode of bottom-fish and sedentary species and is used in so far, as the trawlnet scrapes the bottom very closely, as a matter of fact penetrates into the bottom about one inch. It is further used for placing weirs or sero's, in that other meaning of the term "sedentary fisheries". It is used for laying telegraph cables on it. And now it is used for that new kind of use we are making of the high seas, erecting installations for oil-exploitation. What is the difference with the high seas? In our opinion the sea-bed has much more in common with the high seas than with the subsoil.

We propose to divorce the sea-bed from the subsoil and marry it to the high seas. We have another reason for this, but we will first call for support.

Higgins and Colombos [154], p. 55, say: "A clear distinction must be drawn between the bed of the sea and its subsoil.

As regards the former, the better opinion appears to be that it is incapable of occupation by any State and that its legal status is the same as that of the waters of the open sea above it. The same reasons for maintaining it unappropriated in the interests of the freedom of navigation apply, with equal force, to the bed of the sea. Exceptionally, on grounds based on historical and prescriptive considerations, it has been generally admitted that a limited portion of the bed of the open sea is capable of occupation by individual States for well-defined purposes, and entitled to recognition by other States. This is notably the case of the pearl fisheries ....".

We repeat what has been said in our Section on Sedentary Fisheries that the historic rights of the coastal State to the sea-bed of pearl-banks is an exception and should never be taken as a precedent to build on other kinds of occupation. Occupation of the sea-bed would interfere with the rights of other States, occupation of the subsoil does not interfere with the rights of other States.

The use we make of the sea-bed, however, does not interfere with the rights of other States to use the sea-bed. The sea-bed cannot be consumed. Taking tin-ore from the sea-bed does not interfere with the rights of other States. Once the area has been exploited and left, another (deeper) sea-bed is left and another State can lay its telegraph cable to rest on the new sea-bed. A ship can anchor on that new bed and if the soil is not too rough after dredging, a trawler can again try to catch the bottom-fish which have returned. Laying a cable says Hurst [123], p. 42, is not occupation of the sea-bed. The mere fact that a cable is lying on the sea-bed does not prevent to lay another cable across it.

Gidel [112], p. 500, also gives expression to the opinion that the sea-bed has the same regime as the high seas where he speaks about the exceptional occupation of the sea-bed allowed in the case of sedentary fisheries: "L'admission du principe de la légitimité de l'occupation du fond de la mer comporte nécessairement des conséquences restrictives du principe de la liberté de la mer. A ces conséquences, il n'est pas possible d'échapper, quoi qu'on fasse".

We have also another reason for the proposed divorce. If we keep in mind why we are discussing this problem, i.e. to find the ways to create a system of law which governs the exploitation of the minerals contained in the continental shelf, it is not far-fetched, we believe, if we take a look at the exploitation of minerals ashore. A comparative study of mine law is extremely fertile in this connection. By doing so, one of the first things which strikes us is the difference in principle between, for instance, the European continent and the United States. In French and Dutch mine law (the latter being based on the French one) the owner of the soil is not owner of the minerals contained in the subsoil. In the United States the owner of the soil is owner of the subsoil and everything which is in it (exceptions will be discussed presently) [a].

In France and Holland certain kinds of minerals appertain to the State. The State grants concessions. Exploitation without a concession is forbidden. The concessionaire becomes owner of the mine (i.e. the minerals). This system, dating from the time of Napoleon (the Law is of 10 April, 1810), recognized that it was not in the general interest to leave certain minerals to the owner of the surface. Minerals of the greatest economic importance should only be exploited by way of concessions to those who had the means and the ability for proper exploitation.

The owner of the surface is not allowed to dispose over the minerals. The original rights of the owner of the surface are converted into an allowance, by granting the concession. It is actually a sort of expropriation in the general interest. The

[a] The last principle seems to govern also Islamic law according to el Khoury (S.R. 66[7], p. 21): ".... according to Islamic law, the owner of a territory was also the owner of the air above it and of anything that might be found below the surface of the soil. What was to be found above and below the territory formed a continuous whole ....". With exceptions of certain minerals of great economic importance, it also prevails in the United Kingdom.

State has eminent domain over the most important minerals.

First of all we quote a few articles of the French Law of 1810 as amended several times [223] (p. 652):

Titre II, Article 5. "Les mines ne peuvent être exploitées qu'en vertu d'un acte de concession délibéré en Conseil d'Etat",
and article 6: "Cet acte règle les droits des propriétaires de la surface sur le produit des mines concédées",
and article 7: "Il donne la propriété perpétuelle de la mine, laquelle est dès lors disponible et transmissible comme tous autres biens . . . .".
Finally article 19 (p. 655): "Du moment où une mine sera concédée, même au propriétaire de la surface, cette propriété sera distinguée de celle de la surface, et désormais considerée comme propriété nouvelle . . . .".

As we have said, the Dutch mine law, originally the French one but amended since, is still based on the same principle, although State interference, as everywhere, has increased. The ownership of the mine concerns the minerals only, not the rock in which they are contained (Boekhold [224], p. 86). Before we proceed, it is useful to consider possible analogies with our problem. Would these principles be applicable to the exploitation of the minerals of the continental shelf? If so, it would mean, that the coastal State, in this comparison, would be the owner of the minerals but not the "owner" of the sea-bed (surface), nor the owner of the subsoil. This would more or less be in the line of the United States Proclamation regarding the natural resources of the subsoil and sea-bed as appertaining to the United States. However, the sea-bed would remain under the regime where we believe it ought to remain, i.e. the regime of the high seas. This regime could best be characterized as the general interest of all the States of the world. We escape that way the controversy whether the sea is "res communis" or "condominium" (Gidel [112], p. 213–224). Occupation of the sea-bed is out of the question, just as much as occupation of the high seas.

The whole object of the discussions concerning the continental shelf is to find a solution acceptable to all States for an efficient and proper way of exploiting minerals contained in the continental shelf. This aim is too often lost sight of. The continental shelf theory should not be used for other purposes, i.e. extension of sovereignty, or be mixed with the quite different problem of the fisheries. We believe that the principles of continental mine-law would suffice to attain what we want to attain. That the coastal State should exploit the minerals contained in the shelf adjacent to its coasts is not only reasonable, but probably the only possible way.

Quite a different question is, whether the comparison with the continental mine law should be taken as far as to introduce the principle of the general interest. We can imagine, that times will come when the supply of oil for instance is getting scarce. Only then, it may be necessary to create an organization in order to secure a proper distribution; something on the lines of the U.N.R.R.A. after World War II.

For the time being however, we could say at the most, that the coastal State is entrusted with the task to take its share in the oil-production for the common benefit of mankind. Of course there is an element of general (world) interest in this production. At the same time there is the general (world) interest of the use of the high seas and the sea-bed. If there is a chance that these two sides of the general (world) interest would be contradictory, then an international body would, we believe, be the best way to solve the problem.

Apart from these general principles, there are a few more aspects of continental mine law which may, if necessary, be taken as an example.

Mine law in general, takes into consideration the possibility of damage to the surface, by subsidence. This may become of interest if, as has been suggested (see below) exploitation of the continental shelf should also be undertaken by means of mine-galleries. Subsidence of the sea-bed may cause damage to telegraph cables or pipelines, the latter resulting in oil-pollution of the sea-water. A comparable case was decided by the French "Cour de Cassation", 16 November, 1852 (Boekhold [224], p. 164–165 and p. 194). A gas company had laid pipelines on the surface above a mine and the pipes were damaged through subsidence. The sentence of the first Court was confirmed: "la cour impériale a justement conclu que la compagnie, concessionnaire de la mine, était responsable du dommage éprouvé par la compagnie de l'éclairage au gaz ....", and what is noteworthy, it was said in the judgment: ".... que la circonstance que les travaux de la mine avaient été faits suivant les règles de l'art ne saurait affranchir la compagnie des mines de la responsabilité, par elle encourue; que cette responsabilité existe par cela seul qu'un dommage a été éprouvé et que ce dommage est la conséquence des travaux ou, de l'omission de certaines précautions".

The Netherlands East Indian Mine-law [225] in article 8 and article 20 prohibits mine-exploration and exploitation in areas, where buildings (up to a certain distance thereof), cemetries, *public highways*, *canals* and *railways* are on the surface, or for reasons of the general interest. This may be a principle to keep in mind when drafting a Convention. Application of this principle would mean that a rule should be laid down in the Convention, prohibiting erection of installations which would obstruct or endanger shipping in frequented shipping routes.

It is the principle that the public interest should be protected against the private interest of the concessionaire of the mine. We should think that the same principle ought to govern the relations between States: common interest prevails above the interest of a single State. Shipping should suffer *reasonable* sacrifices, because the oil-company is just as well a user of the high seas and subsoil as the shipowner is, but shipping in general is an interest of a higher order. The situation would change as we have said, if oil supplies would necessitate drastic measures for a reasonable distribution. Oil-production would then rise to the same level of world interest.

The exception we referred to above, to the different principle underlying American, and also English mine law, namely the interference by the Government in the case of certain minerals (coal and oil for instance), has narrowed the gap which existed between the two (Anglo-American and continental) systems. Halsbury [226], p. 2:

".... though prima facie, mines and minerals belong to the owner of the surface over such mines and minerals, the real position is nowadays different in very important respects. Thus mines of gold and silver, and petroleum (including mineral oil or relative hydrocarbons and natural gas) in its natural state in strata in Great Britain, belong to the Crown (Petroleum Production Act 1934, c. 36, s. 1.) .... Further the fee simple in all coal and mines of coal in Great Britain became vested in the Coal Commission on 1st July, 1942 (Coal Act 1938, c. 52, s. 3) and such fee simple ownership in coal and mines of coal, together with the exclusive right to work coal and mines of coal (except for very minor exceptions) became vested on 1st January, 1947, in the National Coal Board (Coal Industry Nationalisation Act 1946, c. 59). Further the Minister of Supply may do on or below the surface of any land such work as he considers necessary for the purposes of discovering whether there is present in or on the land in a natural state or otherwise, any minerals from which may be obtained uranium, thorium, plutonium, neptunium or any compound thereof, or any prescribed substances being such as may be used for the production of atomic energy. (Atomic Energy Act 1946, c. 80, ss 6, 18). In addition, the Minister may compulsorily acquire such substances and any minerals from which such substances can be obtained ....."".

In English law we meet the same provision to protect the surface in the mines (Working Facilities and Support) Act 1923 (13 & 14 Geo 5, c. 20) (Halsbury, p. 3 and 194, 198).

Where oil-exploitation is concerned, we quote Section 2 of the Petroleum Act (Halsbury, p. 217):

"Licenses to search for and get petroleum.
(1) The Board of Trade, on behalf of His Majesty, shall have power to grant to such persons as they think fit licences to search and bore for and get petroleum" [a].

Finally the principle of the general interest concerning certain minerals plays a role in the United States. The Federal Code Annotated, Vol. 9, p. 142 [227]:

"Except where lands have been acquired by the United States for the development of the mineral deposits, by foreclosure or otherwise for resale, or reported as surplus persuant to the provisions of the Surplus Property Act of October 3, 1944 (50 U.S.C. sec. 1611 and the following) (50 appx. 73), all deposits of coal, phosphate, oil, oil-shale, gas, sodium, potassium and sulfur, which are owned or may hereafter be acquired by the United States and which are within the lands acquired by the United States .... may be leased by the Secretary of the Interior .... No mineral deposit covered by this section shall be leased

---

[a] The powers of the Board of Trade have now been transferred to the Ministry of Fuel and Power.

except with the consent of the head of the executive department, independent establishment, or instrumentality having jurisdiction over the lands containing such deposits ....".

This principle (not the law quoted) applies to the resources of the continental shelf according to the Executive Order of 28 September, 1945, in which these resources are reserved, set aside and placed under the jurisdiction and control of the Secretary of the Interior.

The principles we find in mine law and which are, as we believe, worthwhile considering when drafting a Convention are:
1. The public interest (general interest of all the States) prevails above the private interest (the interest of a single State).
2. The distinction made between property rights of the surface and of the minerals.
3. The liability of the mine-owner for damage to the surface.
4. The rules concerning the protection of certain property on the surface, public highways etc.

The principle under 2 is another reason for our proposal to divorce the sea-bed from the subsoil.

That the sea-bed cannot be occupied was according to Waldock [219] the reason that Texas and Louisiana extended their territorial waters. He writes (p. 23) that when these States were "faced with the problem of oil-concessions under the high seas, purported to extend their territorial waters to a distance of 27 nautical miles from shore [a] in order to provide legal cover for their intended oil-concessions. These dubious pretensions were made simply because a coastal State was not then thought to have any natural rights in the sea-bed under the high seas".

We have said above, that occupation of the subsoil is considered legal in International Law. Hyde [177], for instance, writes, p. 467–468: "The subsoil appurtenant to the coast of a State and extending therefrom into an area beneath the high seas is doubtless susceptible to acquisition by that State .... a right of sovereignty therein may be brought into being".

Vorwerk in Strupp [228], p. 190: "Gewisz ist der Meeresboden okkupierbar, jedoch nur vom Lande aus in bergmännischen

---

[a] Texas actually to the edge of the continental shelf.

Verfahren ....". The Chairman (Scelle) said in the 66th Meeting of the International Law Commission (S.R. 66[7], p. 21): "The Commission appeared to be agreed that, where the subsoil could be reached from territorial waters or from the land itself, the right of states to occupy that portion of the subsoil of the high seas was indisputable".

In fact undersea mines (most coal mines) have been driven into the subsoil at several places. Gidel[112], p. 510, mentions in the territorial sea the mines of Dielette on the extreme north-west coast of Cotentin, the mines in Cornwall[a] and Cumberland and gives in a note an impressive list of mines in England, Scotland, Canada (Nova Scotia and Vancouver), Australia, Chile and Japan. He gives the figures for the distances these mines extend from the coast from which it appears that they are all well inside the 3-mile limit. In the Brief for the State of Texas (Supreme Court of the United States, October term 1949)[230] opposite p. 5, is a drawing of a cross-section of the submarine coal-mine of Lota, Chile. In the caption is written "submarine coal-mine running four miles from shore underneath the Pacific Ocean", but measuring the depicted gallery, the scale being 1 : 5000, this gallery does not seem to be longer than 1500 metres. It may be that the depicted gallery is not the longest in the mine. If the figure in the caption is correct it would be the only case of a mine extending beyond the 3-mile limit, being so far (apart from the 200-mile claim) according to Boggs[171], p. 196, the width of the territorial sea of Chile.

The other example usually mentioned is of course the planned Channel Tunnel. In Inclosure 2, the Memorandum of the Report of the Commissioners for the Channel Tunnel and Railway[231], being meant as a basis of the proposed Treaty, the boundary between the two countries was fixed as follows:

"1. The boundary between England and France in the Tunnel shall be half-way between low-water mark (above the tunnel) on the coast of England, and low-water mark (above the tunnel) on the coast of France. The said boundary shall be ascertained

---

[a] The Cornwall submarine mine Act 1858[229], divided the submarine mines between Queen Victoria (those below low-water mark) and the Prince of Wales and Duke of Cornwall (those between high- and low-water marks). It does not appear how far the mines below low-water mark extend.

and marked out under the direction of the International Commission to be appointed, as mentioned in Article 4, before the Submarine Railway is opened for public traffic. The definition of boundary provided for by this Article shall have reference to the Tunnel and Submarine Railway only, and shall not in any way affect any question of the nationality of, or any rights of navigation, fishing, anchoring, or other rights in, the sea above the Tunnel, or elsewhere than in the Tunnel itself".

None of the given examples gives much support to the theory, concerning the possibility to occupy the subsoil.

The mines give at least evidence of State practice concerning the sovereignty of the coastal State over the subsoil underneath its territorial waters. This sovereignty and that over the sea-bed under territorial waters is undisputed.

Several scientific associations drafted articles to that effect, for instance Art. 7 of the draft of Alvarez [232], p. 268; Art. 9 of the Draft Convention of Maritime Jurisdiction [233], p. 43 (where only the sea-bed is mentioned); and Art. 8 of Project No. 10 submitted by the American Institute of International Law to the Governing Board of the Pan American Union on March 2, 1925 [234], p. 319, reading:

"The American Republics exercise the right of sovereignty not only over the water but over the bottom and the subsoil of their territorial sea.
By virtue of that right each of the said Republics can exploit alone or permit others to exploit all the riches existing within that zone".

The replies of the Governments on Point II (Application of the rights of the coastal State to the air above and the sea-bottom and subsoil covered by its territorial waters) Bases of Discussion [126], p. 18–21, were practically unanimously in the affirmative (only 3 Governments did not express a positive opinion).

Art. 2 of Appendix 1 (Acts 1930 [61], p. 213) and the observations read:

"The territory of a coastal State includes also the air space above the territorial sea, as well as the bed of the sea and the subsoil.

Nothing in the present Convention prejudices any Conventions or other rules of international law relating to the exercise of sovereignty in these domains".

"Observations.

It has been thought desirable that a formal provision should be inserted concerning the juridical status of the air above the territorial sea, the bed of the sea, and the subsoil. The text as drafted is on similar lines to the previous article. It therefore follows that the coastal State may also exercise sovereignty in the air space above the territorial sea, and over the bed of the sea and the subsoil. It is important to emphasize that in these domains also sovereignty is limited by the rules of international law. As regards the territorial sea, including the air and the bed of the sea as used in maritime navigation, these limitations are, in the first place, to be found in the present Convention. So far as concerns the air space the matter is governed by the provisions of other Conventions; as regards the bed of the sea and the subsoil, there are but few rules of international law".

SECTION 5. CONTIGUITY AND SECURITY

The wording of some of the instruments we have discussed seems to suggest that they have been inspired by the theory of contiguity. The Proclamation of President Truman and the Royal Pronouncement of the Kingdom of Saudi Arabia use the words: "the contiguous nation". In the Presidential Declaration of Chile is written: "the continental shelf adjacent to their coasts". It may be that this dubious theory has played a role when the instruments were drafted, but we have our doubts. We referred to the theory in Chapter I, Section 6. Although some theories seem to have a tenacious character (like the property of fish born in inland waters) we can hardly believe that a theory which was rejected, would be used again to support a claim.

The theory was rejected by Max Huber in the Award of the Island of Palmas (Miangas) Arbitration.

The following passage from the Memorandum of the United States quoted by Nielsen [235] is noteworthy as it shows the doubts of the United States in relying on this theory, p. 26:

"Perhaps it may be said that definite, comprehensive rules of international law have not been formulated with regard to the rights accruing to a nation by reason of the geographical situation of territory".

Huber said in the Award (Scott [236], p. 111):

".... Although States have in certain circumstances maintained that islands relatively close to their shores belonged to them in virtue of their geographical situation, it is impossible to show the existence of a rule of positive international law to the effect that islands situated outside territorial waters should belong to a State from the mere fact that its territory forms the terra firma (nearest continent or island of considerable size). Not only would it seem that there are no precedents sufficiently frequent and sufficiently precise in their bearing to establish such a rule of international law, but the alleged principle itself is by its very nature so uncertain and contested that even Governments of the same State have on different occasions maintained contradictory opinions as to its soundness .... Nor is this principle of contiguity admissable as a legal method of deciding questions of territorial sovereignty; for it is wholly lacking in precision and would in its application lead to arbitrary results",

and p. 128: "The title of contiguity, understood as a basis of territorial sovereignty, has no foundation in international law".

Why then is this contiguity or being "adjacent" mentioned in the instruments? No State has so far "occupied" the continental shelf adjacent to another State outside its territorial waters (the case in the Persian Gulf is different). It seems that the apprehension that this might happen, has played some role in the action taken.

We believe that in spite of the wording the authors of the instruments do not rely on the theory of contiguity, but were moved by reasons of security. No State would like to see valuable resources so near its coasts to be exploited by another State. No State would like to see foreign installations being built so near its territorial waters. Let us admit immediately that the fear is more of a theoretical than of a practical nature, because the building of installations in front of a foreign coast does not seem practicable in most cases.

The fear for foreign installations in front of one's coast could be compared with the feeling of Judge Scott in the case of the

Anna [237], p. 815, when he argued that the protection of the country had to be reckoned from a number of little mud islands, that they were "the natural appendages of the coast on which they border and from which indeed they are formed" (we note that we have met these arguments also in defence of the claims on the continental shelf or its resources). Scott then says: "Consider what the consequence would be if lands of this description were not considered as appendant to the mainland, and as comprised within the bounds of territory.

If they do not belong to the United States of America, any other power might occupy them; they might be embanked and fortified. What a thorn would this be in the side of America"!

Finally, as far as the Proclamation of President Truman is concerned, there may be a remote influence of the Monroe doctrine on the wording. Wheaton [238] a writes, p. 169:

".... In 1912 the suggested sale by an American company to the Japanese of a large tract of land (400,000 acres), including Magdalena Bay in Mexico, evoked a declaration from the Senate that 'when any harbor or other place in the American continent is so situated that the occupation thereof for naval or military purposes might threaten the communications or the safety of the United States, the Government of the United States could not see without grave concern possession of such harbor or other place by any corporation or association which has such a relation to another government, not American, as to give that government practical power or control for naval or military purposes' ....".

SECTION 6. DIVISION OF A COMMON SHELF

We can be very short about this subject, because we believe that it should be left entirely to the countries concerned.

The situations may differ very much from one case to another and we do not believe that any general rules could be given, not even principles. We think that the Truman Proclamation and Art. 2 of the Iranian Bill (quoted in Chapter I, Section 1) express the idea clearly. Similar wording is used in the instrument of the other countries of the Persian Gulf.

---

a *That is to say: Keith in Wheaton's Elements of International Law* [238].

In the Saudi Arabian Pronouncement for instance "The boundaries of such areas will be determined in accordance with equitable principles by our Government in agreements with other States having jurisdiction and control over the subsoil and sea-bed of adjoining areas". We believe that the Treaty between the United Kingdom and Venezuela [110] would be a good example. We do not share the opinion of the Committee on "rights to the sea-bed and its subsoil", expressed in its Copenhagen Report [24], p. 15, that the International Law Association should have suggested any criteria, such as "the configuration of the coastlines", "the economic value of proven deposits of minerals", etc. We cannot see the necessity of general rules. There are certain peculiarities, which make the problem different from any other division of for instance a river or a narrow street. The Memorandum [22], p. 107 et seq., draws attention to this difference.

The system of the "thalweg" would not apply here. A difficulty would be the division if deposits are in the way. (See Memorandum p. 109). We think the principle should be kept in mind that the dividing line should if it can be avoided not cross an oil "pool".

In granting concessions ashore the following has been found to be a healthy principle: not two straws in one glass. But here again, it is up to the countries concerned to make an agreement. The Memorandum [22] comes to the same conclusion (p. 109).

SECTION 7. DISCUSSIONS INTERNATIONAL LAW COMMISSION AND
PROVISIONAL CONCLUSIONS

In its 69th meeting the International Law Commission discussed the question No. 5 of the First Report of François [10], p. 41, i.e. whether a right of sovereignty was involved or merely rights of control and jurisdiction. In the previous meeting Hudson suggested (p. 9) that control and jurisdiction were exercised by the littoral State for the limited purpose of exploring and exploiting the natural resources (as an alternative suggestion).

We find this idea back in Art. 2 of the proposals of the Commission. Kerno (S.R. 113 [33], p. 19) felt that "with so difficult a problem, it was essential to keep the desired aim clearly in view, and that was to enable, in the general interests, the exploitation of submarine resources. For that purpose it was necessary to accord coastal States a right of control and jurisdiction. There was, however, no question of granting them absolute sovereignty over the continental shelf. It followed that, where working of submarine resources was not possible, no control or jurisdiction was required, and there was, therefore, nothing to regulate".

Division was discussed in the 69th and in the 116th meeting (S.R. 69 [67], p. 7–11 and S.R. 116 [239], p. 3–5). The result of the discussions is Art 7 of the proposals.

Here follow the draft Articles 2 and 7 with the comments (Report 3rd session [36], p. 55–56 and p. 59).

Article 2

"The continental shelf is subject to the exercise by the coastal State of control and jurisdiction for the purpose of exploring it and exploiting its natural resources.

1. In this article the Commission accepts the idea that the coastal State may exercise control and jurisdiction over the continental shelf, with the proviso that such control and jurisdiction shall be exercised solely for the purpose stated. The article excludes control and jurisdiction independently of the exploration and exploitation of the natural resources of the sea-bed and subsoil.

2. In some circles it is thought that the exploitation of the natural resources of submarine areas should be entrusted, not to coastal States, but to agencies of the international community generally. In present circumstances, however, such internationalization would meet with insurmountable practical difficulties, and it would not ensure the effective exploitation of the natural resources which is necessary to meet the needs of mankind. Continental shelves exist in many parts of the world; exploitation will have to be undertaken in very diverse conditions, and it seems impracticable at present to rely upon international agencies to conduct the exploitation.

3. The Commission is aware that exploration and exploitation of the sea-bed and subsoil, which involve the exercise of control and jurisdiction by the coastal State, may to a limited extent affect the freedom of the seas, particularly in respect of navigation. Exploration and exploitation are permitted because they meet the needs of the international community. Nevertheless,

it is evident that the interests of shipping must be safe-guarded, and it is to that end that the Commission has formulated Article 6.

4. It would seem to serve no purpose to refer to the sea-bed and subsoil of the submarine areas in question as *res nullius*, capable of being acquired by the first occupier. That conception might lead to chaos, and it would disregard the fact that in most cases the effective exploitation of the natural resources will depend on the existence of installations on the territory of the coastal State to which the submarine areas are contiguous.

5. The exercise of the right of control and jurisdiction is independent of the concept of occupation. Effective occupation of the submarine areas in question would be practically impossible; nor should recourse be had to a fictional occupation. The right of the coastal State under Article 2 is also independent of any formal assertion of that right by the State.

6. The Commission has not attempted to base on customary law the right of a coastal State to exercise control and jurisdiction for the limited purposes stated in Article 2. Though numerous proclamations have been issued over the past decade, it can hardly be said that such unilateral action has already established a new customary law. It is sufficient to say that the principle of the continental shelf is based upon general principles of law which serve the present-day needs of the international community.

7. Article 2 avoids any reference to 'sovereignty' of the coastal State over the submarine areas of the continental shelf. As control and jurisdiction by the coastal State would be exclusively for exploration and exploitation purposes, they cannot be placed on the same footing as the general powers exercised by a State over its territory and its territorial waters".

Article 7

"Two or more States to whose territories the same continental shelf is contiguous, should establish boundaries in the area of the continental shelf by agreement. Failing agreement, the parties are under the obligation to have the boundaries fixed by arbitration.

1. Where the same continental shelf is contiguous to the territories of two or more adjacent States, the drawing of boundaries may be necessary in the area of the continental shelf. Such boundaries should be fixed by agreement among the States concerned. It is not feasible to lay down any general rule which States should follow; and it is not unlikely that difficulties may arise. For example, no boundary may have been fixed between

the respective territorial waters of the interested States, and no general rule exists for such boundaries. It is proposed therefore that if agreement cannot be reached and a prompt solution is needed, the interested States should be under an obligation to submit to arbitration *ex aequo et bono*. The term 'arbitration' is used in the widest sense, and includes possible recourse to the International Court of Justice.

2. Where the territories of two States are separated by an arm of the sea, the boundary between their continental shelves would generally coincide with some median line between the two coasts. However, in such cases the configuration of the coast might give rise to difficulties in drawing any median line, and such difficulties should be referred to arbitration".

The question of international control will be dealt with in the last Chapter.

Our provisional conclusions could be worded as follows.

1. Unilateral proclamations or national laws can only give an impulse for a new rule of law, nothing more.
2. There are no rights vested "ipso iure" in the coastal State.
3. The sea-bed comes under the same regime as the high seas.
4. Principles of mine law may with advantage be taken into consideration when drafting a Convention.
5. Division of the shelf should be left to the States concerned. Compulsory arbitration would be desirable.
6. A Conference should be convened as early as possible to discuss a draft Convention on the subject.

CHAPTER V

OFFSHORE DRILLING TECHNIQUE AND MINING

SINGLE SECTION

According to Alcorn [240], p. 115: "First drilling in open waters of the Gulf started on October 6, 1937, a joint operation of The Superior Oil Company and the Pure Oil Company, 6000 ft. off shore in 14 ft. of water (mean low tide) known as the Creole field. In 1941 a second one followed. The latest development of the Pure Oil Company is 9 miles off shore from the Louisiana coast-line".

A drilling platform is built on poles driven into the sea-bed. The deck of this structure has to be made at such a height that it cannot be damaged in rough weather. One of the few places where offshore drilling takes place, i.e. the Gulf of Mexico, happens to be a place where hurricanes do occur. The lowest deck of a platform in the Mexican Gulf should be at least 10 metres above the water (the tide should be taken into consideration). Considerable damage was caused in a hurricane in October 1949, by wave action, to a drilling platform, the Ohio Melben "rig" (see Krick [241], p. 154 et seq. and Farley and Leonard [242], p. 136)mainly because the deck was too low above the sea (26 ft.). This particular platform had a smaller one attached to it with a bridge. This small platform carried the crew's quarters. The total length was 333 ft., the width 90 ft. (i.e. 100 × 30 metres). Reading about this damage we thought of the words of Cauchy[243], which he wrote in 1862, p. 34–35: "En vain l'homme essayerait de fonder quelque chose dans les profondeurs de l'abîme: même à quelques pas du rivage, s'il entreprend de construire une digue pour abriter ses vaisseaux, la tempête se lève, et la vague en courroux balaie les assises du roc le plus dur, aussi aisément que l'enfant renverse le château de sable élevé par ses mains".

On the deck(s) we find everything necessary for drilling, the derrick, the engines, the pipes, tanks etc. Sometimes a separate platform is made to serve as a pipe-rack. Alcorn [240], p. 116, gives some further figures about the platform 9 miles off the Louisiana coast: ".... the drilling structure is 50 × 98,5 feet supported by 14 piles, 36 inches in diameter (wall thickness $\frac{5}{8}$ inch). Each pipe is driven to 80 feet of penetration and topped at 25 feet above mean low tide in 38 feet of water. The pipe-rack is 56 × 60 feet and supported by 12 piles, 24 inches in diameter ($\frac{1}{2}$ inch wall thickness)".

The erection of these platforms is of course far more expensive than drilling a well ashore. In an article "The search for oil on the continental shelf" [244], it is said, p. 52: "The most impressive feature of drilling in the Gulf is the tremendously high cost of everything connected with the operations. Ruggedness is essential to everything to withstand the buffeting of 120-mile per hour winds and 32-foot waves of the Gulf hurricanes .... the cost of drilling an offshore well is roughly three times the cost of a comparable land operation. On some locations, high winds and waves or fog cause the loss of more than 20% of the time". The article then draws attention to another reason for extra expense: "Oil produced from most offshore wells has been carried to storage on shore by barges. This method is slow, difficult and expensive. Once again the wind and waves multiply the problems .... the barges must be moored alongside the platforms for several days to receive their load. In bad weather this is extremely difficult and special precautions must be taken to moor the barges to prevent damage to the platforms". Therefore pipelines are being made, but ".... a conservative estimate of the cost of a properly laid and protected eight-inch underwater pipeline, 20 miles long, is from three to four times the cost of a similar line on land".

What is done now, to save costs, is that several platforms are connected with a collecting platform with tanks and separators. The latter is connected with the shore by a pipeline.

Another way to save costs is to drill more wells from one platform. In the article [244], p. 58, a photo is reproduced of Humble's Grand Isle: "7 miles off the Louisiana coast, first and largest self-sustaining platform structure erected in the Gulf.

Three oil-wells have been completed from this structure. Water is 50 feet deep. The upperdeck, 48 feet above the water, is 206 feet by 110 feet. The two decks provide nearly an acre of work space".

Other ways to drill for oil have been tried and engineers are still making and producing new devices. Mc Caslin [245], p. 54, describes, what is called Marine Rig No. 10, an old L.S.T. (landing ship for tanks). The remarkable thing is, that on the place where drilling is going to take place, first of all a pontoon is sunk. This 7½ by 54 feet pontoon fits into the "slot" of the L.S.T. and thus serves as a guide to the L.S.T. as it is pulled on to location. The L.S.T. is supplied with complete equipment for drilling a 15,600 feet well. A derrick has been mounted on it. When the L.S.T. is in position the well is drilled and after completion, the L.S.T. can be used elsewhere.

On these lines completely amphibious contraptions have been designed. For instance a model of a drilling barge, by Briggs [246], p. 108: ".... capable of drilling in sea depths up to 40 or 45 feet. This strange appearing craft has the foundation requirements for drilling and stability in motion while being towed from location to location. The unit derives its turtle-like appearance from sideboards, hinged at the bottom of the barge, and placed on all four sides. In transit these are raised; in operation, these boards carry pontoons for greater stability in setting the rig on bottom; and when this is accomplished, they act as an addition to the foundation for the greater distribution of drilling load. To facilitate the sea-going ability of the barge, the derrick is hinged on two legs and is designed to be laid down across the deck, when not in use or in transit", and p. 112: ".... Upon completion of the well, ... pontoons are refloated; the water expelled from the barge. Cables draw the sideboards to a vertical position, the derrick is lowered and the rig is then sea-worthy". We referred to these amphibious contraptions in Chapter III, Section 4.

We have said something about pipelines above. Denzler [247], p. 128, describes, how after ".... another operator's disastrous blow-out and fire in the Main Pass field in which one well, the structure, drilling barge, and several other barges disappeared beneath the waters, it was resolved that production and drilling must be separated by installing a tank battery . . . ." In the case he described an old platform was used to collect the oil and from there the pipe was laid to the shore. He continues, p. 130: "The California Company has laid to date, in the Gulf of Mexico, 3   4-inch lines with a total length of 47,900 feet and 8   2-inch lines with a total length of 23,800 feet. In the Main Pass the pipelines buried themselves in the soft bottom . . . . the lines in the Bay Marchand field have moved with the current. The bottom is hard blue clay, the lines have not buried themselves. . . . . Marking the Main Pass lines . . . . spar or nun buoys at 1000-foot intervals should suffice to prevent damage from dragging anchors. At Marchand Bay the first line laid has been damaged twice by dragging anchors in spite of the vigilance exercised".

About the results the following: Kastrop [248], p. 59, writes: "1947. First well completed beyond the three-mile limit on the Gulf's continental shelf by Kerr — Mc Gee Oil Industries Inc. in Block 32 Ship Shoal area of Terrebonne Parish Louisiana". This well was made 11 miles off shore in 17 feet of water. ". . . . Since the beginning of offshore activity in 1945, about 40 United States oil-companies have taken to the sea, and combined have spent well over $ 100 million, seeking oil beneath the Gulf's waters. Up to January of this year (1950), value of oil produced amounted to only about $ 340,000. There have been 53 offshore wells drilled through the early part of 1949. 8 of these wells were gas-condensate producers, 12 were oil-wells and 1 was a dry gas-well, with 33 dry holes".

Another way to save on the costs is the dual rig platform described by Kastrop, p. 149: "Unique in drilling operations anywhere, offshore or on land, is the California Company's dual rig platform in the Bay of Marchand field just 5 miles off the coast of Lafourche Parish, Louisiana . . . . By combining two heavy-duty power rigs into a single steel piling structure, and servicing them with only one floating tender, drilling costs are being slashed by 30 to 35 per cent" and p. 152: "Original structures provided for drilling five or six directional wells . . . . The next step in reducing drilling costs was to construct a platform large enough to accomodate two rows of wells, with five per row.

That step involved the same type of structure but utilized two drilling rigs, operating simultaneously ...." and p. 156: "When 10 or 11 wells have been completed, the derrick and substructure, along with other drilling equipment, are transferred to another structure .... At the end of the first year of offshore drilling .... the cost .... was tremendous — several times that of comparable costs for land operations .... The successful completion of 11 wells from one structure by two rigs running simultaneously side by side .... again proves that skill, experience and expert management can beat the sea and the high costs of drilling wells in fields far from shore —".

Ozanne [249], p. 49 et seq., quotes Clayton L. Orn, attorney of the Ohio Oil Company, who: ".... told the senators that his own company and its associates had already spent $ 1,4 million for Gulf of Mexico leases and about the same amount on geophysical exploration in search for oil, plus another $ 5,5 million in drilling 3 dry holes and a small gas-well. Orn pointed out that investment of $ 8,3 million to date has not resulted in the discovery or production of a single barrel of oil ....". He also quotes Walter S. Hallanan (President of the Plymouth Oil Comp. and Chairman of the National Petroleum Council) "who said that offshore operators already had invested $ 234 million in the search for oil in the Gulf of Mexico, but their gross revenues from oil produced so far amounts to only $ 6,3 million. He said the $ 234 million expenditures comprised $ 48 million for geophysical work; $ 60 million for lease bonuses and lease rentals; and $ 126 million for marine equipment, platforms and drilling. Since 1938, he said, 190 wells have been drilled in the Gulf. 77 were wildcats which resulted in the discovery of 14 oil-fields, all off the Louisiana coast. There are 60 wells which now produce from these fields a total of about 11,000 barrels daily. 14 gas-condensate fields and a dry gas field, all of which are now shut in, also have been found. Altogether there have been drilled 66 oil-wells, 22 gas-condensate wells, 3 dry gas-wells and 98 dry holes. Of the 113 development wells drilled, 51 have been dry holes".

Kastrop [143], p. 145: "The location of marine wells and their inacessibility during inclement weather make it necessary to include safety precautions in production practices which are not generally required for similar operations ashore .... With all lines converging on the separator platform, only one large flowline is required to carry the fluid to a distant shore station ... Such was the case with Standard Oil Company of Texas in its Smith Point field of Galveston Bay, Chambers County, Texas. This field consists of 12 producing wells about 10 miles from the Cedar Point tank battery where oil is treated and stored. A separator platform was centrally located in the field to remove gas from the fluids before flowing them through a six-inch line to the shore tank battery (with one well of the East Red Fish field .... making a total of 13 wells producing into this central point)".

We have mentioned several times the expression "directional drilling".

Parks [250], p. 95, writes: "Successful application of high angle directional drilling of unconsolidated sands in Gulf coast offshore drilling operations at Kerr — Mc Gee Oil Industries Inc. 's discovery on Block 32 about 12 miles from the Louisiana shore, marks an important step in future development of such reserves and answers a question which may represent the saving of millions of dollars in future drilling costs .... The tremendous location costs associated with offshore exploratory and development drilling prompted the experiment to see what could be done to drill one or more wells from the same platform with the application of presently developed high angle methods in the loose, unconsolidated sediments of the new discovery .... the objective of 600 feet deviation in 1700 feet of vertical depth with such loose sands as are present in the Block 32 pool had not been attempted, to the author's knowledge, and was an objective well above the tried and accepted practices in the industry. Two wells have been drilled and successfully completed".

Leigh S. Mc Caslin Jr. [251], p. 61, writes: "Magnolia's First Houston. The hazards and costs of offshore drilling were spotlighted this past week when Magnolia Petroleum Company announced its first oil discovery in the Gulf. Magnolia has spent $ 26,000,000 and drilled 16 wells over a 3 year period to bring in a single crude producer. Of the 16 tests, 6 did discover oil or gas-condensate fields. However, marketing of gas and condensate has not yet been attempted from offshore fields. Magnolia's new oiler has the distinction of being the farthest offshore producer in the Gulf of Mexico. It is located 25 miles off St. Mary Parish".

Pratt [37], p. 670, writes about the new oil-reserves on the continental shelf: "But this possible increment to our petroleum-resources, overwhelming in its proportions as it may be, is as yet only of remote interest. It is of no special significance to us as long as we can supply our needs for liquid fuels at lower costs from other sources", and on p. 671: ".... the most practical approach to great stores of petroleum in the sediments of the continental shelves, once we have proved their existence, is not through the waters of the turbulent sea above them, but along the sea-floor beneath these waters; not through a multitude of wells drilled through ocean waters .... but through a few galleries constructed upon or excavated into the ocean-floor from the adjacent land. Into these galleries, hundreds of miles in aggregate length, perhaps, oil would drain from a score of natural reservoirs distributed along their course, through wells drilled downward and outward in appropriate directions; and through suitable pipelines, traversing these galleries, the oil would then flow landward", and on p. 672: ".... these operations do not, however, promise a comprehensive solution of the problem of recovering the petroleum resources of the continental shelves. That solution will not be forthcoming until the time is at hand when our natural resourcefulness fails longer to contrive a preferable source of liquid fuels. In view of the prospect for almost unlimited energy from sub-atomic forces within the next few decades, and the even more imminent promise of liquid fuels from coal at costs fractionally higher than the prevailing cost of gasoline distilled from petroleum, this contingency appears to be remote. We may never find ourselves called upon to attempt the recovery of more than the most available fraction of whatever petroleum the continental shelves may house. But if occasion to develop fully the petroleum resources of the continental shelf ever arises, we may rest assured that the technical problems incident to their development will be readily solved under the spur of an adequate incentive".

Mining may, if conditions are favourable, be a method to produce shale oil as is done in several places in the world.

Of course, even if we have expressed the opinion that the fear for obstruction of shipping by drilling from the surface of the high seas was exaggerated, it needs no explanation that the method of mining would be preferable.

Chapter VI

SOME ASPECTS NOT YET DISCUSSED, SUMMARY AND PROPOSALS

Section 1. Consequences in War Time

Even without the fantasy of H. G. Wells or Jules Verne, some consequences of the offshore oil-production activities in war time may be predicted.

Oil producing wells, storage tanks etc. will certainly be considered as military objectives in war time. Collecting platforms or drilling-platforms are more vulnerable than derricks or tanks on land, because of the possibility of attack by naval forces, especially submarines, and the fact that there is less space available for 'defensive weapons. Although the individual target is usually smaller than similar ones ashore, where they may be placed nearer together, for this very reason there will be little place for an effective anti-aircraft battery for example. Nevertheless they will be armed. They do not fall under the installations envisaged by the American Institute of International Law in their Projects of Conventions of March 2, 1925, Project No. 12 (Jurisdiction) [234], p. 325, Art. 13:

"The American Republics whose coasts are washed by the waters of the sea and which possess a navy or mercantile marine, shall have the right to occupy an extent of the high sea contiguous to their respective territorial sea necessary for the establishment of the following more or less permanent installations, provided they are in the general interest;

1. Bases for non-military airships and dirigibles;
2. Wireless telegraph stations;
3. Stations for submarine cables;
4. Lighthouses;

5. Stations for scientific exploration;

6. Refuge stations for the shipwrecked",

because these installations do not fall under the description given. At any rate they are not in the general interest as meant in that article [a].

Nor would Art. 14 of this draft be applicable:

"It is expressly forbidden to fortify the installations referred to in the preceding article and to use them, even indirectly, as bases of supply, for warships, military airplanes and dirigibles, or for submarines".

These installations will, without any doubt, be considered as falling under the sovereignty of the coastal State, on whose shelf they have been built.

Is the coastal State neutral in a war, then supply of oil from one of its installations to belligerent warships or naval tankers could be considered as a violation of its neutrality [b]. Should we make a provision similar to Art. 19 of the Convention XIII, concerning the rights and duties of neutral Powers in naval warfare, the Hague 1907 (Scott [252], p. 515):

"Belligerent warships may only revictual in neutral ports or roadsteads to bring up their supplies to the peace standard. Similarly, these vessels may only ship sufficient fuel to enable them to reach the nearest port in their own country. They may, on the other hand, fill up their bunkers built to carry fuel, when in neutral countries which have adopted this method of determing the amount of fuel to be supplied", and Article 20 (p. 517):

"Belligerent warships which have shipped fuel in a port belonging to a neutral power may not within the succeeding three months replenish their supply in a port of the same power"?

Another dangerous possibility is the pollution of sea-water

---

[a] The text of this article and the next one is practically the same as the draft rules for maritime communication in peace time presented by Alvarez [331], Art. 17, p. 270, and Art. 18, p. 271.

[b] The possible remark, that oil cannot be used directly for ships is only partly sound. It is of course true for motor vessels, but fuel oil for boilers can sometimes be used directly from the well, as for instance the oil produced in the Tarakan fields, on the island of Tarakan off the east coast of Borneo. Another thing is that refuelling from an installation at sea will often be impossible because of the weather conditions. Besides it will probably be possible only from such installations which have not been connected by pipeline to the shore or a collecting platform.

on a large scale which would result from a torpedo, shell or aerial bomb hitting a producing well or a collecting platform. Damages of this kind will be extremely difficult to repair and even provisional repairs would take considerable time in which the outflowing oil may cause heavy damage to fisheries, sea-birds, beaches and harbours (fires).

The same can be said about a submarine pipeline damaged by a depth charge. François states in his Second Report [26] that, p. 30: "The Commission accepted the principle that all States are entitled to lay submarine telegraph and telephone cables on the high seas, and considered that the same principle should also apply to pipelines".

The analogy, however, in our opinion does not exist in every aspect. When the Institute of International Law in 1879 adopted the following resolutions (Scott [92], p. 24): "1. It would be very advantageous if the several States would agree to declare that the destruction or injury of submarine cables in the high seas is an offence against the Law of Nations ....", submarine pipelines were not existent, and we believe that we may say that if they had been existent, at least the kind used nowadays to connect offshore installations with tanks on land, the Institute would have hesitated to consider the destruction or injury of such a pipeline as an offence against the Law of Nations. Again we may say that such a pipeline misses the characteristic of a submarine cable of being of general interest. The only reason which may be invoked for calling destruction or injury an offence against the Law of Nations, is the pollution of the sea which would result. But the Institute of International Law adopted in 1902 rules regarding submarine cables in war time in which cutting of cables between two belligerents would be allowed anywhere, except in the territorial sea and in the neutralized waters appertaining to a neutral territory (neutralized by treaty or by declaration in accordance with Art. 4 of the Paris resolutions of 1894). A submarine cable connecting two neutral territories would be inviolable, whereas certain rules were also adopted for cables between neutral and belligerent territories, (see Scott [92], p. 162). It is clear that, a fortiori, submarine pipelines would be allowed to be "cut", first of all because they are not of general interest, but rather of national interest, and secondly, because they are, at least the kind we are talking about (other kinds do not exist yet) not connecting two territories, both neutral or at least one of them. On the contrary they connect two "parts" of one country, and if that country is a belligerent it would seem that destruction of the pipeline by an adversary party would not be in contradiction with any existing rule of International Law.

We are afraid that if, in view of the pollution of the sea being the result of such a destruction, rules would be laid down to forbid such operations, such rules would be happily violated because of the "military necessity" or a similar sort of excuse. Besides we do not give acceptance of such rules the slightest chance under the prevailing world conditions.

Another possibility is, that an area where oil-installations occur, will be declared to be a "prohibited zone" for all ships by the coastal State, or a "military area" by the adversary party, which will probably try to make communication with the platforms impossible.

A neutral State (if such a thing still exists) could proclaim a security zone (like the one of the Panama meeting 1939) in which it does not want to suffer hostilities, which are likely to occur near the sources of the "contraband", if enemy destination is taken for granted.

Section 2. Is international control (supervision) possible and desirable?

During the Copenhagen Conference of the International Law Association in 1950, M. A. de La Pradelle said (quoting the provisional minutes of the meeting):

"La communauté de la mer est le principe fondamental, qui doit s'appliquer au droit existant et à son développement future. .... Le contrôle et la juridiction de toutes les utilisations de la mer connues doivent appartenir à une organisme internationale où toutes les nations doivent pouvoir être représentées directement ou indirectement".

M. P. de La Pradelle said (quoting again the provisional minutes): ".... Je comprends que la question ne soit pas mûre pour la création d'une exploitation internationale, mais elle est incontestablement mûre pour l'institution d'un contrôle international". In the first proposal all resources are meant to come under international "contrôle et juridiction", in the second one only the resources of the subsoil.

The nature of this international supervision (contrôle cannot be translated by control in English, because the latter word has a much stronger meaning; supervision seems to be the right translation) would be administrative, authorizing explorations and with a certain say in the matter of exploitations, but without interference with the economic side of the question. The latter could only be realized in the future.

Finally the idea of international control was expressed at Copenhagen by Madame Guldberg (Sweden). She said (provisional minutes): "A mon avis, toutes les richesses marines et sous-marines, du sol et du sous-sol marin doivent être considérées comme le bien commun des habitants de la terre, elles doivent être exploitées par la communauté internationale et pour la communauté internationale". Reference was made to the Schuman-plan and the "International Coöperative Alliance". This proposal goes further than the previous ones.

We will deal with the last one first. We do not believe that this last plan is practicable. Installations cannot be built or operated loose from the nearest shore.

Relief of the men working on the installation, supplies of food and materials has to take place regularly.

Building material has to come from the shore. When drilling operations are going on all the supplies necessary for those activities have to be brought from the land and eventually when the well is completed successfully, the oil must be taken away either by tanker or through a pipeline to the nearest tank battery ashore. We therefore believe that the exploitation can be done best by the coastal State.

Quite a different question is whether some international supervision could be advantageous.

There is something to be said for such supervision. We pointed out that in mine law, the principle of the public interest plays a role. The owner of the surface is not allowed to exploit certain economically important minerals under the surface. The State interferes and grants concessions.

The general interest prevails against the private interest.

Further we mention the 3 law suits: the United States v. California, v. Louisiana and v. Texas. We have quoted short notes about the cases of Louisiana and Texas, but these cases, being mainly an American affair, cannot be dealth with in this work. The gist is that the United States claims to be the owner in fee simple of the minerals underlying the continental shelf outside the inland waters of these States.

Without giving a further opinion about the merits of these cases, we only remark that here the same principle seems to be recognized. The general interest of the greater community of the United States as a whole prevails against the smaller community of one State.

As a third example we mention an Ordinance of the Governor-General of the former Netherlands East Indies of 1903 [253], deciding that local Heads or Rulers recognized by the Government were not entitled to dispose of the revenue on sedentary fisheries. In a letter from the First Secretary of the Government of the former Netherlands East Indies, No. 1591, of 15 July, 1910, to the Resident of the southern and eastern Department of Borneo (Bb. 7295) it was said, that it had been a point of consideration whether the territorial waters adjacent to the self-governing areas belonged to those areas or to the central Government. The Minister for the Colonies decided that only the central Government had these sovereignty rights. The central Government is therefore entitled to regulate fisheries in the marginal belt.

Finally in a governmental guide: "Politiek Beleid en Bestuurszorg 1909 [254]", p. 73, referring to the decision above, it was said that there was still doubt as to who was entitled to grant concessions for tin-ore exploitation of the subsoil off the island of Singkep, the central Government or the self-government of Lingga-Riouw and Dependencies.

Again the Minister decided for the central Government (dépêche 10–5–1905 Lett. A [1]. No. 52 (1458)). In all these examples the same idea of the prevailing interest of the larger community comes to the fore. It is logical that the same thought is applied as between the World community and a single State. The only mistake made here is that the World community has not as yet reached a level of organization comparable with a single State. However, we can imagine that an international body is delegated by the States in a general Convention, to exercise certain powers. Performing that limited task it does represent a facet of the World community already in existence, i.e. the unanimity, or general will, which does exist on that single subject. In the draft of Alvarez [232] such an international body is suggested, Article 3: "Pour veiller à l'observation des dispositions du présent Règlement et pour prévenir et résoudre pacifiquement les conflits entre les Etats à propos des voies de communication maritime, il sera institué une 'Commission Internationale' ....".

The installations we have mentioned in Section 1 of this Chapter had to be approved in advance by this Commission and were put under the supervision of the Commission (Art. 19). Another similar body is the International Waters Office in the Schücking-plan (Report to the Council of the League of Nations[52], p. 38 and p. 58). Art. 3 containing provisions concerning this body, in the original draft was, however, suppressed. We have seen the discussions and proposals of the International Law Commission on this point concerning protection of fish. We believe that a similar organization, as we have proposed in Chapter III, should be given the power to approve the erection of oil-installations in connection with shipping interests and could also be given the power to decide in case of conflicts. We would, however, prefer to have a special Maritime Court to adjudicate conflicts between oil and shipping interests or fishery interests.

Hsu proposed in the 66th meeting of the International Law Commission (S.R. 66 [7], p. 28) "to entrust the development of the continental shelf to the international community" and wanted a joint exploitation of the resources of the continental shelf. This proposal was not accepted by the Commission. We refer to No. 2 of the Comments on Art. 2 in the Annex of the Report of the International Law Commission covering its third session [36], p. 56, which we quoted in Chapter IV. We agree that international exploitation would be difficult to realize, but we do propose a body for the limited purpose described above.

SECTION 3. SUMMARY AND PROPOSALS

We suggested to take the 100-fathom line as delimitation of the shelf and based this proposal on the conviction that it would take a long time before the technique had developed so much as to make exploitation up to that depth possible, and at the same time that exploitation beyond 100 fathoms depth would be extremely unlikely. We may mention a few geologists in support. Carsey [18] wrote that "exploration will proceed in water up to 50 or 100 feet by present methods, but that it will be extremely difficult beyond this point". Emery [255], p. 249, says that 75 feet is the present limit of exploratory activity.

Weaver [256], p. 397, writes: "To date, no one has attempted to drill a well off shore in as much as 60 feet of water, but many engineers consider it feasible. A few seem to feel they would be willing to drill to 100 feet of water". The writer does not consider that, "with the present general methods, anyone will recommend drilling at greater depths".

We believe that we added some evidence to the views held by the International Law Commission that fisheries should be dealt with independently from the continental shelf.

We have seen that there existed some discrepancy between the facts and proposals on international Conferences. We showed the methods adopted by different countries to achieve protection of fish, or to achieve exclusive fishery rights, and we showed the protests which were sent. Expecially the protest of the United States, concerning the Chilian Declaration proclaiming the national sovereignty of Chile over the continental shelf *and* over the seas outside the generally accepted limit of territorial waters, is important, where the first part is concerned also as a means to interpret the United States Proclamation.

We have used our discussions about bottom-fisheries as argument to prove that the sea-bed should come under the same regime as the high seas.

This thesis was further elaborated in Chapter IV, with the object to give another argument to prove that occupation of the sea-bed is illegal.

Where pollution of the sea-water is concerned, we propose that provisions concerning prevention should be laid down in a Convention regulating all the aspects of the exploitation of the mineral resources of the shelf, unless the Economic and Social Council should involve this kind of pollution in its activities.

Regulations concerning seismic operations should likewise be inserted in the Convention.

In the third Chapter we discussed unilateral extensions of territorial waters. We expressed the opinion that the development of unilateral action was undesirable and that an international Conference should be convened to discuss territorial waters, on the basis that uniformity was not strictly necessary.

We propose such a Conference to be convened and believe that there are local circumstances which make a wider belt than the usual 3 miles necessary and acceptable for other countries.

We propose to put on the agenda for the Conference to be convened universal rules for lights and signals for oil-installations. Building installations is a form of use made of the freedom of the seas and sea-bed. Because this sort of use may be dangerous for other users an international body should be empowered to decide whether installations should be built in certain places where hindrance to shipping may occur.

Installations are not artificial islands (because of their temporary character) and should not have a marginal belt, but only a security-zone. The width should be discussed at the Conference.

Territorial waters should not be barred.

Concerning the value of proclamations, declarations and decrees, we have said that they can be considered at the most as giving an impulse to the development of new law. Bingham [257], p. 178, says: "In this case also forehanded action by proclamation is wiser than the slow, tedious, and often disappointing process of negotiation of treaties ·with numerous states". It may be that the American Proclamation concerning fisheries has served a purpose by causing a shock, which was necessary to awaken countries and show the necessity of action. For the continental shelf no such shock was necessary and we believe that this unusual way to realize a desire has caused misunderstanding. According to the unilateral instruments which followed these Proclamations, the States concerned have either read more in the Proclamation than was meant, or used the opportunity to do something they had wanted to do for a long time. The result is that it will be very difficult to repair all this and adopt reasonable rules necessary for the development of the exploitation of the minerals of the shelf.

In an article "Oil under the High Seas: The New Rule" [258], the work of the International Law Commission is criticized, p. 355: "A novel right will, in effect, be granted to States over possibly vast expanses of high seas, without, however, the counterpart of the obligations and responsibilities which accompany normal sovereignty. One needs no reminder to-day that in oil, no less than in other spheres, extreme and irresponsible nationalism is far from serving the 'needs of mankind' which the Commission is seeking to promote".

We are convinced that there is no rule of International Law concerning the use of the continental shelf and hence no State has "ipso iure" rights on the shelf. The right can be housed comfortably under the uses made of the high seas and sea-bed on one side, and the eminent domain over the resources of the nearest subsoil on the other side.

Because of the dangers which may be caused, international discussions are desirable. Some rules of mine law could be used with advantage when drafting the Convention.

Finally we have said that the subject is a fertile field for conflicts. Apart from the type we have mentioned (the Abu Dhabi Award and the fairly similar case of Qatar) we thought of a few possible conflicts.

We may attempt to foresee what sorts of conflicts may occur in connection with oil-exploitation on the continental shelf.

It is not impossible for instance that the boundary of the shelf between two neighbouring States, like the United States and Mexico, gives rise to conflicts, if an installation is built near this boundary-line. Let us suppose that an "International Maritime Court" or an Arbitral Tribunal decides that the installation has been built by nationals of State A on the shelf of State B. How are the financial consequences of such a decision to be settled? Here we think of Art. 5 of the Projects of Conventions prepared by the American Institute of International Law in 1925, Project No. 11 (Rights and Duties of Nations in territories in dispute on the question of boundaries) [234], p. 322: "If an American Republic constructs works or proceeds to carry on work or make contracts in a territorial zone which is disputed by another, and which is assigned to the latter in the settlement of the dispute, the former shall receive from the second an indemnity equivalent to the value of the labor done and the works constructed.

It shall be the same if the works have been constructed and the labor executed while the territory was in litigation".

This Project did not materialize, but in drafting a Convention relating to the continental shelf, a similar provision may be considered.

The same situation would of course occur if a conflict arose concerning the boundary of the shelf between two States situated opposite each other with the shelf between them, like in the Persian Gulf could happen.

Another conflict which will certainly arise is that a well drilled not too far from the boundary-line between the shelf-areas of two neighbouring States produces oil from a pool which is claimed by the neighbouring State, or being made by directional drilling, produces oil from underneath the shelf-area of the neighbouring State.

APPENDIX

Since the manuscript was sent in to the Secretary General of the Institute of International law to be submitted to the jury, some developments took place and some more publications became available.

Following roughly the contents of this work, we start with the Definition of the continental shelf.

We criticized in Chapter I, Section 6, the definition of the International Law Commission in its draft Art. 1 (Third Report [36]*). In the Comments by Governments on Draft Articles on the Continental Shelf and Related Subjects* [259]*, doubt concerning this delimitation is expressed by Brazil (p. 4), the Netherlands (p. 36), wondering "whether a limit of 200 metres in depth would not place the law on a surer foundation and prevent unlimited expansion in the future", and Yugoslavia (p. 48), insisting likewise on a depth limit of 200 metres and drawing attention to the consequences of delimitation on agreements concerning the boundaries on a common shelf (Art. 7 of the draft), and the United Kingdom (Comments, Add. 1, p. 17), suggesting to "Delete 'where the depth of the superjacent waters admits of the exploitation of the natural resources of the sea-bed and subsoil' and substitute 'as far as the 100 fathom line' ".*

Gidel [260] *writes (p. 5) that the extent of the continental shelf would be uncertain and varying. Uncertain, because it would depend on the degree of technical development of a given country and varying, because the extent would change whith the technical developments.*

He finds the delimitation of the International Law Commission difficult to accept and remarks, p. 6: "On ne sait s'il faut l'attribuer à un désir de sa part de limiter actuellement le plus possible le champ du P.C. ou à celui d'en favoriser dans l'avenir la plus large extension".

He touches another point, (also discussed in the same Section 6) i.e. that the Truman Proclamation concerning the natural resources, bases the right of the coastal State on the fact that the continental shelf may be regarded as an extension of the landmass of that State and refers in that connection to the Judgment of the Anglo-Norwegian Fisheries Case, of 18 December 1951 [261]. *The passage referred to reads (p. 138):*

"*Since the mainland is bordered in its western sector by the "skjærgaard" (islands, islets and keys), which constitutes a whole with the mainland, it is the outer line of the "skjærgaard" which must be taken into account in delimiting the belt of Norwegian territorial waters. This solution is dictated by geographic realities*".

We may point out that this passage seems to us to be strictly connected with the particular case the International Court was dealing with and we doubt whether this statement referring to a particular local situation may be considered to be a general directive for other coasts. We come back on this in another connection.

Finally Gidel stands up for maintaining the term "continental shelf", because that would exclude banks not connected with the continent, like the Burwood Banks, south of the Falkland islands and the Saya da Malha Banks in the Indian Ocean.

The United Kingdom (Comments [259], *Add. 1, p. 18) proposes to treat these banks on quite a different footing: "In the opinion of Her Majesty's Government, such submerged plateaux are either* res communis *capable of acquisition by prescription or* res nullius *capable of occupation and exploitation by any State according to the normal law of occupation". Under such banks are comprised "plateaux.... separated from the coast by a channel more than 100 fathoms deep". This is exactly what Norway objects to, because Norway is separated from the North Sea shelf by a long and rather narrow belt of deep water, which would deny Norway participation in the North Sea shelf (Comments, p. 41).*

Fisheries.
Relation between the continental shelf and fisheries.

Iceland (Comments, Add. 1, p. 4) does not agree with the proposal of the International Law Commission to treat fisheries independently from the concept of the continental shelf: "Investigations in Iceland have quite clearly shown that the country rests on a platform or continental shelf whose outlines follow those of the coast itself....

whereupon the depths of the real high seas follow. On this platform invaluable fishing banks and spawning grounds are found upon whose preservation the survival of the Icelandic people depends.... It considers that it is unrealistic that foreigners can be prevented from pumping oil from the continental shelf but that they cannot in the same manner be prevented from destroying other resources which are based on the same sea-bed". (In Annex I is given the Law of April 5, 1948 [153], p. 12; in Annex II, the Regulations concerning conservation of fishing banks off the north coast of Iceland and in Annex III, Regulations concerning conservation of fisheries off the Icelandic coasts).*

Proclamations, Declarations and Decrees.

We can add to Chapter II, Section 3, the following instruments:
1. Proclamation of the President of South Korea, Syngman Rhee, of (presumably) the 19th January 1952. As the official text has not become available to us yet, we quote from the Nippon Times of 25 January, 1952: "The Japanese Government is expected to issue within a few days a statement opposing the action taken last Saturday by the Republic of South Korea in proclaiming Korean sovereignty over the sea 50 to 60 nautical miles from the Korean coasts in an apparent move to shut out Japanese fishermen from these waters.

South Korea last Saturday established a 'Syngman Rhee Line' to replace the 'Mac Arthur Line' which is to lose effect upon coming into force of the Japanese Peace Treaty.... Korean President Rhee in a proclamation said that the Government of the Republic of Korea holds and exercises national sovereignty over seas adjacent to Korea to protect, preserve, and utilize natural resources. The proclamation added, however, that the declaration does not interfere with rights of free navigation on the high seas. The declaration was apparently issued to other countries in general but may be interpreted as prohibiting Japanese fishing operations in that area since only Japanese fishermen are directly interested in the area. The Japanese Government is expected to point out that the Korean proclamation is unprecedented and runs counter to the principle of the freedom on the high seas as recognized by international law".

According to the Nippon Times of 29 January 1952, Japan formally protested against the South Korean action:
"The protest was made in a statement which the Japanese Govern-

ment through the Korean Diplomatic Mission in Japan, sent to the Republic of Korea's Government. The note charged that the Korean proclamation is utterly untenable under any of the accepted ideas of international society and therefore, cannot be acquiesced in by the Japanese Government".

The Nippon Times of 28 January, 1952, gives the Korean answer to the Japanese note. We quote the following passages:

"Pusan Jan. 27 — The Republic of Korea today reasserted its sovereignty over waters adjacent to the peninsula and rejected a Japanese Government charge that such a claim was a violation of freedom of the high seas. The text of the statement follows:

'Variegated analyses are being made as to our recent proclamation of sovereignty over adjacent seas. However, there are well-established international precedents — such as, President Truman's proclamations on coastal fishing and on natural resources on and beneath the shelf, and other proclamations of the same character made by the governments of Mexico, Argentina, Chile, Peru and Costa Rica.... The protective seas as designated in the proclamation bear radically different significance than the Mac Arthur Line. While the Mac Arthur Line limits the area within which Japanese fishing boats are allowed to operate, Korea's protective seas have been established to set limitations on the Koreans as well as Japanese, in order to prevent the exhaustible type of natural wealth in the said area from being exploited.... Proclamation of protective seas does not mean extension of territorial waters into the high seas. The special character of parts of the high seas that at the same time constitute adjacent seas has been recognized by many international bodies including the United Nations Commission on International Law. We do not lack in precedents in the international community which recognize the special status of adjacent seas. Those who still adhere to the 19th century concept of the freedom on the high seas, claiming absolute freedom of fishing on adjacent seas, must be considered as being unaware of the evolution of international law".

Although we should reserve a definite opinion till we have studied the offical text, we get the impression from the quotation given, that here again the Truman Proclamation on coastal fisheries has been misinterpreted.

In view of the fact that at that time a Conference was scheduled to take place on the fisheries question between the two Governments, the *Nippon Times of January 30, 1952, rightly remarked:* "*It is therefore not understandable, as Monday's Japanese Foreign Office statement pointed out, why the Korean Government has issued a proclamation on a subject which was slated for discussion between the two nations. Surely a mutually acceptable formula could have been worked out whereby the open waters under question would be protected from over-fishing*".

The *Nippon Times of 19 March, 1952, reports that at the Conference mentioned above:* "*The Japanese negotiators have said that Japan is ready to enter a pact for conservation of fishing resources and have proposed that the agreement apply to the Japan Sea, the Yellow Sea and the East China Sea*".

2. *Act relating to Fisheries in certain Australian Waters* (assented to *13th March, 1952*) [262].

We quote the following articles or parts thereof:

2. "*This Act shall come into operation on a date to be fixed by Proclamation....*

4. *In this Act, unless the contrary intention appears — 'Australian waters' means —*

    (a) *Australian waters beyond territorial limits; and*

    (b) *the waters adjacent to a territory and within territorial limits;....*

'Proclaimed waters' means Australian waters specified by Proclamation in force under section seven of this Act....

7. *The Governor-General may, by Proclamation, declare any Australian waters to be proclaimed waters for the purpose of this Act.*

8. *The Minister may, by notice published in the Gazette —*

    (a) *prohibit, either at all times or during a period specified in the notice, the taking, from proclaimed waters or from an area of proclaimed waters, of fish or of fish of a species specified in the notice;....*"

Then follow similar provisions concerning the size of fish and methods of fishing, and equipment.

9. "*The Secretary (to the Department of Commerce and Agriculture) or a prescribed authority may grant to a person a licence to take fish in proclaimed waters or in an area of proclaimed waters*"....

Under section 10 of this Act, certain specified officers may board a boat in proclaimed waters, being used or intended to be used for the taking of fish in these waters in order to search the boat for fish or equipment, examine nets etc. and seize, take, detain, remove and secure any fish, boat, net or other equipment, being used or intended to be used in contravention of this Act; and without warrant, arrest a person who has committed an offence against this Act, etc.

The Act further lays down provisions for research and development, penalties etc.

This Act seems to be merely a loaded gun, which is held at the present, in case the development of events would make it necessary to pull the trigger.

The Proclamation mentioned in section 7, would be in contravention of the principle of the freedom of the seas, if an area outside the territorial waters would be declared to be Australian waters. The Act does not mention exclusive fishing rights but the licence system opens the possibility to interfere with foreign fishermen.

In an Article: "Lively Developments in the MOP Industry" [263] *the following comments were given, p. 31:*

"The legislation can only be applied against a foreign vessel after the foreign Government concerned has accepted the legislation. Efforts probably will be made to negotiate agreements with Japan, Philippines, Indonesia, the Netherlands and Portugal. Meanwhile only Australian vessels operating in proclaimed areas outside the 3-mile limit can be interfered with, under International Law".

Of course these comments should be saved up, until the Act would come into operation by virtue of a Proclamation envisaged in section 7.

3. The International Convention for the high seas fisheries of the North Pacific Ocean, signed at Tokyo, 10 May 1952 [264]*, by representatives of the United States, Canada and Japan.*

This Convention results from Art. 9 of the Japanese Peace Treaty, and is based on the principles of the Truman Proclamation on coastal fisheries as far as it lays down cooperative measures to preserve and perpetuate the fish stocks in a certain area (i.e. the North Pacific), but it is also characterized by the following principle:

"The convention introduces a new principle in international conservation practice on the high seas. Under it special treatment is accorded to fisheries already fully utilized and fully conserved. Each

signatory agrees to abstain from exploitation of specified fish stocks which are already exploited to the maximum by one or both of the other parties provided the latter are carrying out programs for the conservation of the stocks and agree to continue to carry out such programs.

The Convention established (Art. II) the International North Pacific Fisheries Commission composed of representatives of the three countries, each with equal vote.

The Commission is empowered to investigate any fish stock of the North Pacific Ocean exploited by two or more of the parties and, when necessary, to recommend joint conservation action to the governments concerned. Excepted from such study and recommendation, however, are fisheries already covered by agreements between the parties, such as the Pacific halibut and sockeye salmon fisheries.

In the special case of fully utilized and conserved fisheries, the Commission may recommend that the country or countries not exploiting the stock refrain from such exploitation. The convention provides, however, that no such abstention should be requested of any parties in waters in which there has been a history of joint conservation activity by such parties, an intermingling of their fleets, and an intermingling of the stocks of fish exploited by their fleets. Accordingly, the convention stipulates that neither the United States nor Canada shall be asked to abstain from any stock in waters off the Pacific Coast of either country from the Gulf of Alaska southward.

To qualify under the abstention proviso, a fishery stock must be fully exploited and be under continuous study and regulation for conservation purposes. The convention recognizes three stocks, salmon, halibut, and herring, off the coasts of North America as meeting these conditions. Accordingly, by the terms of the convention, Japan agrees to abstain from fishing salmon, halibut and herring in specified waters off the coasts of North America; and Canada agrees to abstain from fishing salmon in the Bering Sea east of 175 degrees west longitude.

Also signed today was a protocol to the convention, providing for further study and final determination of the lines bounding the salmon areas. These lines are tentatively set by the convention".

Article I, 2:

"Nothing in this Convention shall be deemed to affect adversely (prejudice) the claims of any Contracting Party in regard to the limits of territorial waters or to the jurisdiction of a coastal state over fisheries".

Article IV provides that "no recommendation (by the Commission) shall be made for abstention by a Contracting Party concerned with regard to .... (2) any stock of fish which is harvested in greater part by a country or countries not party to this Convention".

Article V (2):

"The Contracting Parties recognize that any stock of fish originally specified in the Annex to this Convention fulfills the conditions prescribed in Article IV and accordingly agree that the appropriate Party or Parties shall abstain from fishing such stock and the Party or Parties participating in the fishing of such stock shall continue to carry out necessary conservation measures".

Article VI:

"In the event that it shall come to the attention of any of the Contracting Parties that the nationals or fishing vessels of any country which is not a Party to this Convention appear to affect adversely the operations of the Commission or the carrying out of the objectives of this Convention, such Party shall call the matter to the attention of other Contracting Parties. All the Contracting Parties agree upon the request of such Party to confer upon the steps to be taken towards obviating such adverse effects or relieving any Contracting Party from such adverse effects".

It remains to be seen how this article will work out as against third States, but we hope, that the country concerned will be invited to join the Convention and will be willing to do so.

Article IX lays down rules for violation of the provisions i.e. concerning the arrest and trial of the offender (limited of course to fishing vessels of a Contracting Party).

We see in this Convention a happy development of International Law. No State can be forced to abandon rights based on the principle of the freedom of the seas, in this case the right to fish. But a State can, as a wise precaution to forestall conflicts and with an open eye for the imperative necessity to preserve the stock of marine resources, voluntarily waive these rights in a certain area of the high seas and for a certain period. In the willingness to sacrifice rights for the common interest and peace we see a proof of growing consciousness of the solidarity of the denizens of this planet and a real hope for a favourable development and approaching rule of International Law.

It may serve as an example for other areas, where several States have fishery interests, although it may be said that the circumstances were particularly favourable in this case.

Juji Enomoto writes in his paper for the Madrid Conference of July, 1952, of the International Bar Association [265], *p. 13: "Consider-*

ing that the object of conservation cannot be attained without cooperation among the countries concerned, it may be said to be both proper and desirable that the state to which a new participant belongs should strive in a spirit of cooperation and fairness to conclude an agreement with the coastal state concerned".

In this connection we refer to a Resolution of the Institute of International Law of 1937 [266], *p. 271: „En tout état de cause, l'Institut est d'avis qu'un Etat manquerait à ses devoirs internationaux s'il négligeait de prendre les mesures appropriées pour empêcher les pratiques qui, à la lumière de la science, sont notoirement contraires à l'exploitation et à la protection rationelles des richesses de la mer".*

In other words: fishing regardless of the necessary protection should be considered as a misuse of the high seas.

This Convention does not interfere with the rights of third States [a].

As we mentioned several times the Anglo-Norwegian Fisheries Case, a subject partly belonging to Chapter II, Section 6 and partly to Chapter III, Section 2, we will quote the Judgment of the International Court of Justice of December 18th, 1951 [261], *p. 143:*

"The Court, rejecting all submissions to the contrary, finds by ten votes to two, that the method employed for the delimitation of the fisheries zone by the Royal Norwegian Decree of July 12th, 1935, is not contrary to international law; and by eight votes to four, that the base-lines fixed by the said Decree in application of this method are not contrary to international law".

It is here not the place to discuss this judgement in detail. The only questions which follow logically on what we have said before, are in the first place, what are the possible consequences of this decision, and secondly, are any generally applicable rules of International Law given, bearing on our subject.

Before we answer these questions we draw attention to art. 59 of the Statute of the Court stating that "the decision of the Court has no binding force except between the parties and in respect of that

---

[a] *The same opinion is expressed by Bishop* [267], *p. 718: "Such a Treaty would in no way impair the legal rights of any third nation to fish wherever, and in whatever manner, it now has a right to fish".*

For further comments on this convention see Allen [268], *p. 319–323, and Selak* [269], *p. 323–330.*

particular case". Furthermore, as Johnson [270] *says, p. 180: "The judgement is also not a precedent in the strict sense for the reason that the Court went out of its way to stress the exceptional features of the case even to the extent of making those exceptional features one of the bases of its decision".*

Nevertheless, the decision may encourage countries, who believe to be in the same, or nearly the same position as Norway and trust that in case of conflicts the Court will decide in the same sense. We believe that the Icelandic Provisional Act of 19 March, 1952, and the Regulations concerning conservation of fisheries off the Icelandic coasts, 19 March, 1952 may be mentioned as an example.

In the Comments by Governments on the Draft Articles of the International Law Commission [259], *the Government of Chile (p. 10) expresses the view that the sovereignty of a coastal State extends to the continental shelf and to the superjacent high seas.*

On page 13, the Chilean Government criticizes the articles concerning the resources of the sea and sedentary fisheries and reaffirms its views that the coastal State should have the right to establish an exculsive hunting and fishing zone 200 sea-miles wide. The Chilean Declaration of 1947 was motivated by the particular geographical situation of Chile and by economic considerations. These motives played a role in the Anglo-Norwegian Fisheries case (see Judgment [261], *p. 128, 129, 131, 133, 134, 135, 139 and 142).*

However the geographical conditions which were accepted as warranting a deviation from the low-water mark rule, which as the Judgment states "has generally been adopted in the practice of States" (p. 128), will not be found in many countries.

Even more cautiously should be handled the argument based on economic interests of the coastal State. Although the importance of the fisheries for the coastal population was mentioned p. 128: "In these barren regions the inhabitants of the coastal zone derive their livelihood essentially from fishing", this argument was only turned into a legal consideration through a qualification. The Judgment mentioning the economic interests p. 133, adds: " . . . .the reality and importance of which are clearly evidenced by a long usage".

Again, where the Judgment referred to the Norwegian claim on a historic title to the waters of Lopphavet; it was said (p. 142) that such rights (traditional rights), "founded on the vital needs of the population and attested by very ancient and peaceful usage may legitimately be taken into account in drawing a line which moreover, appears to the Court to have been kept within the bounds of what is moderate and reasonable".

This qualification will lessen considerably the chances of relying on the argument of economic interests, which in its pure unqualified form does not differ much from the right of conservation promulgated by Fauchille [70], *but not accepted in International Law.*

The claims of some countries on stretches of the high seas can hardly be called moderate, or reasonable.

We are not afraid that this judgment will give rise to further extravagant claims nor do we believe that existing claims would find much foothold on this decision to rely on in case of conflicts.

We are firmly convinced that such countries, instead of relying on unilateral declarations, which cannot be enforced against third States, would find more reliable promotion of their economic interests if they followed the example of the Convention for the high seas fisheries of the North Pacific Ocean of May 10, 1952.

In this connection the comments of Iceland are of interest (Comments, Add. 1, p. 4). "The Icelandic Government considers itself entitled and indeed bound to take all necessary steps on a unilateral basis to preserve these resources.... The Government of Iceland does not maintain that the same rule should necessarily apply in all countries. It feels rather that each case should be studied seperately and that the coastal State could within a reasonable distance from its coasts determine the necessary measures for the protection of its coastal fisheries in view of economic, geographic, biological and other relevant considerations".

It is not surprising that the United Kingdom takes the opposite view (Comments, Add. 1, p. 20): "Her Majesties Government wish to place on record their emphatic opposition to the proposal contained in note 5 (on Art. 2 of the draft on the resources of the sea). In the opinion of Her Majesties Government no State has the right to enforce conservation measures against the fishing vessels of other States outside its territorial waters except by international agreement. Unilaterally declared conservation zones outside territorial waters are illegal as being in contravention of the principle of the freedom of the seas".

The other question, concerning the Judgment, will be dealt with later.

The Regime of the High Seas and Navigation. Territorial waters.

We mention the Decree of the Presidium of the National Assembly concerning the Territorial and Inland Waters of the People's Republic of Bulgaria of October 23, 1951, modified, November 9, 1951 [271], p. 67–69.

The territorial waters shall extend 12 miles from the water mark of the sea coast.... and the line of the inland waters (section 1).

In section 2 this line is further described as a straight line connecting certain headlands.

Zones of the territorial waters may be closed to all navigation for security reasons (section 3) and the ports of Stalin and Sozopol shall be closed to all foreign ships (section 7).

Measurement of depth around the vessel (in the territorial and inland waters) may be allowed only when the vessel has run aground. (a rather useful provision from a nautical point of view!) [a]

We further refer to François, Report on the Regime of the Territorial Sea [273], where a list is given on p. 11, with the breadth of the territorial waters of different States, specifying between fishing, neutrality, customs etc.

In Art. 4, a maximum breadth of 6 marine miles is suggested. Admitting that "it seems very doubtful that a compromise on the six-mile rule can easily be achieved", François stresses the necessity to solve the problem, "since if each State were left absolutely free to determine the breadth of its territorial sea itself, the principle of the freedom of the seas would suffer to an inadmissible extent" (loc.cit. p. 18 and 20).

In Art. 5 he suggests, as an exception to the general low-water mark rule, in case a coast is deeply indented and cut into, or bordered by an archipelago, base-lines, joining appropriate points on the coast, but following the general direction of the coast.

Finally, in Art. 15 he suggests the same provision as contained in the Report of the Codification Conference of the Hague in 1930, concerning innocent passage: "A coastal state may put no obstacle in the way of the innocent passage of foreign vessels in the territorial sea".

Regarding contiguous zones the United Kingdom shows a certain amount of relaxation as compared with its strong opposition against

---

[a] *See on this Decree also the article of Pundeff [272], p. 330–333.*

these zones during the Codification Conference in the Hague in 1930. "Her Majesties Government are satisfied, however, that on the basis of established practice, the article proposed by the Commission (Art. 4, Part II) is acceptable provided that.... this Article is read in conjunction with another Article stating that the territorial waters of a State shall not extend more than three miles from the coast unless in any particular case a State has an existing historic title to a wider belt".

Islands on the continental shelf.

The passage of the Judgment in the Anglo-Norwegian Fisheries case concerning the "skjærgaard" "which constitutes a whole with the mainland" (quoted above) could be misused by the sponsors of the thesis that islands on the continental shelf belong to the coastal State.

We believe that this would be a misuse, because this statement is typically limited to the case under consideration.

Pipelines.

François in his Third Report on the Regime of the High Seas [274], *p. 11 and 12, treats the pipelines on the same footing as submarine telegraph cables and suggests the following provisions (we leave the telegraph cables out).*

Art. 1 "All States may lay.... pipelines on the bed of the high seas" (this seems to include not only pipelines connecting oil-installations with the shore, but also those which may be layed over the continental shelf of another State).

Art. 2 "The breaking or injury.... outside of the territorial waters, done wilfully or through culpable negligence.... of a submarine pipeline, shall be a punishable offence" (cases of "force-majeur" excepted).

Art. 3 "The owner of.... a pipeline outside the territorial waters who, by the laying or repairing of that.... pipeline shall cause the breaking or injury of another.... pipeline, shall be required to pay the cost of the repairs...."

Art. 4 "All fishing gear used in trawling shall be so constructed and so maintained as to reduce to the minimum the danger of fouling submarine.... pipelines on the sea-bed".

Finally as regards this Chapter we mention Menzel [275], *who in*

*his paper prepared for the Conference of the International Bar
Association in Madrid (July 1952) fails to distinguish claims on the
mineral resources and on sea-areas. He writes for instances (to
mention only one example) on p. 23: „Die Übertragung der Sou-
veränitätsrechte der Küstenstaaten auf die "Sockel"-Zone würde
praktisch bedeuten, dass allein nach der Truman-Proklamation von
1945 nicht weniger als 10 Millionen Quadratmeilen dem bisherigen
Bereich der freien Schiffahrt entzogen würden".*

Mineral resources.

*In the Judgment of the Anglo-Norwegian Fisheries Case, we find
the following passages, p. 132:*

*"The delimitation of sea areas has always an international
aspect; it cannot be dependent merely upon the will of the coastal
State as expressed in its municipal law. Although it is true that the
act of delimitation is necessarily a unilateral act, because only the
coastal State is competent to undertake it, the validity of the delimitation with regard to other States depends upon international law".*

*We believe this phrase, to be a general statement, independent of
the special merits of the case, and take it therefore as an answer on
the second question we put above in connection with this Judgment.*

*We feel that this statement largely strenghtens our thesis that
unilateral acts are not the proper means to create new rules of
International Law and gives at the same time an answer to our
question:*

"What is the value of proclamations, declarations and national
decrees in International Law".

*It is a confirmation of the general principle of International Law,
that rights and obligations cannot be altered unilaterally by passing
domestic legislation.*

*In this Section we also discussed whether law existed on this
subject, and we came to the conclusion that it did not.*

*Our thesis and our often repeated urge that a Conference be
convened to draft a Convention is strenghtened by the comments of
Sweden (Comments Add. 3, p. 3):*

*"The Swedish Government feels bound to regard any proposal
to grant rights over the continental shelf to coastal States as being* de
lege ferenda *and considers that such a proposal could only be put
into effect by an international convention providing for certain*

concessions to coastal States which are in a position to exploit the continental shelf. The conclusion of such a convention is a matter of expediency", and by the comments of the Netherlands (Comments, p. 35):

"In this Government's opinion it is absolutely essential that rules of international law should be established on this subject so as to put an end to the present practice by which States issue regulations unilaterally".

Chile on the contrary objects to Art. 2 of the draft of the International Law Commission and suggests a wording by which sovereignty, dominion and jurisdiction are vested "ipso iure" in the coastal State (Comments, p. 9).

Green [276], *p. 79 says: "..... mere proclamations and unilateral declarations can amount to no more than inchoate titles requiring some measure of occupation or exploitation to perfect them".*

His premise is, however, different from ours. He says (same page): "It has been suggested throughout this paper that title to the continental shelf and its resources depends on effective occupation....". We explained our objections to this notion in this connection and believe that title to the resources can only be obtained from a Convention to be concluded.

The meaning of control and jurisdiction and other rights claimed.

As to the nature of the rights which a coastal State can claim on the resources of the continental shelf we mention the comment of the Government of the United States (Comment, Add. 2): "This Government wonders, accordingly, whether it would not be advisable to make it clear, at least in the commentaries, that control and jurisdiction for the purpose indicated in the draft articles mean in fact an exclusive, but functional, right to explore and exploit". This would not mean that foreign nationals are ipso iure excluded. A coastal State may lease a "block" to a foreign company, but may of course also refuse an application for such a lease.

Insertion of the word "exclusive" in Art. 2 of the draft is also suggested by the Government of Brazil (Comments, p. 4–5), but at another place and for another reason. Brazil suggests to insert it before the word "purpose", to emphasize the object of the provision that "control and jurisdiction over the continental shelf should be exercised solely for the purpose stated".

Thirdly the word "exclusive" is suggested to be inserted in Art. 2 of the draft by Denmark (Comments, p. 15), but here with quite a different meaning: "The Danish authorities would find it appropriate that the right of the coastal State as set out in part I, article 2, be expressly characterized as an exclusive right since that would preclude any idea of expansion of the territory of the State concerned". What is meant here is that the article purports to give only this right and nothing more.

Also the Swedish Government (Comment, Add. 3, p. 4): "thinks it proper that the rights of coastal States, in respect of the continental shelf, should be confined to the purposes stated".

On the contrary some countries prefer to describe the right of the coastal State as "sovereignty": Chile (Comments, p. 8–9); Ecuador (Comments, p. 22–23, actually only stating that the national law of Ecuador does not limit the State's jurisdiction to the exploration and exploitation of the shelf's natural resources); Israel (Comments, p. 30–31), and the United Kingdom (Comments, Add. 1, p. 18), reasoning that "If the expression 'sovereignty' were used, there would be no doubt that a crime committed in a tunnel under the continental shelf would come within the jurisdiction of the coastal State. . . .".

We believe that using the word "sovereignty" is not advisable. It is one of the most misused words of the human language. We believe that the notion is "devaluating" rapidly. The tendency of International Law is to diminish sovereignty, not to extend it, or as Lauterpacht [218] *says, p. 391: "In fact through the entire branch of International Law relating to state territory and territorial sovereignty there runs as a constant theme the phenomenon of limitation of sovereignty in various spheres and directions". Lauterpacht, however, argues that for this reason it is not improper to call the right sovereignty, because sovereignty is a limited right anyhow. We feel, that the coastal State could easily misuse its sovereignty, especially if the right would extend to the shelf, for instance by probihiting trawling on "their" shelf.*

We should like to forestall this possibility and to limit the right strictly to the minerals and consider the coastal State as owner of these minerals, with the right to exploit them in a way compatible with the rules of International Law.

As to the remarks by Hurst concerning a horizontal delimitation of sovereignty, a remark which we tried to refute, Vallat [277] *writes, p. 3: "This horizontal division is a comparatively new idea, but there does not seem to be any compelling logical or practical reason why the earth, the sea and the surrounding space should not be divided horizontally as well as vertically. Indeed.... it is very likely that scientific progress will in time require a horizontal limitation on the extent of the space above the territory of a State which is subject to its sovereignty".*

Sea-bed and subsoil.

Gidel [260], *p. 12–13, discusses the thesis of giving the sea-bed the same legal status as the high seas. He objects that the idea of the continental shelf theory is to penetrate the sea-bed to extract the minerals from the subsoil. But this penetration is only a form of use. Moreover it is tiny in dimensions and temporary in character.*

Division of a common shelf.

Several countries in their comments expressed the view that Art. 7 of the draft, leaving the division of a common shelf to the countries concerned, is not satisfactory and that some rules should be laid down internationally, rules which may thereafter guide an Arbitration Tribunal in its decisions. Doubt is expressed as to the advisability of arbitration "ex aequo et bono": Denmark, Comments, p. 15–18; Israel, p. 32–34; the Netherlands, p. 37; Norway, p. 40–41 and the United Kingdom (Comments, Add. 1, p. 19). Yugoslavia, p. 48–49, criticizes draft Art. 7 because of the faulty definition of the continental shelf in draft Art. 1 and suggests the geometric middle as a boundary-line (which of course would only be applicable in the case of States lying opposite each other on the same shelf as for instance in the Persian Gulf). Furthermore the United States (Comments, Add. 2, p. 2) and Sweden (Comments, Add. 3, p. 5–6), suggesting that previous arbitration cases may possibly provide useful material for rules concerning division, drawing attention to the Hague arbitral award of 1909 on the maritime frontier between Sweden and Norway (meant is the Grisbadarna arbitration).

Gidel, on the contrary agrees with the draft articles, that it is difficult to give general rules (p. 17).

We end with a few remarks. The Comments we mentioned several times, only contain the answers of a limited amount of States. Many countries have not sent in their comments yet.

The Conference of the International Bar Association in Madrid, in July, 1952, adopted a resolution to the effect that the Conference agreed with the draft articles from Art. 2 onwards, but reserved its opinion on Art. 1.

We stressed throughout this work our opinion that a Conference should be convened to lay down the rules which do not exist yet; and Gidel (p. 22) is of the same opinion. We have tried to give some suggestions about what we think these rules ought to contain. Therefore this work has to be consulted in exactly the opposite sense as the Supreme Court did in the case of the "Paquete Habana" and the "Lola" (8 January 1900) when it considered the case in the light of the authority of jurists and commentators and is was pointed out "that such works were resorted to by judicial tribunals, not for the speculations of those authors concerning what the law ought to be, but for trustworthy evidence of what the law really was".

BIBLIOGRAPHY

(sequence of the text)

1. Dr. Otto Krümmel (Prof. Geography Kiel Univ.) Handbuch der Ozeanographie, Band I, Stuttgart, 1907.
2. J. H. F. Umbgrove (Prof. Geology, Delft), The Pulse of the Earth, The Hague, 2nd edition, 1947.
3. De Zeeën van Nederlandsch Oost-Indië, published by the Royal Netherlands Geographical Society, Leiden, 1922 (Prof. Dr. G. A. F. Molengraaff wrote Chapter VI on Geology).
4. J. H. F. Umbgrove (as under 2) Origin of continental shelves, Bulletin of the American Association of Petroleum Geologists, vol. 30, Part I, 1946.
5. Ph. H. Kuenen (Prof. Geology, Groningen Univ.) Marine Geology, New-York, London, 1950.
6. XIII, Department of State Bulletin, September 30, 1945.
7. Summary Record of the 66th meeting, International Law Commission 2nd session, A/CN.4/SR.66, 12 July 1950.
8. Richard Young, Saudi Arabian offshore legislation, The A.J.I.L. [a], July 1949.
9. Richard Young, Further claims to areas beneath the high seas, The A.J.I.L., October 1949.
10. J. P. A. François, Report on the High Seas, International Law Commission, Second Session, A/CN.4/17, 17 March, 1950.
11. J.Y.B(rinton). Jurisdiction over sea-bed resources and recent developments in Persian Gulf Area, Revue Egyptienne de droit international, vol. 5, 1949.
12. Richard Young, The Legal Status of submarine areas beneath the high seas, The A.J.I.L., April 1951.
13. Dr. Edgar Dacqué (Privatdozent Univ. München), Grundlagen und Methoden der Paläogeographie, Jena, 1915.
14. Report of the International Law Commission covering its second session, 5 June — 29 July, 1950. United Nations, General Assembly official records, Fifth session, Supplement No. 12 (A/1316) 1950.
15. Report of the Committee on Coastal Waters and Appurtenant Subsoil, Third International Conference of the Legal Profession. International Bar Association, London, July 1950.
16. Handbuch der Geophysik, Band II, Aufbau der Erde, Herausgegeben von Dr. B. Gutenberg (Prof. für Geophysik und Meteorologie am California Institute of Technology, Pasadena); Berlin, 1933 (Dr. E. Kossinna, Berlin, wrote Abschnitt VI, die Erdoberfläche).

---

[a] The A.J.I.L.: The American Journal of International Law.

17. PAUL WEAVER, Variations in history of continental shelves, Bulletin of the American Association of Petroleum Geologists, Vol. 34, Number 3, March 1950.

18. J. BEN CARSEY (Chief Geologist, Humble Oil and Refining Company), Geology of Gulf Coastal Area and Continental Shelf, Bulletin of the American Association of Petroleum Geologists, Vol. 34, Number 3, March 1950. ,

19. CARLOS MORALES, Plataforma submarina, Revista de Derecho y Legislacion, Año XXXVIII (1949), Tomo Trigesimo Octavo, Caracas.

20. Dr. PH. H. KUENEN (as under 5). The Formation of the Continental Terrace, The Advancement of Science, Vol. VII, Number 25, 1950.

21. JACQUES BOURCART (Prof. à la Sorbonne), Géographie du Fond des Mers, Paris, 1949.

22. Memorandum on the Regime of the High Seas, prepared by the Secretariat of the United Nations, A/CN.4/32, 14 July 1950.

23. RENÉ JOSUÉ VALIN, Nouveau Commentaire sur l'Ordonnance de la Marine du mois d'août, 1681, Tome Second, A La Rochelle, MDCCLX.

24. Introductory Report by the Committee (on Rights to the Sea-bed and its Subsoil), International Law Association, Copenhagen Conference, 1950.

25. Summary Record of the 67th meeting, International Law Commission, Second Session, A/CN.4/SR.67, 13 July 1950.

26. J. P. A. FRANÇOIS, Second Report on the High Seas, International Law Commission, Third Session, A/CN.4/42, 10 April 1951.

27. FRANCIS P. SHEPARD, Submarine Canyons: A joint product of rivers and submarine processes, Science, Vol. 114, Number 2949, July 6, 1951.

28. Prof. OREN F. EVANS (Department of Geology, University of Oklahoma), Shoreline Processes of the Continental Shelf, World Oil, June 1949.

29. International Law Association, Branche Française, Propriété du Sous-Sol Marin, Rapport de la Commission, signed: Le Président James Paul Govare (this report was distributed at the Conference of the International Law Association at Copenhagen, 1950).

30. Le baron DE PUFENDORF, Le droit de la Nature et des Gens, Tome Premier, traduit du Latin par Jean Barbeyrac, Amsterdam, 1734.

31. THOMAS WEMYSS FULTON (Lecturer on the scientific study of Fishery problems, Univ. of Aberdeen), The Sovereignty of the Sea, Edinburgh, 1911.

32. JOSÉ LUIS DE AZCÁRRAGA Y BUSTAMENTE, Los Derechos sobre la plataforma submarina, Revista Española de Derecho Internacional, Vol. II, Núm. 1, 1949.

33. Summary Record of the 113th meeting, International Law Commission, 3rd Session, A/CN.4/SR.113, 11 August 1951.

34. Summary Record of the 117th meeting, International Law Commission, 3rd Session, A/CN.4/SR.117, 11 August 1951.

35. Summary Record of the 123rd meeting, International Law Commission, 3rd Session, A/CN.4/SR.123, 24 August 1951.

36. Report of the International Law Commission Covering its Third Session, 16 May — 27 July, 1951, A/CN.4/48, 30 July 1951.
37. WALLACE E. PRATT, Petroleum on Continental Shelves, Bulletin of the American Association of Petroleum Geologists, Vol. 31, Number 4, April 1947.
38. The Science of the Sea, prepared by the Challenger Society, originally edited by G. Herbert Fowler, B. A., Ph. D., second ed. edited by E. J. Allen D. Sc., F.R.S., Oxford, 1928 (Sir John Murray wrote the Chapter on the Sea Floor).
39. Geological Nomenclator (in Dutch, German, English and French), by W. E. Boerman; G. van Dijk; B. G. Escher; H. F. Grondijs; J. A. Grutterink; G. A. F. Molengraaff; P. Kruizinga; K. Oestreich; L. Rutten; G. Schouten; edited by L. Rutten, The Hague, 1929.
40. GÉRARD DE RAYNEVAL, Institutions de droit de la Nature et des Gens, Seconde Edition, Paris, An. XII, 1803.
41. EMORY CLARK SMITH, The character and scope of the rights asserted by States in the Western Hemisphere over coastal waters and appurtenant subsoil, presented to the third International Conference of the International Bar Association, London, July 1950.
42. L. OPPENHEIM, M.A., LL.D., International Law, Vol. I, sixth edition, edited by H. Lauterpacht, M.A., LL.D., 1947.
43. L. E. VISSER, De territoriale zee, Utrecht, 1894.
44. The International Law Association, Report of the forty-third Conference, Brussels, 1948.
45. MIGUEL RUELAS, La Cornisa Continental Territorial, Revista de Derecho Internacional, Organo del Instituto Americano de Derecho Internacional, Habana, Año. IX, Tomo XVII, Enero–Junio 1930.
46. G. M. LEES, Nature of continental shelves, Bulletin of the American Association of Petroleum Geologists, Vol. 35, Number 1, January 1951.
47. PAUL WEAVER, Continental Shelf of Gulf of Mexico, Bulletin of the American Association of Petroleum Geologists, Vol. 35, Number 2, February 1951.
48. Annual Report of the Secretary of the Interior, 1945, United States Department of the Interior, Harold L. Ickes, Secretary, United States Government Printing Office, Washington, 1945.
49. Summary Record of the 68th meeting, International Law Commission, Second Session, A/CN.4/SR.68, 14 July 1950.
50. Summary Record of the 130th meeting, International Law Commission, Third Session, A/CN.4/SR.130, 27 August 1951.
51. Summary Record of the 134th meeting, International Law Commission, Third Session, A/CN.4/SR.134, 28 August 1951.
52. League of Nations, Committee of Experts for the progressive Codification of International Law, Report to the Council of the League of Nations on the questions which appear ripe for international regulation. (Questionnaires 1–7) adopted by the Committee at its third session held in March–April 1927, C 196 M. 70, 1927 V, C.P.D.I. 95(2), Geneva, 1927. (Public. of League of Nations V Legal 1927, V 1).
53. JAMES JOHNSTONE D. Sc. (Prof. Oceanography Liverpool Univ.),

Andrew Scott, A.L.S. and Herbert C. Chadwick, A.L.S., The Marine Plankton, London, 1934.

54. H. U. SVERDRUP (Prof. Oceanography, Univ. California) Martin W. Johnson (Ass. Prof. of Marine Biology, Univ. of California) and Richard H. Fleming (Ass. Prof. of Oceanography, Univ. of California).
The Oceans, Their Physics, Chemistry and general Biology, New York, 1946.

55. MARGARETHA BRONGERSMA-SANDERS, D.Sc. The Importance of upwelling water to vertebrate Paleontology and Oil Geology, Verhandelingen der Koninklijke Nederlandse Academie van Wetenschappen, afd. Natuurkunde, Tweede Sectie, Deel XLV, No. 4, Amsterdam, 1948.

56. Fishery resources of the United States, 79th Congress, 1st Session, Senate, Document No. 51, Letter of the Secretary of the Interior transmitting, persuant to Law, a Report on a survey of the Fishery Resources of the United States and its Possessions, March 1, 1945. (U.S. Government Printing Office, Washington, 1945).

57. Dr. J. L. KASK (Chief Fisheries Biologist of the United Nations Food and Agriculture Organization). Pacific can produce more fish, Fisheries Newsletter, Vol. 8, No. 2, April, 1949.

58. Sidney Shapiro (Fishery research biologist, Fish and Wildlife Service, Washington), The Japanese long-line fishery for tunas, Commercial Fisheries Review, Washington, Vol. 12, No. 4, April, 1950.

59. Report No. 104, The Japanese Tuna Fisheries, Natural Resources Section, General Headquarters, Supreme Commander for the Allied Powers, Tokyo, March, 1948, Reproduced in Fishery Leaflet 297, Fish and Wildlife Service, Washington, April, 1948.

60. Report of M. JOSÉ LEÓN SUÁREZ, Supplement to The A.J.I.L., Vol. 20, Special Number, July, 1926, p. 231–241.

61. Acts of the Conference for the Codification of International Law, held at the Hague from March 13th to April 12th, 1930, Vol. III, Minutes of the second Committee Territorial Waters, League of Nations, C 351 (b), M 145 (b), 1930 V, Geneva, August 19, 1930 (Series of League of Nations Publications V. Legal, 1930, V 16).

62. The A.J.I.L., Vol. 40, January, 1946, Official Documents, p. 45–48.

63. WALTER M. CHAPMAN, United States Policy on High Sea Fisheries XX Department of State Bulletin, January 16, 1949.

64. Report No. 31, The Japanese Salmon Industry, Natural Resources Section, General Headquarters, Supreme Commander for the Allied Powers, Tokyo, 25 April, 1946.

65. CHARLES B. SELAK Jr., Recent developments in high seas fisheries jurisdiction under the Presidential Proclamation of 1945, The A.J.I.L., October, 1950.

66. K.B., Japanese Fishery Control and Bristol Bay, Far Eastern Survey, American Council, Institute of Pacific Relations, Vol. VI, No. 9, April 28, 1937.

67. Summary Record of the 69th meeting, International Law Commission, 2nd Session, A/CN.4/SR.69, 17 July 1950.

68. WILLIAM W. BISHOP Jr. (Prof. of Law, Univ. of Michigan Law School), The exercise of jurisdiction for special purposes in high seas areas beyond the outer limit of territorial waters (e.g. conservation etc.). Committee X, Theme 2, Territorial Waters and Ocean Fisheries, Sixth Conference (Paper prepared for the North American Bar Association, 1949).

69. CHARLES G. FENWICK (Prof. Political Science Bryn Mawr College, 1918–1940, Director Dept. International Law and Organ., Pan American Union), International Law, Third Ed., New York, 1948.

70. PAUL FAUCHILLE, Traité de Droit International Public, 8e Ed., Tome 1er, Deuxième Partie, Paris, 1925.

71. W. E. HALL, M.A., A Treatise on International Law, Eighth Edition, Edited by Pearce Higgins, C.B.E., K.C., LL.D., London, 1938.

72. Naval War College, International Law Documents, 1948–1949, Navpers 15031, Vol. XLVI, United States, Gvt. Print. Off., Washington, 1950.

73. RICHARD YOUNG, Recent developments with respect to the continental shelf, The A.J.I.L., October, 1948.

74. The A.J.I.L., Vol. 41, January, 1947, Official Documents, p. 11–12.

75. ERWIN SCHWEIGGER, Dr. PHIL., Pesqueria y Oceanografia del Peru y Proposiciones para su Desarrollo Futuro, Informe elevado a la Compañia Administradora del Guano, Lima, 1943.

76. MICHAEL SCULLY, Peru goes fishing, "Rediscovery" of the Humboldt Current yields rich new food source, Américas, published by the Pan American Union, Vol. 3, Nr. 8, August, 1951.

77. Year book of Fisheries Statistics, 1948–1949, Food and Agriculture Organization of the United Nations, Washington-Rome, 1950.

78. F.A.O. Fisheries Bulletin, Vol. 4, Number 3, Rome, May/June, 1951.

79. The International Law Quarterly, Vol. 2, 1948, Declaration by the President of the Republic of Chile regarding Chilian territorial claims, June, 1947; Peruvian Decree regarding national sovereignty and jurisdiction over the continental and insular shelf.

80. Dr. J. P. A. FRANÇOIS (Prof. Economic Highschool, Rotterdam), Handboek van het Volkenrecht, Eerste Deel, Zwolle, 1949.

81. Anglo-Norwegian Fisheries Case, Letter dated 19th September, 1951, from the Agent of the Government of the United Kingdom to the Registrar of the International Court of Justice.

82. Replies from Governments to Questionnaires of the International Law Commission, A/CN.4/19, 23 March 1950.

83. XXIV, Department of State Bulletin, January 1, 1951.

84. Anglo-Norwegian Fisheries Case, International Court of Justice, Reply submitted by the Government of the United Kingdom of Great Britain and Northern Ireland, Vol. II, Annexes.

85. Anglo-Norwegian Fisheries Case, International Court of Justice, Verbatim Report, Public Sitting, October 18th, 1951, C.R. 51/26.

86. Summary Record of the 63rd meeting, International Law Commission, 2nd Session, A/CN.4/SR. 63, 7 July 1950.

87. STUART A. MOORE, F.S.A., A History of the Foreshore, and Hall's Essay on the Rights of the Crown in the Sea-shore, London, 1888.

88. Anglo-Norwegian Fisheries Case, International Court of Justice, Verbatim Report, Public Sitting, October 19th, 1951, C.R. 51/27.

89. Summary Record of the 119th meeting, International Law Commission, 3rd Session, A/CN. 4/SR. 119, 21 August 1951.

90. EDWARD W. ALLEN, The Fishery Proclamation of 1945, The A.J.I.L., Vol. 45, January, 1951.

91. HUGO GROTIUS, De Jure Praedae Commentarius, Translation of the original manuscript of 1604 by Gwladys L. Williams with the collaboration of Walter H. Zeydel, Carnegie Endowment for International Peace, London, 1950.

92. JAMES BROWN SCOTT, Resolutions of the Institute of International Law, dealing with the Law of Nations, New York, 1916.

93. Annuaire de l'Institut de Droit International, Treizième Vol., Session de Paris, mars 1894.

94. L. LARRY LEONARD, International regulation of fisheries, Carnegie Endowment for International Peace, Washington D.C., 1944.

95. Summary Record of the 118th meeting, International Law Commission, 3rd Session, A/CN. 4/SR. 118, 22 August 1951.

96. STEFAN A. RIESENFELD (Assistant Prof. of Law, Univ. Minnesota), Protection of coastal fisheries under International Law, Carnegie Endowment for International Peace, Washington, 1942.

97. Regime of the High Seas, Questions under study by other Organs of the United Nations or by Specialized Agencies (Memorandum presented by the Secretariat), A/CN.4/30, 23 June 1950.

98. Summary Record of the 132nd meeting, International Law Commission, 3rd Session, A/CN.4/SR. 132, 28 August 1951.

99. Convention for the Regulation of Aerial Navigation, Paris, October 13, 1919, H.M. Stat. Off. London, Treaty Series 1922, No. 2, Cmd. 1609.

100. Protocols regarding the Amendment of the Convention relating to the Regulation of Air Navigation of October 13, 1919, Brussels, June 1, 1935, H.M. Stat. Off., Miscellaneous No. 7 (1936) Cmd. 5332.

101. International Civil Aviation Conference, Part 1, Final Act and Appendixes I–IV, Chicago, 7th December, 1944, H.M. Stat. Off. London, Miscellaneous No. 6 (1945) Cmd. 6614.

102. Convention for the protection of fur seals in the waters of the North Pacific, Washington, 7 July, 1911, Nouveau Recueil Général de Traités, Continuation du Grand Recueil de G. Fr. de Martens (p. Heinrich Triepel) Troisième Série, Tome V, p. 720.

103. Convention revising the Convention of May 9, 1930, for the preservation of halibut fishery of northern Pacific Ocean and Behring-Sea, Ottawa 29, 1937. Supp. The A.J.I.L., Vol. 32, 1938, Official Documents, p. 71–76.

104. Final Act of the International Whaling Conference with International Convention for the Regulation of Whaling and Protocol of the 1947–48 Season, Washington, 2nd December, 1946, H.M. Stat. Off., London, Miscellaneous No. 3 (1947), Cmd. 7043.

105. Final Act of the International Fisheries Conference, London, 22nd October, 1943, H.M. Stat. Off. London, Miscellaneous No. 5 (1943), Cmd. 6496.

106. International North-west Atlantic Fisheries Conference, Final Act and Convention, Washington, 8th February, 1949, H.M. Stat. Off. London, Miscellaneous No. 4 (1949), Cmd. 7658.

107. Establishment of the Indo-Pacific Fisheries Council, Formulated at Baguio, February 26, 1948, Entered into force November 9, 1948, Treaties and other International Acts Series 1895, Department of State Publication 3473, United States Gvt. Print. Off., Washington, 1949.

108. Mexico-United States, Convention for the establishment of an International Commission for the scientific investigation of tuna, signed at Mexico City, January 25, 1949; ratifications exchanged at Washington, July 11, 1950, The A.J.I.L., Official Documents, p. 51 et seq., April 1951.

109. Summary Record of the 114th meeting, International Law Commission, 3rd Session, A/CN.4/SR. 114, 11 August 1951.

110. Treaty between His Majesty in respect of the United Kingdom and the President of the United States of Venezuela relating to the Submarine Areas of the Gulf of Paria, Caracas, February 26, 1942 (Ratifications exchanged in London, September 22, 1942), H.M. Stat. Off. London, Treaty Series No. 10 (1942), Cmd. 6400.

111. EDGAR THURSTON (C.M.Z.S. etc.) Pearl and Chank Fisheries of the Gulf of Manaar, Madras, 1894.

112. GILBERT GIDEL (Prof. à la Faculté de droit de l'Univ. de Paris et à l'Ecole libre des sciences politiques), Le Droit International Public de la Mer, Tome I, Chateauroux, 1932.

113. An Act to carry into effect the Convention between His Majesty and the King of the French concerning the fisheries in the seas between the British Islands and France, 22 August, 1843, 6 & 7 Victoria, cap. 79, The Statutes at Large, Anno Regni Victoriae, Brittanniarum Reginea, Sexto & Septimo.

114. Parliamentary Debates, Fifth Series No. 163, House of Commons, Fourth Vol. of Session 1923.

115. The Pearl-shell and Bêche-de-mer Fishery Act 1881, 45 Vict. No. 2, The Queensland Statutes, Vol. I, Brisbane, 1911.

116. PHILIP C. JESSUP, LL. B., Ph.D., The Law of territorial waters and maritime jurisdiction, New York, 1927.

117. Official Yearbook of the Commonwealth of Australia No. 15, 1922, Commonwealth Bureau of Census and Statistics, Prepared by Chas. H. Wickens, F.I.A., F.S.S., Melbourne.

118. An Act to carry into effect a Convention between Her Majesty and the Emperor of the French concerning the Fisheries in the seas adjoining the British Islands and France and to amend the Laws relating to British sea fisheries, 13th July, 1868, 31 & 32 Vict., cap. 45, The Law Reports, the Public general Statutes, passed in the 31st and 32nd years of the reign of Her Majesty Queen Victoria, Vol. III, 1868.

119. Ordonnantie van 15 Januari 1905, regelende de parelvisserij in de Assistent Residentie Zuid-Nieuw-Guinea, Ned. Indisch St.bl. 1905, No. 50.

120. Ordonnantie van 29 Januari 1916, Algemene regelen voor het
     vissen van parelschelpen, paarlemoerschelpen, tripang en sponzen,
     binnen de afstand van niet meer dan 3 zeemijlen van de kusten
     van Nederlands Indië, Ned. Indisch St.bl. 1916, No. 157.
121. JOHN WESTLAKE, K. C., LL.D. (Late Whewell Prof. of Inter-
     national Law, Univ. Cambridge), International Law, Part I,
     Cambridge, 1910.
122. Prof. Dr. L. OPPENHEIM, Der Tunnel unter dem Ärmelkanal
     und das Völkerrecht, Zeitschrift für Völkerrecht und Bundes-
     staatsrecht, herausgegeben von Prof. Dr. Jozef Kohler, II Band,
     Breslau, 1908.
123. Sir CECIL J. B. HURST, K.C.B., K.C., Whose is the bed of the
     sea? The British Yearbook of International Law, 1923–1924.
124. Correspondence relating to an Agreement for Lease of Pearl
     Fisheries on the coast of Ceylon, May, 1906, H.M. St. Off., London,
     Cmd. 2906.
125. Parliamentary Debates, Fifth Series No. 164, House of Com-
     mons, Fifth Volume of 1923.
126. League of Nations, Conference for the Codification of Inter-
     national Law, Bases of Discussion, drawn up for the Conference
     by the Preparatory Commission for the Conference, Vol. II
     Territorial Waters, C. 74. M. 39. 1929. V, Geneva, 1929.
127. Summary Record of the 120th meeting, International Law
     Commission, Third Session, A/CN.4/SR. 120, 21 August 1951.
128. M. F. LINDLEY, LL.D., B. Sc. (London), The acquisition and
     government of backward territory in International Law, London,
     1926.
129. Bijblad op het Staatsblad van Ned. Indië, Deel LXV, 1928
     (Bb, 11416).
130. Summary Record of the 65th meeting, International Law Com-
     mission, Second Session, A/CN.4/SR. 65, 11 July 1950.
131. W. R. BISHOP, LL.D., Oil in navigable waters, The International
     Law Association, Report of the 33rd Conference, held at Stock-
     holm, September 8th to 13th, 1924, London, 1925.
132. International Conference at Washington, June, 1926, Oil Pollution
     of Navigable Waters, Final Act, signed 16th June, 1926, and
     Annex, H.M. Stat. Off. London, 1926, Cmd. 2702.
133. United Nations, Economic and Social Council, Transport and
     Communications Commission, Fourth Session, 27 March, 1950,
     Item 10 (h) on the provisional agenda, Pollution of seawater by
     oil, Note by the Secretary General, E/CN.2/68, 20 December,
     1949.
134. Société des Nations, Journal Officiel, Supplement spécial No. 127,
     Genève, 1934.
135. Société des Nations, Journal Officiel, XVIIe Année No. 11,
     novembre 1936.
136. United Nations, Economic and Social Council, Transport and
     Communications Commission, Fifth Session, 19 March, 1951,
     Item 8 on the provisional agenda, Pollution of sea-water, Note
     by the Secretary General, E/CN.2/100, 9 January, 1951.
137. HENRY LEENHARDT, De l'action du mazout sur les coquillages,
     Conseil Permanent International pour l'Exploration de la Mer,

Rapports et Procès-Verbaux des Réunions, Volume XXXV, Copenhague, janvier 1925.

138. C. H. ROBERTS, B.A., Oil Pollution, The effect of oil pollution upon certain forms of aquatic life and experiments upon the rate of absorption, through films of various fuel oils, of atmospheric oxygen by sea-water, Conseil Permanent International pour l'Exploration de la Mer, Journal du Conseil, Volume I, No. 3, Copenhague, août 1926.

139. An Act to vest in the Government of the Colony the property in Petroleum and Natural Gas within the Colony and to make provision with respect to the searching and boring for and getting of Petroleum and Natural Gas and for the purposes connected with the matters aforesaid, 3 April, 1945, Statutes No. 1, Bahamas Acts passed in the year 1945, Nassau, 1945.

140. Rules and Regulations governing geological, geophysical and other explorations in areas within tidewater limits belonging to the State of Texas as promulgated by Bascom Giles, Commissioner of the General Land Office, under the provisions of House Bill No. 665 passed by the 51st Legislature, June 16, 1949.

141. Rules and Regulations governing drilling and producing operations in coastal waters (of Texas), given under signature and seal of office, 6 April, 1948, by Bascom Giles, Commissioner of the General Land Office (persuant to the provisions of Article 5366 of the Revised Civil Statutes of Texas, 1925).

142. Louisiana Oil and Gas Conservation Law, Revised Statutes, Title 30 Section 1–20 (inclusive).

143. J. E. KASTROP, Unique Hookup for testing marine wells, World Oil, August, 1949.

144. Summary Record of the 133rd meeting, International Law Commission, Third Session, A/CN.4/SR. 133, 28 August 1951.

145. JOHN E. FITCH and PARKE H. YOUNG, Use and effect of explosives in California coastal waters, California Fish and Game, Vol. 34, Number 2, San Francisco, April 1948.

146. Explosieven in de visserij, Berita Perikanan (Visserijnieuws), No. 4, Juni 1950, extract from a note by Boon Indrambarya B.Sc., (Head of the Fishery-service in Thailand), called: Note on the effect of explosives on fish in Siamese coastal waters.

147. M. EWING, J. L. WORZEL, N. C. STEENLAND, FRANK PRESS, Geophysical investigations in the emerged and submerged Atlantic coastal plain, Bulletin of the Geological Society of America, Vol. 61 September, 1950.

148. Summary Record of the 115th meeting, International Law Commission, Third Session, A/CN.4/SR. 115, 11 August 1951.

149. Final Act and Convention of the International Overfishing Conference, H. M. Stat. Off. London, Miscellaneous No. 7 (1946), Cmd. 6791.

150. XX, Department of State Bulletin, June 12, 1949.

151. PHILIP C. JESSUP, The Pacific Coast Fisheries, The A.J.I.L., Vol. 33, 1939.

152. Dr. J. HATSCHEK (Prof. Univ. Göttingen), Völkerrecht als System Rechtlich Bedeutsamer Staatsakte, Leipzig, 1923.

153. United Nations Legislative Series, Laws and Regulations on the

Regime of the High Seas, Vol. I, 1. Continental Shelf —2. Contiguous Zones — 3. Supervision of Foreign Vessels on the High Seas, New York 1951 (ST/LEG/SER.B/1, 11 January 1951).

154. HIGGINS and COLOMBOS, The International Law of the Sea, Second Revised Edition by C. John Colombos LL.D., London 1951.

155. Draft Conventions and comments on .... territorial waters, prepared by the Research in International Law of the Harvard Law School, Supplement to The A.J.I.L., Vol. 23, Special Number, April 1929.

156. H. LAUTERPACHT, The Function of Law in the International Community, Oxford 1933.

157. ALISON REPPY (Prof. of Law, New York Univ.), The Grotian Doctrine of the Freedom of the Seas reappraised, International Bar Association, London Conference, 1950.

158. HUGO GROTIUS, De Jure belli ac pacis, Libri Tres, Translation by Francis, W. Kelsey, Carnegie Endowment for International Peace, London 1925.

159. M. THÉODORE ORTOLAN (Capitaine de Vaisseau), Règles Internationales et Diplomatie de la Mer, Quatrième Edition, Tome Premier, Paris 1864.

160. ORVILLE A. PARK Jr., Extension of territorial waters in the Western Hemisphere, Third World Conference of the Legal Profession (International Bar Association), London, July 19–26, 1950.

161. M. D. A. AZUNI, Droit Maritime de l'Europe, Tome Premier, A Paris, XIII, 1805.

162. General Sir HOWARD DOUGLAS, BART., Translated in Dutch by H. A. Gobius, Verhandeling over de zee-artillerie, Vlissingen 1839.

163. General Sir HOWARD DOUGLAS, BART., G.C.B., G.C.M.G., D.C.L., F.R.S., A Treatise on Naval Gunnery, Fifth Edition, Revised, London 1860.

164. H. J. PAIXHANS, Translated from French. Proefnemingen, gedaan door de Franse Marine omtrent de Bombe-kanons, 's-Gravenhage 1826.

165. J. LAFAY, Aide-Mémoire d'Artillerie Navale, Paris, 1850.

166. FREDERICK LESLIE ROBERTSON, The Evolution of Naval Armament, London 1921.

167. WYNDHAM L. WALKER, M.A., LL.B., Territorial waters: The Cannon Shot Rule, The British Yearbook of International Law, 1945.

168. EDWIN BORCHARD, Resources of the Continental Shelf, The A.J.I.L., Vol. 40, January 1946.

169. GILBERT GIDEL, Le Droit International Public de la mer, Tome III, Chateauroux, 1934.

170. Rapport de M. le Professeur OPPENHEIM (sixième Commission, La mer territoriale) Annuaire de l'Institut de Droit International, Vingt-sixième Volume, Session d'Oxford, août 1913, Paris 1913.

171. S. WHITTEMORE BOGGS (special Adviser on Geography to the Department of State), National claims in adjacent seas, Geographical Review, published by the American Geographical Society of New York, April, 1951.

172. S. WHITTEMORE BOGGS, Delimination of seaward areas under national jurisdiction, The A.J.I.L., Vol. 45, April, 1951.

173. JACOB VAN DER LEE, Divergencies in International Law with special reference to the law of territorial waters, Dissertation submitted for the M.Litt. Degree in the University of Cambridge, June, 1951.

174. Report on the meeting of the Ministers of Foreign Affairs of the American Republics, Panama, September 23–October 3, 1939, Congress and Conference Series No. 29, Pan American Union, Washington, 1939.

175. JOHN BASSETT MOORE, LL.D., A Digest of International Law, Vol. I, Washington, Gvt. Print. Off., 1906.

176. Naval War College, International Law Situations, with solutions and notes 1939, United States Gvt. Print. Off., Washington, 1940.

177. CHARLES CHENEY HYDE (Prof. Int. Law and Diplomacy, Columbia Univ.), International Law, Vol. I, Boston, 1945.

178. United Nations Textbook, texts of important U.N. Documents etc. compiled by the "Professor Telders" Study Group for International Law at Leyden University, assisted by Dr. F. M. Baron Van Asbeck (Prof. of Int. Law and Int. Polit. History at Leyden Univ.) and Dr. J. H. W. Verzijl (Prof. of Int. Law at Utrecht Univ. and of the History of Int. Law at Leyden Univ.), Leiden, 1950.

179. General MAC ARTHUR's Address before Congress, United States Information Service, Press Section, Daily News Bulletin, The Hague, No. 76/# 96, April 19, 1951.

180. Recent Statutes, International Law, Power of a State to extend its boundary beyond the three-mile limit, Columbia Law Review, Vol. XXXIX, New York, 1939.

181. United States Supreme Court Reports, from beginning of Vol. 139 to end of Vol. 142, Cases argued and decided in the October terms 1890, 1891, ed. by Stephen K. Williams LL.D., Book XXXV, The Lawyers Co-operative Publishing Company, New York, 1891, 1892.

182. Territorial Waters of Saudi Arabia, Decree No. 6/4/5/3711, May 28, 1949, The A.J.I.L., Official Documents, p. 154–156, July ,1949.

183. Lieutenant Commander B. H. BRITTIN, U.S. Navy, International Law Aspects of the Acquisition of the Continental Shelf by the United States, United States Naval Institute Proceedings, Vol. 74, No. 550, December, 1948.

184. GEORGES SCELLE (Prof. à la Faculté de Droit de Paris), Cours (Manuel) de Droit International Public, Paris, 1948.

185. Convention and Statute on Freedom of Transit, Barcelona, April 20, 1921, League of Nations Treaty Series, Vol. VII, Numbers 1, 2 and 3, 1921–1922.

186. International Court of Justice, Reports of Judgments, Advisory Opinions and Orders, The Corfu Channel Case (Merits), Judgment of April 9th, 1949, Leyden.

187. FRANZ VON LISZT, Das Völkerrecht, 12e Aufl., bearbeitet von Dr. Max Fleischmann, Berlin, 1925.

188. H. A. Smith, D.C.L. (Prof. Emeritus Univ. London), The Law and Custom of the Sea, London, 1948.

189. L. Oppenheim, M.A., LL.D., International Law, Vol. II, sixth edition, revised, edited by H. Lauterpacht, M.A., LL.D., London, 1944.

190. Antonio Sanchez de Bustamante y Sirven (Juge à la Cour Perm. de Justice Int., Prof. de Droit Int. Univ. de la Havane), La Mer Territoriale, Paris, 1930.

191. A. D. Belinfante, Wetgevingspolitiek, Rechtsgeleerd Magazijn Themis — 1950, Afl. 1.

192. David Rowland, A Manual of the English Constitution, London, 1859.

193. Thomas Pitt Taswell-Langmead B.C.L., English Constitutional History, Tenth Edit., revised and enlarged by Theodore F. T. Plucknett, London, 1946.

194. William Stubbs, D.D., Hon. LL.D., D.C.L., (Bishop of Oxford), Select Charters and other illustrations of English Constitutional History, Oxford MDCCCCV.

195. Green Haywood Hackworth, Digest of International Law, Vol. II, Washington, 1941.

196. United States Code, 1946 Ed., Containing the General and Permanent Laws of the United States, in force on January 2, 1947, Vol. IV, U.S. Gvt. Print. Off., Washington, 1948.

197. John Bassett Moore (Hamilton Fish Professor of Int. Law and Diplomacy, Columbia Univ.), History and Digest of the International Arbitrations to which the United States has been a party, Vol. I, Washington, 1898.

198. D. H. N. Johnson, Artificial Islands, The International Law Quarterly, Vol. 4, No. 2, April, 1951.

199. State of Louisiana, Department of Conservation, Minerals Division, New Orleans, Louisiana, State-wide Order governing the drilling for and producing of oil and gas in the State of Louisiana, Order No. 29–B, July 19, 1943.

200. V. L. Lakhtine, Rights over the Arctic Regions, Moskva (Moscow) 1928.

201. C. H. M. Waldock, C.M.G., O.B.E., B.C.L., M.A. (Chichele Prof. Publ. Intern. Law, Univ. Oxford), Disputed Sovereignty in the Falkland Islands Dependencies, The British Yearbook of International Law, 1948.

202. British and Foreign State Papers, 1883–1884, Vol. LXXV, London, 1891.

203. Summary Record of the 124th Meeting, International Law Commission, third session, A/CN.4/SR. 124, 24 August 1951.

204. Summary Record of the 125th Meeting, International Law Commission, third session, A/CN.4/SR. 125, 24 August 1951.

205. Royal Pronouncement concerning the Policy of the Kingdom of Saudi Arabia with respect to the subsoil and sea-bed of areas in the Persian Gulf contiguous to the coasts of the Kingdom of Saudi Arabia, May 29, 1949, The A.J.I.L., Official Documents, Vol. 43, July 1949.

206. Iran, Bill relating to Persian Gulf subsea resources approved by the Council of Ministers and submitted to Majlis, May 1949,

Revue Egyptienne de Droit International, Vol. 5, Documents (Territorial Waters), 1949.

207. Bahrain Government, Proclamation No. 37/1368, June 5, 1949, The A.J.I.L., Official Documents, Vol. 43, October, 1949.

208. The Bahamas (Alteration of Boundaries) Order in Council 1948, No. 2574, 26 November, 1948, Statutory Instruments 1948, Vol. I, Part. I, H.M. Stat. Off., London, 1948.

209. The British Honduras (Alteration of Boundaries) Order in Council 1950, No. 1649, 9 October, 1950, and The Falkland Islands (Continental Shelf) Order in Council 1950, No. 2100, 21st December, 1950, Statutory Instruments 1950, Vol. I, H.M. Stat. Off., London, 1951.

210. The Jamaica (Alteration of Boundaries) Order in Council 1948, No. 2575, Statutory Instruments 1948, Vol. I, Part II, H.M. Stat. Off., London, 1949.

211. FRED K. NIELSEN, Is the jurisdiction of the United States exclusive within the three-mile limit? Does it extend beyond this limit for any purpose?
Proceedings of the American Society of International Law at its seventeenth annual meeting, held at Washington D.C., April 26–28, 1923.

212. HERBERT W. BRIGGS, The Law of Nations, Cases, Documents and Notes, New York, 1938.

213. British and Foreign State Papers 1896–1897, Vol. LXXXIX, London, 1901.

214. Sir CECIL HURST, G.C.M.G., K.C.B., K.C., The Continental Shelf, The Grotius Society, Transactions for the year 1948, Vol. 34, 1949.

215. Statute of the International Law Commission and other Resolutions of the General Assembly relating to the International Law Commission, A/CN.4/4, 2 February, 1949.

216. Award of the Umpire, The Right Hon. Lord Asquith of Bishopstone in the Matter of an Arbitration between Petroleum Development (Trucial Coast) Limited and His Excellency Sheikh Shakhbut Bin Sultan Bin Za'id, Ruler of Abu Dhabi and its Dependencies.

217. JURAJ ANDRASSY, Epikontinentalni Pojas, Jugoslavenska Akademija Znanosti i Umjetnosti, Zagreb, 1951.

218. H. LAUTERPACHT, K. C., F.B.A. (Whewell Prof. Int. Law, Univ. of Cambridge), Sovereignty over submarine areas, The British Yearbook of International Law, 1950.

219. C. H. M. WALDOCK (Chichele Prof. Int. Law and Diplomacy, Univ. Oxford), The Legal Basis of Claims to the Continental Shelf, read before the Grotius Society on April 5th, 1950 (paper circulated by the courtesy of the Society to assist the Copenhagen Conference of the International Law Association, 1950).

220. F. A. VALLAT, The Continental Shelf, The British Yearbook of International Law, 1946.

221. GEORG COHN (Minister, dr. jur.) Den Kontinentale Sokkel, Jus Gentium, Nordisk Tidsskrift for Folkeret og international Privatret, Vol. II, Fasc. 1–2, København, 1950.

222. Regeerings Almanak Nederlandsch Indië 1930, I, Bijlage III, Concessiën tot mijnontginning, p. 812, No. 52.

223. Loi du 21 avril 1810, Mines, minières, carrières; Dalloz, Code administratif, Paris, 1949.
224. B. E. BOEKHOLD, Mijnrecht, Den Haag, 1912.
225. Wet van 23 Mei 1899 (Ned. Staatsblad No. 124), De wetboeken en Verordeningen van Indonesië, Engelbrecht, 1950.
226. Halsbury's Statutes of England, Second Ed., Vol. 16, 1950.
227. United States, Federal Code Annotated, Vol. 9, 1944 (replacement volume), Cumulative Pocket Supplement, Vol. 9, 1950.
228. Dr. KARL STRUPP, Wörterbuch des Völkerrechts und der Diplomatie, Erster Band, Berlin und Leipzig, 1924.
229. The Cornwall submarine mine Act, 1858, 21 & 22 Victoria, Cap. CIX.
230. Supreme Court of the United States, October term, 1949, No. 13 Original, United States of America (Plaintiff) v. State of Texas (Defendant), Brief for the State of Texas in opposition to motion for judgment.
231. Report of the Commissioners for the Channel Tunnel and Railway, Presented to both Houses of Parliament by Command of Her Majesty, 1876 (H.M. Stat. Off.), C. 1576.
232. Projet d'une réglementation des voies de communications maritimes en temps de paix, présenté par M. le Professeur A. Alvarez, The International Law Association, Report of the 33rd Conference, held at Stockholm, September 8th–13th, 1924, London, 1925.
233. Draft Convention, Laws of maritime jurisdiction in time of peace, International Law Association, Report of the 34th Conference, held at Vienna, August 5th–11th, 1926, London, 1927.
234. Supplement to the American Journal of International Law, Vol. 20, Special Number, October 1926.
235. Report of FRED. K. NIELSEN, The Island of Palmas Arbitration before the Permanent Court of Arbitration under the special Agreement concluded between the United States of America and the Netherlands, January 23, 1925, Washington, May 2, 1928.
236. JAMES BROWN SCOTT, The Hague Court Reports, Second Series, Carnegie Endowment of International Peace, New York, 1932.
237. The English Reports, Vol. CLXV, Ecclesiastical, Admiralty and Probate and Divorce V, London, 1923, (5.C.Rob(inson) 373).
238. WHEATON's Elements of International Law, sixth English ed. by A. Berriedale Keith, D.C.L., D.Litt., Vol. I, London, 1929.
239. Summary Record of the 116th meeting, International Law Commission, 3rd session, A/CN.4/SR. 116, 11 August 1951.
240. I. W. ALCORN, The Pure Oil Company's Tideland Development, World Oil, May, 1949.
241. Dr. IRVING P. KRICK, Preliminary Report on Oceanographic and Meteorologic Factors, involved in damage to offshore drilling structures, The Oil and Gas Journal, February 23, 1950.
242. R. C. FARLEY and J. S. LEONARD, Gulf Hurricane and consequent damages, Oil and Gas Journal, February 23, 1950.
243. EUGÈNE CAUCHY, Le Droit Maritime International, Tome premier, Paris, 1862.

244. The search for oil on the continental shelf, World Oil, December 1949.
245. LEIGH S. Mc CASLIN Jr., Marine Rig No. 10, a new approach to the reduction of high costs of offshore drilling, The Oil and Gas Journal, January 12, 1950.
246. FRANK BRIGGS, A model design of a new type drilling barge, World Oil, February 1, 1950.
247. H. E. DENZLER Jr., JOHN W. Scott and James M. West, Installation of offshore flow lines, World Oil, February 1, 1950.
248. J. E. KASTROP, 50 Years of Gulf Coast Oil; and Dual Rig Platform, reduces offshore drilling costs, World Oil, June, 1950.
249. HENRY OZANNE, Federal State Dispute slows Continental Shelf development, World Oil, September, 1950.
250. GEORGE B. PARKS, Directional Drilling in offshore operations, World Oil, June, 1949.
251. LEIGH S. Mc CASLIN Jr., Along the Gulf, The Oil and Gas Journal, April 13, 1950.
252. JAMES BROWN SCOTT, The Hague Peace Conferences of 1899 and 1907, Vol. II, Documents, Baltimore, 1909.
253. Besluit van den Gouverneur-Generaal van 15 September 1903, No. 4, Ned. Indisch St.bl. 1903, No. 318.
254. Politiek Beleid en Bestuurszorg in de Buitenbezittingen, Tweede Gedeelte A, Batavia, 1909.
255. K. O. EMERY, Continental Shelf — Southern California, Bulletin of the American Association of Petroleum Geologists, Vol. 35, No. 2, February, 1951.
256. PAUL WEAVER, Continental Shelf of Gulf of Mexico, Bulletin of the American Association of Petroleum Geologists, Vol. 35, No. 2, February, 1951.
257. JOSEPH WALTER BINGHAM, The continental shelf and the marginal belt, The A.J.I.L., Vol. 40, January, 1946.
258. Oil under the High Seas: The New Rule, Petroleum Press Service, Vol. XVIII, No. 11, November, London, 1951.
259. Regime of the High Seas, Comments by Governments on Draft Articles on the Continental Shelf and Related Subjects, International Law Commission, Fourth Session, A/CN. 4/55, 16 May 1952.
260. Gilbert Gidel, Le Plateau Continental, Fourth International Conference of the Legal Profession, International Bar Association, Madrid, July 1952.
261. International Court of Justice, Reports of Judgments, Advisory Opinions and Orders, Fisheries Case (United Kingdom v. Norway), Judgment of December 18th, 1951.
262, The Commonwealth of Australia, Fisheries, No. 7 of 1952, An Act relating to Fisheries in certain Australian Waters (assented to 13th March, 1952).
263. Lively Development in the MOP Industry, Pacific Islands Monthly, published by Pacific Publication PTY, Ltd., Sydney. Vol. XXII, No. 9, April, 1952.
264. International Convention for the high seas fisheries of the North Pacific Ocean, Tokyo, 10 May, 1952, Department of State for the Press, No. 370, May 9, 1952.
265. Juji Enomoto, Character and Scope of rights asserted and exercised

over coastal waters and appurtenant subsoil, Paper for the Madrid Conference of the International Bar Association, July, 1952 (April 1952).

266. *Annuaire de l'Institut de Droit International, 1937.*

267. *Wm. W. Bishop Jr., The need for a Japanese Fisheries Agreement, The A.J.I.L., Vol. 45, October, 1951.*

268. *Edward W. Allen, A new concept for fishery Treaties, The A.J.I.L., Vol. 46, April, 1952.*

269. *Charles B. Selak Jr., The proposed international convention for the high seas fisheries of the North Pacific Ocean, The A.J.I.L., Vol. 46, April, 1952.*

270. *D. H. N. Johnson, The Anglo-Norwegian Fisheries Case, The International and Comparitive Law Quarterly, Vol. 1, Part 2, April, 1952.*

271. *Decree of the Presidium of the National Assembly concerning the Territorial and Inland waters of the People's Republic of Bulgaria, The A.J.I.L., Official Documents, Vol. 46, April. 1952.*

272. *Marin Pundeff, Bulgarian Decree on Territorial Waters, The A.J.I.L., Vol. 46, April, 1952.*

273. *J. P. A. François, Report on the Regime of the Territorial Sea International Law Commission, Fourth Session, A/CN.4/53, 4 April 1952.*

274. *J. P. A. François, Third Report on the Regime of the High Seas, International Law Commission, Fourth Session, A/CN.4/51, 29 February 1952.*

275. *Dr. Eberhard Menzel, (Privat Dozent, Forschungsstelle für Völkerrecht und ausl. öfftl. Recht der Univ. Hamburg) Die Lehre von „kontinentalen Sockel" (continental shelf) und ihre Bedeutung im modernen Völkerrecht, paper for the Madrid Conference of the International Bar Association, July, 1952.*

276. *L. C. Green, The Continental shelf, Current Legal Problems, 1951, edited by George W. Keeton and Georg Schwarzenberger, Vol. 4, London, 1951.*

277. *F. A. Vallat, Continental Shelf, paper for the Madrid Conference of the International Bar Association, July, 1952.*

INDEX

The small figures refer to the Bibliography. A figure between brackets indicates that the idea comprehended by the word in question will be found on that page, although the word itself does not appear.